When Computers Were Human

When Computers Were Human

David Alan Grier

PRINCETON UNIVERSITY PRESS

PRINCETON AND OXFORD

LIBRARY OF CONGRESS CATALOGING-IN-PUBLICATION DATA

Grier, David Alan, 1955 Feb. 14–
When computers were human / David Alan Grier.
p. cm.
Includes bibliographical references.
ISBN 0-691-09157-9 (acid-free paper)
1. Calculus—History. 2. Science—Mathematics—History. I. Title.
QA303.2.G75   2005
510′.92′2—dc22   2004022631

British Library Cataloging-in-Publication Data is available

This book has been composed in Sabon

Printed on acid-free paper. ∞

www.pupress.princeton.edu

Printed in the United States of America

10 9 8 7 6 5 4 3 2 1

FOR JEAN

Who took my people to be her people and my stories to be her own without realizing that she would have to accept a comet, the WPA, and the oft-told tale of a forgotten grandmother

Contents

When Computers Were Human

A Grandmother's Secret Life

> After a while nothing matters . . . any more than these little
> things, that used to be necessary and important to forgotten
> people, and now have to be guessed at under a magnifying glass
> and labeled: "Use unknown."
>
> Edith Wharton, *The Age of Innocence* (1920)

IT BEGAN with a passing remark, a little comment, a few words not understood, a confession of a secret life. On a cold winter evening, now
many years ago, I was sharing a dinner with my grandmother. I was
home from graduate school, full of myself and confident of the future. We
sat at a small table in her kitchen, eating foods that had been childhood
favorites and talking about cousins and sisters and aunts and uncles.
There was much to report: marriages and great-grandchildren, new
homes and jobs. As we cleared the dishes, she became quiet for a moment, as if she were lost in thought, and then turned to me and said, "You
know, I took calculus in college."

I'm certain that I responded to her, but I could not have said anything
beyond "Oh really" or "How interesting" or some other empty phrase
that allowed the conversation to drift toward another subject and lose
the opportunity of the moment. In hindsight, her statement was every bit
as strange and provocative as if she had said that she'd fought with the
Loyalists in the Spanish Civil War or had spent her youth dealing bac-
carat at Monte Carlo. Yet, at that instant, I could not recognize that she
had told me something unusual. I studied with many women who had
taken calculus and believed they would have careers in the mathematical
sciences like my intended career. I did not stop to consider that only a few
women of my grandmother's generation had even attended college and
that fewer still had ever heard of calculus.

My grandmother's comment was temporarily ignored, but it was not
lost. It came rushing back into my thoughts, some six or seven years later,
as I was sitting in a mathematics seminar. Such events are often conducive
to reflection, and this occasion promised plenty of opportunity to think
about other subjects. The speaker, a wild-haired, ill-clad academic, was
discussing a new mathematical theory with allegedly important applications that were far more abstract than the theory itself. As I was helping

myself to tea and cookies, a staple of mathematical talks, I caught a remark from a senior professor. I had always admired this individual, for he had the ability to sleep during the boring parts of seminars and still catch enough of the material to ask deep and penetrating questions during the discussion period. This professor, who had recently retired, was describing his early days at the university during the Great Depression of the 1930s. Having just arrived in the United States from his native Poland and knowing only rudimentary English, he was assigned to teach the engineering calculus course. "This," he stated with a flourish, "was the first time that calculus was required of engineering students at the university."

As I listened to his story, I heard my grandmother's phrase from that night long before. "You know, I took calculus in college." I did not know when she had attended college, but having heard my mother's stories of the Depression, I was certain it would have been before 1930. As I settled into my chair, I started to ponder what my grandmother had said. For the next hour, I was lost in my own thoughts and oblivious to the talk, which proved to be the best seminar of the term. During the discussion period, I was asking myself the questions I should have raised at that dinner years before: Where had my grandmother attended college? What courses had she taken? What had she hoped to learn from calculus?

By then, it was too late to ask these questions. My grandmother was gone, and no one knew much about her early life. My mother believed that my grandmother had studied to be an accountant or an actuary. My uncle thought that my grandmother had taken some bookkeeping classes. Our family genealogist, a distant cousin who seemed to know everything about our relations, expressed her opinion that my grandmother's family had been too poor to send her to college. Still bothered by that one phrase, I decided to see what I could learn. My grandmother had been raised in Ann Arbor, the home of the University of Michigan. So one day, I called the college registrar and asked if she had a transcript for my grandmother. I tried to use a tone of voice to suggest that it was the most natural thing in the world for a grandson to review his grandmother's college grades, rather than the other way around. With surprisingly little hesitation, the registrar agreed to my request and left the phone. In a few minutes, she returned and said, "I have her records here."

Catching my breath, I asked, "When did she graduate?"

"Nineteen twenty-one," the registrar responded.

"What was her major?" was my next question.

After a moment of shuffling paper, she replied, "Mathematics."

Three weeks later, I was sitting at a long library table with a little gray box that contained the university's record of my grandmother's life. As I worked through her transcript and the course record books, I was surprised but pleased to see that she had taken a rigorous program of study.

1. Calculus class, University of Michigan, 1921. Author's grandmother is right-most woman

In all, she had taken about two-thirds of the mathematics courses that I had taken as an undergraduate, and she had studied with several well-known mathematicians of the 1920s. The professors' record books were particularly intriguing, for they contained little notes that hinted at the activity and turmoil outside the classroom. One mentioned the male students who had left for the First World War; another recorded that he had devoted part of the term to analyzing ballistics problems; a third mentioned that two students had died in the influenza epidemic.[1]

Perhaps the most surprising revelation was the fact that my grandmother was not the only female mathematics student. Of the twelve students who had taken a mathematics degree in 1921, six of them, including my grandmother, were women. The University of Michigan was more progressive than the Ivy League schools, but its liberalism had limits. About a quarter of the university student body was female, but the school provided no dormitory for women and barred them from the student union building, as it was attached to a men's residence hall. University officials also discouraged women from studying medicine, business, engineering, physics, biology, and chemistry. For women with scientific interests, the mathematics department was about the only division of the school that welcomed them. Much of this welcome was provided by a single professor, James W. Glover (1868–1941), who served as the advisor to my grandmother and most of her female peers.[2]

Glover was an applied mathematician, an expert in the mathematics of finance, insurance, and governance. He had been employed as an actuary for Michigan's Teacher Retirement fund, had held the presidency of Andrew Carnegie's Teachers Insurance and Annuity Association, and, in the early years of the century, had served as a member of the Progressive Party's "brain trust," the circle of academic advisors to the party leader, Robert La Follette.[3] Within the University of Michigan, Glover was an advocate for women's education, though he was at least partly motivated by a desire to increase enrollments in mathematics courses. He welcomed women to his classes, encouraged them to study in the department lounge, prepared them for graduate study, and helped them search for jobs. He pushed the women to look beyond the traditional role of schoolteacher and consider careers in business and government. At a time when clerical jobs were still dominated by men, Glover helped his female students find positions as assistant actuaries and human computers, the workers who undertook difficult calculations in the days before electronic computers. At the end of his career, he recorded that he had advised nearly fifty women and that only "one-third have married and have retired from active business life."[4]

Of the six women who graduated in 1921, only one, my grandmother, never worked outside the home. The remaining five had mathematical careers that lasted into the 1950s. One was a human computer for the United States Army and prepared ballistics trajectories. A second did calculations for the Chemical Rubber Company, a publisher that sold handbooks to engineers and scientists. Another compiled health statistics for the state of Michigan. The fourth worked for the United States Bureau of Labor Statistics and eventually became the assistant director of a field office in Baton Rouge. The last female mathematics major of 1921 became an actuary, moved to New York City, and operated her own business.[5]

Though my grandmother's hidden mathematical career held a special emotional appeal to me, it was the story of the other five women that captured my interest. What kind of world did they inhabit? What were their aspirations? What did they do each day? At the ends of their careers, what had they accomplished? Rather than restrict my scope to the five women who had known my grandmother or even the women mathematics graduates of the University of Michigan, I decided to look at the history of scientific computers, the workers who had done calculations for scientific research.

Scientific computation is not mathematics, though it is closely related to mathematical practice. One eighteenth-century computer remarked that calculation required nothing more than "persevering industry and attention, which are not precisely the qualifications a mathematician is most anxious to be thought to possess."[6] It might be best described as "blue-collar science," the hard work of processing data or deriving predictions from scientific theories. "Mental labor" was the term used by the English mathematician Charles Babbage (1791–1871).[7] The origins of scientific calculation can be found in some of the earliest records of human history, the clay tables of Sumeria, the astronomical records of ancient shepherds who watched over their flocks by night, the land surveys of early China, the knotted cords of the Inca.[8] Its traditions were developed by astronomers and engineers and statisticians. It is kept alive, in a sophisticated form, by those graduate students and laboratory assistants who use electronic calculators and computer spreadsheets to prepare numbers for senior researchers.

Though many human computers toiled alone, the most influential worked in organized groups, which were sometimes called computing offices or computing laboratories. These groups form some of the earliest examples of a phenomenon known informally as "big science," the combination of labor, capital, and machinery that undertakes the large problems of scientific research.[9] Many commentators identify the start of large-scale scientific research with the coordinated military research of the Second World War or the government-sponsored laboratories of the Cold War, but the roots of these projects can be traced to the computing offices of the eighteenth and nineteenth centuries.[10]

It is possible to begin the story of organized computing long before the eighteenth century by starting with the great heavenly *Almagest*, the charts of the planets created by Claudius Ptolemy (85–165) in classical Egypt. As the connection between the ancient world and its modern counterpart is sometimes tenuous, we will begin our story just a few years before the opening of the eighteenth century with two events: the invention of calculus and the start of the Industrial Revolution. Both events are difficult to date exactly, but that is of little concern to this narrative. Identifying the inventors of specific ideas is less important than understanding how these ideas developed within the scientific community. Calculus gave scientists new ways of analyzing motion. Most historians of mathematics have concluded that it was invented independently by Isaac Newton (1642–1727) and Gottfried Wilhelm Leibniz (1646–1716) in the 1680s. It was initially used in astronomy, but it also opened new fields for scientific research. The Industrial Revolution, the economic and social change that was driven by the factory system and the invention of large machinery, created new techniques of management, developed public journalism as a means of disseminating ideas, and produced the modern factory.[11] Most scholars place the start of the Industrial Revolution at the end of the eighteenth century, but this change was deeply influenced by the events of Newton's time. "It is enough to record that by 1700 the foundations of modern technology have been laid,"[12] concluded historian Donald Caldwell.

By starting with the invention of calculus, we will overlook several important computational projects, including the *Arithmetica Logarithmica* by Henry Briggs (1561–1630), the ballistic trajectories of Galileo Galilei (1564–1642), and the planetary computations in the *Rudolphine Tables* by Johannes Kepler (1571–1630). Each of these projects contributed to the development of science and mathematics. Briggs gave science one of its most important computing tools, the logarithm table. Galileo and Kepler laid the foundation for calculus. However, none of these projects is an example of organized computation, as we define it. None of these scientists employed a staff of computers. Instead, they did the computations themselves with the occasional assistance of a student or friend.

The story of organized scientific computation shares three themes with the history of labor and the history of factories: the division of labor, the idea of mass production, and the development of professional managers. All of these themes emerge in the first organized computing groups of the eighteenth century and reappear in new forms as the story develops. All three were identified by Charles Babbage in the 1820s, when he was considering problems of computation. These themes are tightly intertwined, as mass production clearly depends upon the division of labor, and the appearance of skilled managers can be seen as a specific example of divided and specified labor. However, this book separates these ideas and treats them individually in an attempt to clarify and illuminate the different forces that shaped computation.

The first third of this book, which deals with computation from the start of the eighteenth century up to 1880, treats the first theme, the division of labor. During this period, astronomy was the dominant field of scientific research and the discipline that required the greatest amount of calculation. Some of this calculation was done in observatories for astronomers, but most of it was done in practical settings by individuals who used astronomy in their work, most notably navigators and surveyors. It was a period when the borders of scientific practice were not well defined and many a scientist moved easily through the learned disciplines, scanning the sky one night, navigating a ship the next, and perhaps, on the night following, designing a fortification or preparing an insurance table. The great exponent of divided labor, the Scottish philosopher Adam Smith (1723–1790), wrote *The Wealth of Nations* during this period. Smith discussed the nature of divided labor in scientific work and even commented briefly on the nature of astronomy. The astronomers of the age were familiar with Smith's ideas and cited them as the inspiration for their computing staffs.

The second third of the book covers the period from 1880 to 1930, a time when astronomy was still an important force behind scientific computation but was no longer the only discipline that required large-scale calculations. In particular, electrical engineers and ordnance engineers started building staffs to deal with the demands of computation. The major change during this period came from the mass-produced adding and calculating machines. Such machines have histories that can be traced back to the seventeenth century, but they were not commonly found in computing offices until the start of the twentieth century. While these machines decreased the amount of time required for astronomical calculations, they had a greater impact in the fields of economics and social statistics. They allowed scientists to summarize large amounts of data and to develop mathematical means for analyzing large populations. With the calculating machines came other ideas that we associate with mass production, such as standardized methods, generalized techniques, and tighter managerial control.

The final third of the book discusses computation during the Great Depression, the Second World War, and the early years of the Cold War. It was at this time that human computers attempted to establish their work as an independent discipline, distinct from the different fields of scientific research and even from mathematics itself. This activity required human computers to create a literature of computation, define formal ways of training new computers, and create institutions that could support their work. Historians discuss such subjects under the topic of "professionalization," a term that suggests independence, societal respect, and control of one's activities. In the case of the human computer, professionalization produced no independence, little respect, and nothing that could be characterized as self-governance. Professionalization came just as human computers were being replaced by computing machines that were built with tubes, powered by electricity, and controlled by a program.

The story of the human computer is connected to the development of the modern electronic computer, but it does not provide the direct antecedent of the machines that were built for scientific and business calculation in the last half of the twentieth century. To be sure, the two stories twist about each other, touching at regular points and sharing ideas with the contact. The developers of electronic computers often borrowed the mathematical techniques of hand calculation and, from time to time, asked human computers to check some number that had been produced by their machines; however, few human computers contributed to the invention of electronic computing equipment, and few computing offices were connected to machine development projects. It is best to view the human computing organizations as the backdrop against which the story of electronic computers unfolds. Human computers plugged away at their calculations with little influence over those engaged in machine design. Most computers were intrigued with the electronic computing machines and looked forward to using these devices, but they would prove to be the secondary characters in the narrative, the Rosencrantz and Guildenstern instead of the Hamlet and Ophelia. The human computers occupied a small corner of the stage, somewhat unsure of their role, as engineers developed electronic replacements for the computing laboratories and their large staffs of workers.

This book attempts to invert the history of scientific computing by narrating the stories of those who actually did the calculations. These stories are often difficult to tell, as the vast majority of computers left no record of their lives beyond a single footnote to a scholarly article or an acknowledgment in the bottom margin of a mathematical table. Furthermore, the few surviving human computers often failed to appreciate the full scope of what they did. As often as not, they would deflect inquiries with remarks like "It was nothing" or "You should have asked my supervisor about that." The stories unfolded in unusual ways from unlikely sources. There was a bound volume of correspondence in the Library of Congress, the cassette tape that had been carefully guarded by a family, a scrapbook that had been long filed away, the box of records with the confusing label on the shelves of the National Archives, the correspondence from an obscure university official, and the four-hour telephone conversation with a man on a hospital bed. Each of these stories illustrated a different aspect of the human computer, but each, in its own way, returned to the statement of a grandmother, "You know, I took calculus in college."

PART I

Astronomy and the Division of Labor 1682–1880

> If your wish is to become really a man of science and not
> merely a petty experimentalist, I should advise you to apply to
> every branch of natural philosophy, including mathematics.
> Mary Shelley, *Frankenstein* (1818)

CHAPTER ONE

The First Anticipated Return: Halley's Comet 1758

> When they come to model Heaven
> And calculate the stars, how they will wield
> The mighty frame . . .
>
> John Milton, *Paradise Lost* (1667)

OUR STORY will begin with a comet, a new method of mathematics, and a seemingly intractable problem. The comet is the one that appeared over Europe in August 1682, the comet that has since been named for the English astronomer Edmund Halley (1656–1742). This comet emerged in the late summer sky and, according to observers at Cambridge University, hung like a beacon with a long, shimmering tail above the chapel of King's College. To that age, comets were mysterious visitors, phenomena that appeared at irregular intervals with no obvious explanation. Their origins, substance, and purpose were matters of pure speculation. Some thought that they were wayward stars. Others suggested that they might originate in the atmosphere, each a burning piece of Helios's chariot, perhaps, that had been caught between the earth and the moon.

The only aspects of the 1682 comet that could be studied with certainty were its position against the fixed stars of night and the length of its tail. The young Edmund Halley recorded both measurements on at least seven distinct nights that summer. He was a gentleman of private life, possessed of an independent income and a new house in a prosperous village just north of London. His collection of scientific instruments included a sextant, a small telescope mounted on an arc of a circle, which allowed him to measure the distance of the comet's head from nearby stars. His measurements were not in miles or meters or light-years but in degrees of an angle. His home marked the joint of that celestial angle. One leg stretched from the earth to the head of the comet. The second leg reached to a star, the end of the tail, or some other reference point. The work required patience and a steady hand. By the time the comet vanished, Halley had traced its path across the sky and recorded the advance and retreat of the tail. At the time, it was not entirely clear what Halley might do with these measurements. If they had been the measurements of a planet, he might have computed an orbit, but few believed that comets moved in ellipses around the sun as the

2. Halley's comet over Cambridge, 1682

planets did. Halley had other interests to pursue, so he put his comet data away for future use.

The new method of mathematics was calculus, a subject then known in England as fluxions. Calculus is the mathematics of physical activity, the mathematics of change. It probes the nature of movement by dividing it into smaller and smaller steps and then reassembling these tiny units into surprisingly elegant and simple expressions. The techniques of calculus had their origin in an attempt to explain the motion of the planets by physical laws rather than by the arbitrary actions of superhuman beings. The

English proponent of calculus was Isaac Newton (1642–1727), who developed the method while he was writing his masterwork, *Philosophiae Naturalis Principia Mathematica* (*The Mathematical Principles of Natural Philosophy*), a book commonly called *Principia*. In *Principia*, Newton explained that he was attempting to analyze "the motions of the planets, the comets, the moon, and the sea," the last term referring to the movements of the tides.[1] In the central part of the book, Newton considered the motion of two objects under the influence of a single universal force, which he called gravity. The two objects might be the moon and the earth, a planet and the sun, or even a comet and some other celestial object. In these circumstances, Newton argued that gravity impels the bodies to follow certain kinds of paths: the gentle bend of the hyperbola, the tight hairpin of a parabola, and the cyclical orbit of an ellipse.

The intractable problem appeared when the calculus of Newton met the comet data of Halley. Halley called upon Newton in 1684, when *Principia* was nothing more than a collection of notes. He helped Newton prepare the final manuscript for publication in 1687 and promoted Newton's ideas at the Royal Society, the central organization of seventeenth-century English science. Though he frequently thought about the problems of comets and astronomy, he let thirteen years pass after his initial observations in 1682 before he undertook a serious analysis of his data. During those intervening years, he had other problems to keep him busy. He served as clerk to the Royal Society and as the editor of its journal, *Philosophical Transactions*. He also studied a number of other scientific problems, such as the design of diving bells and the mathematics of finance.

In September 1695, Halley returned to his comet data and attempted to validate the statements that Newton had made about comets in *Principia*. Newton had speculated that comets moved in parabolas around the sun, narrow curves that started at a distant point in the universe, sped past the earth, turned sharply at the sun, and then rushed back to the void whence they came. It seemed a plausible theory, but he had never done the analysis to verify it.[2] Halley spent about a month working with the measurements from four different comets, trying to identify the path that each object made through the solar system. From an individual comet, he would select three observations, each recorded on a different day. From these numbers he computed the parameters of a parabolic curve. Newton had done this sort of work with graphs, but after a little practice Halley could report, "I am now become so ready at the finding a Cometts orb by calculation."[3] Once he had calculated the parabola, he adjusted the curve by comparing it to the other observations of the comet. If he found that all of the observations were close to the parabola, he would conclude that he had found the proper path. If he discovered that some of them fell at a distance from the curve, he would attempt to adjust the parameters in order to bring the parabola closer to the observations.

The procedure worked well for the first three comets: one observed by Newton in 1664, a second that Halley had observed just before the 1682 comet, and a third that had appeared shortly after.[4] Each of these objects seemed to followed a parabolic curve. When Halley began to work on the 1682 comet, the comet that he had observed from his home, he altered his methods. He chose to fit the data to a closed ellipse rather than an open-ended parabola. Halley's biographer has noted that this idea did not come from calculation but was "based upon somewhat inspired insight."[5] Halley had noted that the 1682 comet followed a path that had been traversed by two earlier comets, one observed in 1531 by the German astronomer Peter Apian (1495–1552) and a second recorded in 1607 by Johannes Kepler. With his 1682 data, Halley computed the values for an elliptical orbit and then compared the curve to the earlier observations. Pleased with the results, he wrote to Newton, "I am more and more confirmed that we have seen that Comett now three times since ye Year 1531."[6]

Though he was certain that the 1682 comet orbited the Sun, Halley recognized that his calculations did not prove his claim. His work did not address a substantial inconsistency in his data. Seventy-six years separated Apian's observations from those of Kepler. Only seventy-five years passed between Kepler's sighting and Halley's data from 1682. The analysis suggested that the comet should have a fixed period, that it should return without fail every seventy-five years. Halley speculated that the discrepancy might be caused by the gravitational pull of the outer planets, forces which could easily disturb the orbit of the comet and change the date of its return. Writing to Newton, he asked, "When your more important business is over, I must entreat you to consider how far a Comet's motion may be disturbed by the Centers of Saturn and Jupiter, particularly in its ascent from the Sun."[7]

Newton responded quickly, but his reply was vague and unhelpful. "How far a comet's motion may be disturbed," he wrote to Halley, "cannot be affirmed without knowing the Orb of ye Comet & times of its transit through ye Orbs of [the two planets]."[8] Once Saturn and Jupiter became part of the equations, the calculations were no longer straightforward and could not be handled by a single astronomer in his spare minutes and hours. The Sun, Saturn, and Jupiter form a three-body system, three objects moving through space, each exerting an influence upon the other two. Newton had been unable to find a simple expression that described the motion of such a system, even though he had been able to find solutions for two bodies in motion. In his best effort, he had devised an approximation that crudely described the movement of three bodies, but this expression was not precise enough to explain the variation in the comet's period.[9]

The lack of a simple solution to the three-body problem stymied Halley's calculations, but it did not shake his faith. He freely discussed his ideas in public and published his theory of comets in *Astronomiae Cometicae Synopsis* (*A Synopsis of the Astronomy of Comets*).[10] In this book, he claimed that he could "undertake confidently to predict the return" of the comet in 1758. Some scholars noted a lack of mathematical rigor in Halley's analysis and questioned this claim. Responding to the criticism, Halley weakened his statements, claimed that the comet might return at any time within a 600-day period that began in 1757, and replaced his confident prediction with a sentence that began, "I think, I may venture to foretell" the return of the comet.[11]

From time to time, Halley tried to improve his predictions for the 1758 return. He made little progress, as he was unwilling, or perhaps unable, to refine his estimates into a detailed computation. His final effort occurred in about 1720, just before he became Astronomer Royal and director of the Royal Observatory in Greenwich. For this calculation, he had a new approximate solution for the three-body problem of Saturn, Jupiter, and the Sun. From this solution, he deduced that the comet was pulled farther from the Sun after its 1682 return and hence would require more time to traverse its path. It was one more crude estimate, but it would stand as his final word on the subject. In his last revision of his *Astronomiae Cometicae*, which was published after his death in 1742, he announced that his comet would return "about the end of the year 1758, or the beginning of the next."[12] With this opinion on the subject, he bequeathed the comet to future generations. "Having touched upon these things," he wrote, "I shall leave them to be discussed by the care of posterity, after the truth is found out by the event."[13]

Posterity made the return of Halley's comet a test for Newton's theory of gravitation. Newton's "followers have, from his principles, ventured even to predict the returns of several [comets]," wrote the Scottish philosopher Adam Smith (1723–1790), "particularly of one which is to make its appearance in 1758." If scientists could predict the date of return, they would take the agreement between prediction and observation as evidence that Newton's ideas on gravity were correct. If the predicted date did not coincide with the actual date, then they would conclude that other forces were at work in the universe. Smith believed that Newton's analysis was probably correct. "His system," he stated, "now prevails over all opposition, and has advanced to the acquisition of the most universal empire that was ever established in philosophy." However, Smith was not willing to accept the prediction for Halley's comet without a proper test. "We must wait for that time before we can determine, whether his philosophy corresponds as happily to [comets] as to all the [planets]."[14]

A thorough test of the gravitational theory required computational techniques beyond the mathematics that Halley had used for his initial analysis of the comet. Newton's calculus would never provide a simple way to describe the motion of three or more bodies and hence would never give an accurate date for the comet's return. The only way to determine the comet's orbit was to substitute brawn for brain, to divide the comet's progress into tiny steps, analyze the forces pulling on the comet, and then combine these steps into a serviceable whole through the tedious process of summation. "What immense labor," wrote one astronomer, "what geometrical knowledge did not this task require?"[15] Among the astronomers that followed Edmund Halley, few even considered undertaking the labor. Only one, a French mathematician named Alexis-Claude Clairaut (1713–1765), made a serious attempt to predict the date of the 1758 return, an attempt that required both a computational technique beyond those developed by Newton and the means of dividing the work among computing assistants.

Clairaut was described by his contemporaries as "ambitious," "vivacious by nature," and "successful in society."[16] He had already made a reputation as a mathematician by extending Newton's calculus and developing a computational method of handling the three-body problem. He had used his method to find solutions to other problems in astronomy, but the challenge of Halley's comet had a special appeal. The comet was well known to astronomers in France and England. Observers in both countries were scanning the sky for a fuzzy speck of light that would be the first sign of the comet. If he could predict the point of the comet's first appearance, Clairaut would become famous indeed. Such a calculation had many practical problems, not the least being the ability of local weather to obscure the night sky. Instead of predicting the first observation of the comet, Clairaut computed the date of the perihelion, the date the comet made its closest approach to the Sun.[17]

Clairaut decided to undertake the calculation sometime in the spring of 1757. He may have been encouraged by two friends, Joseph-Jérôme Le Français de Lalande (1732–1807) and Nicole-Reine Étable de la Brière Lepaute (1723–1788), with whom he divided the computations. Joseph Lalande was a young astronomer, a scientist near the start of his career. He had studied in a seminary to become a Jesuit priest and had nearly taken orders, but, as the historian Ken Alder has noted, he "had an insatiable thirst for fame."[18] He left the clerical life to make his reputation as an astronomer. By the time he was twenty-one, he had undertaken a major astronomical project that combined the efforts of the Paris and Berlin observatories. He had also been elected to the French scientific society, the

3a. Computers for first return of Halley's Comet: Alexis-Claude Clairaut

3b. Computers for first return of Halley's Comet: Joseph-Jérôme Le Français de Lalande

3c. Computers for first return of Halley's Comet: Nicole-Reine Étable de la Brière Lepaute

Académie des Sciences. Nicole-Reine Lepaute came from the thin stratum of wealthy French bourgeoisie.[19] She had been born in the Palais Luxembourg and had been educated by her parents. "In her early childhood, she devoured books, [and] passed nights at readings," wrote Lalande, who also claimed that she was "the only woman in France who [had] a true knowledge of astronomy." When she was twenty-five, she had married the royal clockmaker, Jean André Lepaute (1709–1789). It seems to have been a generous marriage, one that allowed Mme Lepaute some freedom to exercise her scientific skill. Lalande recorded that "she observed, she calculated, and she described the inventions of her husband."[20]

Joseph Lalande and Nicole-Reine Lepaute had an unconventional relationship that reflected the unconventional work in which they were engaged. Lalande never married; "he was a supremely ugly man," wrote Alder, yet he "loved women, especially brilliant women, and promoted them in both word and deed."[21] The two met when Lalande, acting as a representative of the Académie des Sciences, called upon Lepaute's husband and asked to examine the royal clocks. Lalande had been asked to prepare a report on clockmaking, and he returned regularly to the Lepaute apartment. He soon came to appreciate that Mme Lepaute had substantial mathematical talent and that her "astronomical tables were next to her household account books." He asked Nicole-Reine Lepaute to prepare a table for his report on the royal clocks, a request that started a thirty-year collaboration. "She endured my faults," wrote Lalande of his partner, "and helped to reduce them."[22]

Clairaut's computing plan was a step-by-step process that treated Jupiter and Saturn as if they were the hour and minute hands of a giant clock, ticking their way around the Sun. Each step of the calculation would advance the two planets a degree or two in their orbits. The procedure then required an adjustment of the orbits based upon the gravitational pull between the two planets and the sun. Saturn might be pulled a bit closer to the Sun, or Jupiter might be pushed forward in its orbit. On this celestial clock, the comet was a second hand that circled on a long, skinny ellipse in the opposite direction of the planets. The calculation of this orbit could be done after the calculations for Jupiter and Saturn, for though the two planets pulled on every step of the comet, the comet had no substantial effect on the planets. This part of the computation would advance the comet a small distance along the ellipse, compute the forces from the two giant planets, and adjust the orbit accordingly.[23] Lalande and Lepaute handled the first part of the computation, the three-body problem involving Saturn and Jupiter, while Clairaut took the orbit of the comet itself.

The three friends began their computations of Halley's comet in June 1757. They worked at a common table in the Palais Luxembourg using goose-quill pens and heavy linen paper.[24] Lalande and Lepaute worked at one side of the table and handed their results to Clairaut. Presumably, they wore the formal court dress of palace residents: coats and knee breeches for the men, a dress with underskirts for Lepaute, and powdered wigs for all. They would have begun their work late in the morning, as palace residents were rarely early risers. The computations would take them through the early afternoon and into the evening, when dimming lights and fatiguing hands discouraged further efforts. At times, they would compute while they ate their meals. When servants appeared with food, the three would simply push their papers and ink pots aside to make room for the dishes and continue with their work.[25]

In addition to stepping Halley's comet through its orbit, Clairaut took the responsibility of checking all of the calculations for errors. With three people contributing to the task, the project offered many opportunities for arithmetical mistakes. Identifying the source of such mistakes was difficult, and even a small error might be compounded in the calculations and ultimately render the final result meaningless. Critics of scientific research had pointed to such errors as a weakness of mathematical methods and suggested that numerical techniques were too fallible to capture the nature of the universe. Such critics included the English writer and social commentator Jonathan Swift (1667–1745), who attacked, in his satire *Gulliver's Travels*, both the general validity of mathematics and the specific accuracy of Edmund Halley's calculations.

Gulliver's Travels appeared about thirty-five years before Clairaut began his computations. At the time of its publication, Halley had just become the Astronomer Royal of England. One section of the book describes an astronomer-king, much like Halley, who governs a mythical land of scientists called Laputa. In Laputa, the residents decorate their clothes with astronomical symbols and keep one eye turned always to the heavens. This astronomer-king eats food cut into geometrical shapes and gets so absorbed in calculation that he requires a servant to arouse his attention. His subjects are absent-minded, selfish, and contemptuous of practical issues. They "are dexterous enough upon a piece of paper," the fictional Gulliver notes, "yet in the common actions and behaviour of life, I have not seen a more clumsy, awkward, and unhandy people." They are deeply afraid of a comet, easily recognizable as Halley's, which they believe will destroy the earth in its next passage.[26]

As Swift mocks the community of scientists, he directly attacks scientific calculation by equating the work of computation with the craft of a foolish and obstinate tailor. This tailor is asked to prepare a new suit of clothes for Gulliver and uses the full range of astronomical tools to create the pattern. Gulliver reports that the tailor "first took my altitude by a quadrant," a device slightly larger than Halley's sextant, "and then, with a rule and compasses, described the dimensions and outlines of my whole body." When Gulliver returns to accept the new clothes, he records that the tailor "brought my clothes very ill made, and quite out of shape, by happening to mistake a figure in the calculation." The tailor makes no effort to correct the mistake, and Gulliver notes that "such accidents [occur] very frequently and [are] little regarded."[27]

In the summer of 1757, French translations of *Gulliver's Travels* could be easily found at Paris booksellers, though it does not seem to be the sort of book that Alexis Clairaut would have enjoyed. He had a methodical nature, one that would have carefully sought to avoid producing a suit of "clothes very ill made" or a cometary orbit ill calculated. He devised procedures that would check every step of the computation. Some values were double-computed; others were checked with graphs.[28] The equations themselves were checked with two preliminary calculations. Starting with observations from the 1531 return, the three workers computed a full orbit and compared the final positions of the comet with the data from 1607. After adjusting the equations, Clairaut, Lalande, and Lepaute repeated the process. Beginning with the 1607 observations, they calculated a second orbit and compared the final values with Halley's observations of the 1682 return.[29] These computations tripled the amount of work, but the process gave Clairaut, Lalande, and Lepaute a measure of confidence in their results.

The daily routine of computation, the effort to advance the comet eight or ten degrees in its orbit, tested the stamina of the three. "It is difficult to comprehend the courage which was demanded by this enterprise," Joseph Lalande later wrote. He complained that "as a result of this hard work, I acquired an illness which, for the rest of my life, shall be with me." By contrast, Nicole-Reine Lepaute carried her share of the burden without complaint. "Her ardor is surprising," remarked Clairaut. Lalande confirmed this opinion, stating that "we would not, without her, have dared undertake this enormous work."[30]

The "enormous work" kept Clairaut, Lalande, and Lepaute busy at their computing table through late September. By then, astronomers were searching the night sky, expecting to catch the first gleam of the comet. Already several bodies had been falsely identified as the returning comet of 1682. Each passing day brought more pressure upon Clairaut, until he decided that he would have to increase the speed of the calculations. As the group began computing the last segment of the orbit, the part that came closest to the sun and planets, Clairaut simplified the equations by removing the terms that accounted for the gravity of Saturn. He believed that the influence of Saturn was small in that part of the orbit and that the revised calculation would still give a reasonable approximation of the comet's motion.[31]

On the fourteenth of November, Clairaut presented the results to the Académie des Sciences. He predicted that the comet would reach its perihelion, the point closest to the sun, in mid-April. His calculations showed that the date would be April 15, 1758, but he knew that it was unlikely that the calculation captured perfectly the movement of the comet.[32]

"You can see with what caution I make such an announcement," he said, "since so many small quantities, which must be neglected in methods of approximation, can change the time by a month."[33] His results suggested that the comet might round the sun as early as March 15 or as late as May 15.

For six weeks, the computation stood as a major accomplishment while astronomers watched and waited for the first observation of the comet. No sign of the comet was found in November and none in the first weeks of December, but on Christmas Day 1757, a German astronomer sighted the fuzzy glow of the comet nucleus. By the middle of January, enough observations had been made of the approaching body to show that the calculation had overestimated the date of the comet's arrival. The comet reached its perihelion on March 13, a few days outside of the interval that Clairaut had presented to the Académie des Sciences.[34] As the news of the discrepancy began to spread, astronomers started to assess the calculation and debate its meaning. The most public critic of the computation was French mathematician Jean Le Rond d'Alembert (1717–1783). D'Alembert was no social commentator in the manner of Jonathan Swift, but an established astronomer and an editor of the *Encyclopédie*, the French Enlightenment's grand catalog of knowledge. He argued that Clairaut's work was not based upon well-defined principles of scientific inquiry but was only a rough and unsubstantiated approximation.[35] He denounced what he called the "spirit of calculation" and claimed that the computation was more "laborious than deep."[36]

D'Alembert's criticisms sparked an extended argument among French scientists, an exchange that was fueled in no little part by the egos and ambitions of the two protagonists.[37] From the tenor of the debate, it seems clear that even Clairaut had more faith in Newton's theory of gravitation than in his own calculation of the comet's perihelion. Clairaut never suggested that Newton's theory needed to be adjusted to fit the prediction; instead, he returned to his figures and attempted to find some mistake that had caused the discrepancy. Under the best of circumstances, this is a dangerous endeavor, for there is a tremendous temptation to bend the equations in order to fit the data. With the perspective of modern astronomy, we know that Clairaut did not account for the influences of Uranus and Neptune, two large planets that were unknown in 1757. Without these planets, the equations could produce the correct date for the perihelion only if they misstated the influence of Jupiter and Saturn. Clairaut never took his adjustments that far, though he did produce a better prediction. A modern analysis of the result concluded that there were still many problems in the computation and that Clairaut benefited from the "fortuitous cancellation of opposing errors."[38] After two years of public rancor, Clairaut decided that he was tired of the debate, and he ended the controversy by begging d'Alembert, who clearly had no interest in large calculations, to "leave in peace those who did."[39]

In the most charitable light, Clairaut had improved Halley's work by a factor of ten. By the time Halley had completed his research, he had recognized that the comet had a variation in its return of approximately two years, about 600 days. Clairaut's calculation missed the actual date of return by 33 days. As he might have undershot the date of perihelion by an equal amount, the rough accuracy of his calculations was twice 33, or 66 days. Beyond the simple accuracy of his result, Clairaut's more important innovation was the division of mathematical labor, the recognition that a long computation could be split into pieces that could be done in parallel by different individuals. In spite of d'Alembert's criticism, the astronomers of the mid-eighteenth century recognized that Clairaut's division of labor was an important contribution to astronomical practice.[40] Clairaut never undertook another calculation of equal complexity, but his two assistants, Joseph Lalande and Nicole-Reine Lepaute, were involved with computation for the rest of their careers. In 1759, Lalande became the director of the *Connaissance des Temps*, an astronomical almanac published by the Académie des Sciences, and he appointed Nicole-Reine Lepaute as his assistant. The two of them prepared tables for the *Connaissance des Temps* that predicted the positions of the stars, the sun, the moon, and the planets. These tables were easier to calculate than the orbit of Halley's comet, as they were based on a long history of data and as they could be corrected from observations taken throughout the year. Likewise, the division of labor for these calculations was simpler than it had been for Halley's comet. Lalande prepared the computing plans and checked the results, while Lepaute calculated the values for the tables.

Lalande called Lepaute his "assistant without equal,"[41] but of course, she was anything but equal to him. Lalande was able to advance from his position and would eventually be appointed a professor of astronomy and director of the Paris Observatory. Lepaute had no such opportunities and would spend fifteen years as a computer for the *Connaissance des Temps*. However, even with its limitations, the appointment to the *Connaissance des Temps* had certain advantages for Lepaute. It gave her an official standing among French scientists, a rare accomplishment for a woman. During her fifteen-year career, she found that she could occasionally do some astronomical work outside of the *Connaissance des Temps*. In 1764, she published a map under her own name that predicted the extent and duration of an upcoming solar eclipse. She was apparently quite proud of that work, for she asked that it be included in a portrait that she gave to Lalande.[42]

"No scientific discovery is named after its original discoverer," wrote the historian Stephen Stigler.[43] Even if Clairaut, Lalande, and Lepaute were the first astronomers to divide the labor of scientific calculation, their names did not travel with their contribution. Others would rediscover the idea without knowing of Halley's comet or the computations that were done at a table in the Palais Luxembourg. Still, the work of those three scientists during the summer and fall of 1757 identified a pattern that touched three or four generations of human computers, a pattern that divided calculations into independent pieces, assembled the results from each piece into a final product, and checked that result for errors.

CHAPTER TWO

The Children of Adam Smith

> Even in the quieter professions, there is a toil and a labour of
> the mind, if not of the body. . . .
>
> Jane Austen, *Persuasion* (1818)

"THE SUPERIOR GENIUS and sagacity of Sir Isaac Newton," wrote the philosopher Adam Smith, "made the most happy, and, we may now say, the greatest and most admirable improvement that was ever made in philosophy." Smith turned his generous praise on Newton's calculus and stated that new discoveries would come from "more laborious and accurate calculations from these principles."[1] At least one scholar has found Smith's praise insincere and has suggested that the philosopher distrusted any science that rested "primarily upon mathematics, rather than easily visualized phenomena, common to the mind of all men."[2] Smith was more interested in things of earth than in things of heaven, the movement of goods and services rather than the cycles of planets and comets. At the time of the 1758 return, he was collecting the material that formed the basis for his book *An Inquiry into the Nature and Causes of the Wealth of Nations*. In this work, he sought a fixed principle that explained economic behavior as well as Newton's gravity explained celestial motion. He found this principle in the marketplace, "the propensity to truck, barter, and exchange one thing for another," though he was not certain that the rules of the market equaled those of celestial motion. "Whether this propensity be one of those original principles in human nature," he speculated, "or whether, as seems more probable, it be the necessary consequence of the faculties of reason and speech, it belongs not to our present subject to inquire. It is common to all men, and is to be found in no other race of animals."[3]

Smith claimed that the market encouraged people to specialize, to produce those goods that gained them the most profit. Butchers did not make shoes, nor did cobblers slaughter their own animals. This specialization was one part of a more general idea that Smith identified as the division of labor. "The greatest improvements in the productive powers of labour," he wrote, "seem to have been the effects of the division of labour." He claimed that there were three benefits to be gained from such division. First, it led to the "increase of dexterity in every particular workman." Laborers could focus on a small number of tasks and thus gain skill and efficiency. Second, divided labor made workers more productive by reducing "the time which is commonly lost in passing from one species of work to another." Finally, the division of labor encouraged workers to improve their tools, to invent "a great number of machines which facilitate and abridge labour, and enable one man to do the work of many."[4]

Later generations would explore Smith's market principles and divided labor with the calculus of Isaac Newton. Smith was content to simply describe how his laws touched different aspects of economic behavior. He claimed that his ideas applied equally to manufacture and to "natural philosophy," the term he used to describe scientific research. The "subdivision of employment in philosophy," he wrote, "improves dexterity, and saves time." Smith conceded that philosophers might not be motivated by the traditional economic forces of profit and loss, yet he argued that they desired to extract the greatest results from their limited resources, so that "more work is done upon the whole, and the quantity of science is considerably increased by it."[5] The work of Alexis Clairaut, Nicole-Reine Lepaute, and Joseph Lalande was an early example of this observation. Without the division of labor, Clairaut could not have completed the calculations before the comet's reappearance and could not have devoted so much effort to checking the results.

At the time that Adam Smith was writing *The Wealth of Nations*, the British Admiralty, the executive office of the English navy, was organizing a new computing office and taking a further step in the division of labor. The Admiralty created this office in order to produce a nautical almanac, a volume of tables that gave the position of the sun and the moon, the planets and the stars. The founder of this office was the new Astronomer Royal, Nevil Maskelyne (1732–1811). Maskelyne was the successor, twice removed, of Edmund Halley, the fifth scientist to oversee the Royal Observatory at Greenwich. This appointment was not purely a scientific honor, as it carried a practical responsibility for the country's fleet of naval and merchant ships. Anyone who accepted the king's warrant for astronomy was required to develop methods of celestial navigation, particularly techniques for the "finding out of the longitude of places."[6] As if to emphasize this charge, the Greenwich Observatory sat on the high bank of the River Thames in the midst of a royal estate. From his desk, Maskelyne could view the ocean traffic as it moved between the London docks and the open waters of the North Sea.

The *Nautical Almanac* was the outgrowth of a competition between two methods for finding longitude, one computational and the other mechanical. The two methods were nearly identical and differed only on a single point: the means of determining the time at Greenwich. The time at Greenwich was important because it allowed a navigator to compare two observations of a single star. The first measurement would be taken by the navigator in the dim moments before dawn or in the dusky hour of twilight, when the thin line of the horizon was visible from the ship and at least a few bright stars could be seen in the violet sky. After determining the position of a star, the navigator would turn to a nautical almanac and find the position of the same star as it would be viewed at the identical moment from the observatory at Greenwich. The difference between these two positions, properly adjusted with a dozen steps of calculation, was the longitude of the ship.

In the 1760s, there were two possible ways of determining the time at Greenwich, both with advantages and drawbacks. The simpler way used a mechanical clock set to the time at Greenwich. This solution was problematic, as no common clock could guarantee sufficient precision under shipboard conditions. The roll of the waves disrupted pendulums. Variations in heat and humidity caused springs to expand and contract. A good clock might lose or gain four minutes a day, enough time to allow the earth to spin a full degree in its rotation. In the middle latitudes, a four-minute error could translate into a deviation of fifty miles. A navigator relying on such a clock could easily calculate a longitude that placed his ship at a safe distance from the shore when, in fact, the vessel was about to strike coastal rocks. In the early 1760s, English inventors strove to develop a precision clock that could record the time accurately under shipboard conditions. Of the timekeeping devices presented to the British Admiralty, one created by John Harrison (1693–1776) was the most promising.[7]

The second approach to determining the time at Greenwich used the moon as a timekeeper. This technique was known as the lunar distance method. The moon moves twelve degrees across the sky each night, passing neighboring stars as if they were marks on a watch dial. That motion is enough to allow a skilled navigator to compute the time at Greenwich with sufficient accuracy, though the calculations are admittedly lengthy and require a special table that predicts the moon's position. The lunar distance method had been developed in the early eighteenth century but had been dismissed by most navigators because of the difficulties in predicting the position of the moon. Like the calculation of the perihelion for Halley's comet, the prediction of lunar position required the solution of a three-body problem. In this case, the three-body system involved the moon, the earth, and the sun. An acceptable solution to this particular system appeared only in the late 1750s, when the German astronomer Tobias Mayer (1723–1762) published a detailed table of lunar positions.[8] Astronomers praised Mayer's work as "the most admirable masterpiece in theoretical astronomy," and in 1761, the *Connaissance des Temps* published an article that showed how Mayer's tables could be used in navigation.[9]

In popular accounts of the competition between Harrison's clock and the lunar distance method, Nevil Maskelyne has been portrayed as a villain, a powerful scientist who undercut a valid technology for personal reasons. His alleged villainy came when he was asked by the British Admiralty to compare Harrison's clock with the lunar distance method. Some writers have charged that Maskelyne was a prejudiced evaluator of the two techniques because he had publicly stated his admiration of Mayer's lunar tables before the trial began and was known to favor the techniques of astronomical calculation.[10] His conclusions from a test voyage certainly confirmed his opinion that the lunar distance method was a practicable means of determining longitude, and he dismissed Harrison's clock.[11] From a modern perspective, precision clocks, now called chronometers, clearly provide the easiest way of determining the time at Greenwich, but such a conclusion may not have been so clear in the 1760s. The historian Mary Croarken has noted that Harrison's clock was an immature technology and was "much too expensive to be taken to sea by the majority of [English] navigators."[12]

On the trial voyage from England to Barbados and back, Maskelyne had required four hours to make a single computation of longitude with the lunar distance method.[13] "It is rather to be wished," he wrote, "that such parts of the computations as conveniently can, were made previously at land by capable persons."[14] Those parts of the computations that could be done in advance took the form of a set of tables that gave the distance from the moon to easily recognizable stars in a simplified form. These tables needed to be prepared and published annually, as the position of the moon varied from year to year. With such tables, a navigator could compute the time at Greenwich with a handful of operations and determine a ship's longitude with only thirty minutes of work.[15] Maskelyne wanted to include these tables as part of a general nautical almanac, as such values could be used for purposes beyond the problem of finding the time at Greenwich. They could even be used to check the settings of a chronometer in the middle of the ocean or guide a ship back to land should the chronometer fail.

In February 1765, the British Admiralty approved Maskelyne's plan for an almanac, gave him a staff of five computers, and told him to begin work on celestial tables for 1767. To all involved with the project, including Maskelyne and the members of the Admiralty, this assignment must have seemed quite reasonable. Under normal circumstances, the almanac staff would have to produce a new publication every year. Maskelyne had fully twenty-two months before the start of 1767, almost twice the time he should have needed, but the process of recruiting and training his computing staff proved to be more challenging than he had anticipated. He organized the computing staff as a cottage industry, a form of production that was still common in England, even though it was starting to be eclipsed by the factory. In cottage production, the workers labored in their own homes. Their materials, instructions, and often their tools were provided by the company or individual for whom they worked. In the clothing industry, cottage workers might receive carded wool and spin it into yarn. For the *Nautical Almanac* computers, Maskelyne provided paper, ink, and instructions that were called "computing plans." Maskelyne wrote these plans on one side of a heavy sheet of folded stationery. The instructions, scrawled in a slightly disheveled hand, summarized each step of the calculation. Occasionally, he would illustrate the computations with a hasty sketch of an astronomical triangle. On the other side of the paper he drew a blank table, ready for the computer to complete.[16]

The computers produced tables that tracked the motion of a planet or the sun, tables that were called ephemerides in the plural (or an ephemeris in the singular). Most of these ephemerides were double-computed, prepared by two independent computers working from the same plan. Each computer would send Maskelyne a version of the ephemeris. Maskelyne would forward the two ephemerides to a third computer, who had the title of comparator. The comparator would search the two ephemerides for discrepancies and correct the mistakes. The only tables that were not double-computed were those of lunar motion. These tables were divided in half. One computer would calculate the moon's position at noon. The other would compute the position at midnight. The comparator would merge the two tables and make sure that the two sets of calculations were consistent.[17]

Initially, Maskelyne assigned two computers to prepare the 1767 volume of the almanac. A third acted as the comparator, and the remaining two were put to work on the 1768 volume. From what we know of his staff, all of them came from the second tier of astronomical talent. Most commonly, they had demonstrated some skill at astronomy but lacked the resources or the connections to acquire one of the prestigious scientific appointments at Cambridge or Oxford. The first computer of the 1767 volume, William Wales (1734–1798), came from a poor family in the north of England. The second computer, Israel Lyons (1739–1775), was a Jew and was unwilling to make the profession of belief that might gain him a place at the church-centered universities. The comparator, Richard Dunthorne (1711–1775), had shown the greatest ability to advance himself as a scientist. He, too, was born to a lower-class family but had demonstrated his mathematical prowess by analyzing the motion of

4a. Computing sheet of Nevil Maskelyne

the moon. This work had given him a minor reputation as an astronomer and had connected him to a wealthy patron who provided Dunthorne with a regular income.[18]

Whenever possible, Maskelyne attempted to reduce the amount of calculation by borrowing tables from other sources, such as the *Connaissance des Temps*. He also simplified some of the calculations by employing the method of interpolation. Interpolation expands a table by esti-

4b. Computing instructions prepared by Nevil Maskelyne

mating intermediate values rather than by calculating these numbers from the original equations. It is a mathematical means of connecting the dots. The computers would link the moon's position at noon to its location at midnight using a polynomial, a mathematical expression that is the sum of terms such as x, x^2, and x^3. With this polynomial, the computers estimated the moon's location at three-hour intervals without having to calculate new values from Mayer's tables.[19]

Even with Maskelyne's attempts to minimize the workload, the 1767 almanac fell behind schedule. Wales and Lyons needed time to learn the new computing procedures and develop the skill that would get the work done most efficiently. By the spring of 1766, Maskelyne recognized that his two computers would not be able to finish their computations in time for publication the following fall. Both were engaged in at least one other job and could not devote extra time to completing the calculations. Lyons worked as a surveyor and did other work for the British Admiralty, while Wales was involved in a number of astronomical projects. Fortunately, Maskelyne had a reserve pool of labor, the two computers who were preparing the second issue of the almanac. He told this pair to put aside their calculations and assist Wales, Lyons, and Dunthorne with the first issue. Together, the five computers finished the tables in late fall. The 1767 issue of the almanac appeared only six days after the start of the year.[20]

As the computers moved to the second and third almanacs, they were able to claim the first benefit that Smith had ascribed to divided labor, the increase in dexterity and speed. After three years of calculations, the almanac staff had completed all of the almanacs through the 1773 volume and were beginning the calculations for 1774. By 1780, they were creating tables six years in advance, and Maskelyne was able to reduce the number of computers from five to four.[21]

The division of labor for Maskelyne's first *Nautical Almanac* offered no innovation beyond the methods commonly applied in English commerce. The only difference between the computers and the carders and weavers of the cloth industry was the fact that the computers' product, the ephemerides, could be folded into a neat packet and sent through the mails. A more radical approach to the division of labor was found at the French Bureau du Cadastre. The bureau, a civil mapping agency, was a product of the French Revolution and hence embraced notions about labor and organization that were far more radical than those employed by Maskelyne at the British *Nautical Almanac*. The bureau prepared maps for governance, taxation, and land transactions. Initially, it had no computing division beyond a few assistant surveyors. It assembled a staff of almost one hundred computers when it became involved with the standardization of weights and measures that produced the metric system.

The metric system grew out of an attempt by the National Assembly to gain control of the French economy. In March of 1790, barely eight months after the storming of the Bastille prison, the National Assembly debated a proposal to discard "the incalculable variety in our weights and measures and their bizarre names" and adopt a unified measurement system based upon scientific principles.[22] At that time, each region was free to establish its own set of measures. Local officials easily manipulated these measures to their own advantage in a number of ways. Commonly, they could keep a large measure to collect taxes of grain and produce but reserve smaller measures for the payment of their own debts.[23]

The Académie des Sciences agreed to create the new system of weights and measures. They quickly stipulated several basic principles for the new system. They agreed that the standards of weight and length should be beyond the control of any political organization and that the units for area, volume, and even weight should be related to the unit for length. In one of their final discussions, the Académie stated that the new measures should form a decimal system. All units should be related through multiples of ten. For example, the meter, the standard measure of length, would be divided by ten to produce the decimeter, which in turn could be divided to produce the centimeter, the millimeter, and the micrometer. The liter, the gram, and the dyne, the standard units of volume, mass, and force, could also be divided or expanded in decimal multiples. The members of the Académie argued that this same principle should govern all standard units, including those that measured angles. Under their proposal, a right angle would no longer have 90 degrees. Instead, it would be split into one hundred new units called grades.[24]

The proposal for the decimal measurement of angles produced a major computational problem that led to the creation of a computing office at the Bureau du Cadastre. The principal users of angle measure, navigators and surveyors, did their work with sines, cosines, and other trigonometric functions. Without trigonometric tables prepared for the decimal grades, the new standard for angle measure would be unused. No surveyor or navigator, even one ardently committed to the revolutionary cause, would measure angles in grades if he had to convert his numbers into degrees in order to use a sine table. Openly or surreptitiously, they would measure their angles in degrees and use the trigonometric tables of the ancien régime to calculate their position on the globe or the area of a piece of land.

The director of the Bureau du Cadastre was Gaspard Clair François Marie Riche de Prony (1755–1839), a civil engineer with the country's elite Corps des Ponts et Chaussées, the Corps of Bridges and Highways. De Prony came from a family of "modest but ancient title" in the province of Beaujolais. His mathematical skill had brought him to the attention of the corps and gained him entrance to the corps's preparatory school in Paris. He graduated at the top of his class from the school and proudly accepted the uniform of a corps officer, which came with royal fleur-de-lys buttons.[25]

Following his graduation, de Prony had hoped to live in the capital and pursue scientific research, but the conventional assignment for young engineers was a term of service in the field. He fulfilled his duty and found a way back to Paris by taking an appointment as an inspector, an officer who critiqued the mathematical analyses of other engineers and verified calculations. Though the assignment involved no independent work, he found it a pleasing activity, reporting that "happy circumstances then put me in contact, with the most distinguished savants of the capital." It also brought him into contact with foreign savants, including Nevil Maskelyne. De Prony met Maskelyne when the corps decided that it needed to remeasure the difference in longitude between Paris and London. Beyond the intellectual challenge of this work, an accurate measurement of the difference would allow French surveyors to use the British *Nautical Almanac* in their work and enable the Académie des Sciences computers to compare their tables with the output of Maskelyne's computers. As part of the effort to measure the difference between the two capitals, de Prony traveled to Greenwich, met Maskelyne, and inspected the work of the almanac computers.[26]

De Prony had become leader of the Bureau du Cadastre because of a cautious and uneasy relationship between the corps engineers and revolutionaries. The revolutionaries were wary of the corps, their uniforms, and their fleur-de-lys buttons, yet they needed the services of corps officers and could not easily dismember the group. In an attempt to weaken the authority of the corps, the revolutionary government tried to disperse the officers, most of whom resided in Paris, and have them take positions in the countryside.[27] In early 1791, the government ordered de Prony to move to southwest France but found that he was unwilling to take the assignment. "I have received your letter, informing me that I am appointed engineer-in-chief for the departement of the Pyrenees," he wrote to his superintendent. "In these circumstances, I beg you to permit me to remain in Paris." He argued that he was working on two reference books that would be difficult to complete without the resources found in the capital. Claiming that the research had "already cost me many sleepless nights," he stated that he was willing to forgo a formal assignment so long as he could stay in Paris and continue with his research. After a brief confrontation, a senior officer intervened and gave de Prony the job at the Bureau du Cadastre.[28]

In ordinary times, de Prony would have been responsible for organizing surveying teams and dispersing them across the country, but 1791 was no ordinary time. The dangers of civil unrest forced de Prony to keep the survey teams in Paris and occupy them with "tasks of public usefulness."[29] Under these circumstances, he sought or accepted the task of preparing the trigonometric tables for the decimal grade system of angle measure. De Prony wrote that the assignment "over-burdened" him.[30] The Académie des Sciences wanted the work done quickly and required that the final tables leave "nothing to be desired in accuracy."[31] As he contemplated the work before him, he devised a plan based upon the first chapter of *The Wealth of Nations*. A French commentator has suggested that de Prony approached the book haphazardly, the way one might flip open a Bible to search for an inspiring verse. "He opened it at random and his eye fell on the first chapter, which was called *A Treatise on the Division of Labor*, and which cited the example of making pins."[32] It is a dramatic rendition of the story, but it overlooks the long-standing influence of *The Wealth of Nations* upon the engineers in the Corps des Ponts et Chaussées. The book had circulated among the officers during the early 1780s and was probably an old and familiar friend to de Prony.[33]

De Prony recalled that the pin example provided the insight he needed. "Suddenly," he wrote, "I conceived how I might apply the same method to the work which had burdened me." Referring to one of the tables he needed to compute, he recorded that his bureau "could manufacture logarithms as easily as one manufactures pins."[34] His recollection of the event ignores a great deal of hard work and gives his accomplishment a patina of false confidence. The structure of de Prony's computing office cannot be easily seen in Smith's example. His computing staff had two distinct classes of workers. The larger of these was a staff of nearly ninety computers. These workers were quite different from Smith's pin makers or even from the computers at the British *Nautical Almanac* and the *Connaissance des Temps*. Many of de Prony's computers were former servants or wig dressers, who had lost their jobs when the Revolution rendered the elegant styles of Louis XVI unfashionable or even treasonous.[35] They were not trained in mathematics and held no special interest in science. De Prony reported that most of them "had no knowledge of arithmetic beyond the two first rules [of addition and subtraction]."[36] They were little different from manual workers and could not discern whether they were computing trigonometric functions, logarithms, or the orbit of Halley's comet. One labor historian has described them as intellectual machines, "grasping and releasing a single piece of 'data' over and over again."[37]

The second class of workers prepared instructions for the computation and oversaw the actual calculations. De Prony had no special title for this group of workers, but subsequent computing organizations came to use the term "planning committee" or merely "planners," as they were the ones who actually planned the calculations. There were eight planners in de Prony's organization. Most of them were experienced computers who had worked for either the Bureau du Cadastre or the Paris Observatory. A few had made interesting contributions to mathematical theory, but the majority had dealt only with the problems of practical mathematics.[38] They took the basic equations for the trigonometric functions and reduced them to the fundamental operations of addition and subtraction.

From this reduction, they prepared worksheets for the computers. Unlike Nevil Maskelyne's worksheets, which gave general equations to the computers, these sheets identified every operation of the calculation and left nothing for the workers to interpret. Each step of the calculation was followed by a blank space for the computers to fill with a number. Each table required hundreds of these sheets, all identical except for a single unique starting value at the top of the page.

Once the computers had completed their sheets, they returned their results to the planners. The planners assembled the tables and checked the final values. The task of checking the results was a substantial burden in itself. The group did not double-compute, as that would have obviously doubled the workload. Instead the planners checked the final values by taking differences between adjacent values in order to identify miscalculated numbers. This procedure, known as "differencing," was an important innovation for human computers. As one observer noted, differencing removed the "necessity of repeating, or even of examining, the whole of the work done by the [computing] section."[39]

The entire operation was overseen by a handful of accomplished scientists, who "had little or nothing to do with the actual numerical work." This group included some of France's most accomplished mathematicians, such as Adrien-Marie Legendre (1752–1833) and Lazare-Nicolas-Marguerite Carnot (1753–1823).[40] These scientists researched the appropriate formulas for the calculations and identified potential problems. Each formula was an approximation, as no trigonometric function can be written as an exact combination of additions and subtractions. The mathematicians analyzed the quality of the approximations and verified that all the formulas produced values adequately close to the true values of the trigonometric functions.

Joseph Lalande visited the Bureau du Cadastre in 1794, after the computers had been working for nearly two years. He probably saw only the office of de Prony and his planners. No record has been found of a centralized computing floor for the former hairdressers.[41] Given the size of the computing staff and the tradition of cottage work, it is likely that the computers did their calculations at home. Lalande, who had become the éminence grise of French astronomy, reported that de Prony's computers were "producing seven hundred results each day."[42] At that pace, they could have duplicated Clairaut's calculations for Halley's comet in about a week. Perhaps more to the point, if Lalande, Lepaute, and Clairaut had been asked to prepare de Prony's decimal trigonometry tables, they would have spent a century and a half sitting at their table in the Palais Luxembourg.

Lalande's visit seems to have marked a high point of the cadastral computers. In 1795, the revolutionary government instructed de Prony to prepare the tables for publication "at the expense of the nation,"[43] but by the time de Prony completed the work, the nation was not all that interested in paying for the new trigonometry. Decimal angle measure was not included in the law implementing the new metric system, which was passed by the National Assembly on August 1 of that year.[44] "In the face of popular indifference and hostility," wrote historian Ken Alder, "the government began to lower its sights."[45] Officials had difficulty enforcing the use of metric measures in Paris, and there is little evidence that it penetrated into the countryside.

De Prony kept his project in operation through 1800 or 1801, even though it appears that most of the work had been completed by 1796. Even before the typesetters began work on the tables, de Prony's publisher began to promote the new trigonometric functions with the story of their creation. According to the advertisement, de Prony had created a new "process of manufacturing" that was "strange in the history of science, where there is no other example." The publisher argued that the tables would never have been created "if M. de Prony had not had the fortunate idea of applying the powerful method of division of labour," but such words could not bring the tables into print.[46] The publisher was bankrupted during a national fiscal crisis. The French government, then led by Napoleon, had no interest in completing the work. De Prony retained the nineteen-volume manuscript and made occasional, though fruitless, efforts to publish it.[47]

The story of Gaspard de Prony and his computers at the Bureau du Cadastre would have been little more than an odd footnote to the history of economics were it not for the attention of Charles Babbage. Babbage is generally remembered as a mathematician and as a designer of early computing machinery, but he was a broad and eclectic scholar whose interests ranged from mathematics and astronomy to economics and railroad construction. To a certain degree, he fit the stereotype of the Victorian gentleman scientist. He had been educated at Cambridge in the canonical works of Newton and Halley; he possessed a comfortable, though not extravagant, income that allowed him to pursue his own interests; he lived in London and mingled freely with the country's political and intellectual elite. In some ways, he was more interested in the organizations and institutions of science than he was in the science itself.[48]

Babbage arrived in London in 1814, finished with his Cambridge studies and newly married. He applied to be a computer at the Royal Greenwich Observatory, but friends encouraged him to direct his talents elsewhere.[49] Looking for ways to establish a reputation as a scientist, Babbage decided to give a series of public lectures on astronomy. His mathematical training was deeply connected to astronomical problems,

5. Charles Babbage

but that training did not make him an expert on stars and comets and planets. In preparing the lectures, he relied on the advice of more accomplished colleagues, notably a college friend, John Herschel (1792–1871), and Herschel's aunt, Caroline Herschel (1750–1848). The two were members of England's premier astronomical family. William Herschel (1738–1822), the father of John and the brother of Caroline, had discovered the planet Uranus in 1781. Caroline Herschel had served as her brother's assistant before she became recognized for discovering comets and cataloguing nebulae.[50]

Babbage's lectures brought him to the attention of a group of businessmen and amateur astronomers who were organizing a society for "the encouragement and promotion of Astronomy." This group, originally called the Astronomical Society of London, invited Babbage and John Herschel to their first meeting. This meeting was held in January 1820 at a tavern situated among the business houses of central London.

Most of the founders had some connection to ocean trade and the problems of celestial navigation. They were merchants, currency traders, stockbrokers, and business teachers.[51] Though many of them were comfortable behind the lens of a telescope or computing the orbit of a planet, they described astronomy as if it were another commercial endeavor. The language of Adam Smith and divided labor permeated their words and sentences. They stated that the society would coordinate the "labours of insulated and independent individuals" and that they were "ready and desirous to divide at once the labour and the glory" of celestial observation. Their descriptions of the society suggest that they had some first-hand experience with the problems of management, for they wrote of the need "to preserve a perfect unity of design" while simultaneously preventing the loss of effort.[52]

In 1821, Babbage and Herschel agreed to undertake one of the first projects sponsored by the society, a set of mathematical tables that would augment the material in the British *Nautical Almanac*. The two college friends prepared a computing plan and hired a pair of computers to produce two independent versions of the table. Once the computers had finished their work, Babbage and Herschel compared the results. Sitting together, one of them read aloud the values from his version of the table. The other held the second table and confirmed each number. "Finding many discordancies," Babbage later wrote, "I expressed to my friend the wish that we could calculate by steam."[53] He would point to this work as the start of his study of computing machines, but he would repeat at least two inconsistent versions of this story. A second narrative placed his moment of insight during his student days in Cambridge. Both versions suggested that the idea to design a computing machine grew out of a desire to improve the accuracy of calculation. As Babbage would acknowledge, he was neither a trained engineer nor a skilled machinist.[54] His design for a computing engine would be based upon the division of labor, though not upon Maskelyne's ideas but upon those of de Prony.[55]

Babbage knew of de Prony and the cadastral computers when he and Herschel prepared their tables in 1821. He may have seen the manuscript tables when he visited Paris with his wife in 1819, or he may have learned of the computing effort when de Prony asked a wealthy English physician to help publish the tables.[56] Babbage was little interested in the decimal trigonometry tables but clearly understood the benefits of de Prony's organizational plan. He wrote that de Prony's experience showed that the division of labor was not restricted to physical work but could be applied to "some of the sublimest investigations of the human mind," including the work of calculation. After his own attempts at calculation, Babbage turned to de Prony's analysis and took the division of labor to its next logical step, the invention of a machine to "facilitate and abridge" the work.[57]

When he began work on his calculating machine, Babbage was following a path that was already well marked. Inventors in both England and the United States had built machines based upon Adam Smith's example of the divided labor in pin manufacture. Their machines followed each step that Smith had identified in *The Wealth of Nations*. They cut a roll of wire into fixed lengths, sharpened one end of each wire segment, affixed a head to the other, and placed the finished pin in a paper holder. Babbage took the opportunity to study one of these pin-making machines and reported that "it is highly ingenious in point of contrivance," especially interesting "in respect to its economical principles."[58]

Babbage designed a machine that might be considered more flexible than the pin-making machines. Rather than analyze the equations that had been used to create a specific table, such as the decimal trigonometry tables, he considered a single computational technique that could be applied to many kinds of calculation. The technique that he chose was a process of mathematical interpolation known as the finite difference method. The finite difference method is one way of computing intermediate values of a table, such as the intermediate positions of the moon that Nevil Maskelyne's computers prepared for the British *Nautical Almanac*. It is especially amenable to the division of labor because it reduces the entire process into the simple operation of addition. A simple application of this method can compute a list of the squared integers (4, 9, 16, etc.) without performing a single multiplication. First, one computes a list of the odd integers: 1, 3, 5, 7, 9, etc. This can be done by starting with the number 1 and successively adding 2: $1 + 2 = 3$, $3 + 2 = 5$, $5 + 2 = 7$. Once this has been completed, one can sum the list of odd integers to get the list of squares: $1 + 3 = 4$, $4 + 5 = 9$, $9 + 7 = 16$, and so on.

Though Babbage's machine would be far more complicated than the pin-making device, it was simpler in one respect. The pin machine needed to perform four different fundamental operations. Babbage's machine would only need to do one: addition. Babbage started with a geared adding mechanism originally developed by Blaise Pascal (1623–1662) in 1642,[59] improved the design, and cascaded the devices so that the results of one addition would be fed to the next. To create a list of squared integers, one mechanism would repeatedly sum the number 2 to create new odd numbers. The next mechanism would sum the odd numbers to create the squares. By the spring of 1822, Babbage had completed a demonstration model of his machine, which he named the "Difference Engine."

The London of 1822 was wholly unprepared for Babbage's machine. It was a world of gaslights and horse-drawn carriages, of servants and walking sticks. Most residents had not yet seen a steam locomotive, as the city's first railroads were still under construction.[60] Though the idea of the adding machine was one hundred and eighty years old, there was

6. Difference engine constructed from the original plans of Charles Babbage

none to be bought or sold. The first commercial machine, which would be produced in France, existed only as a crude prototype.[61] Babbage anticipated that his machine might be met with disbelief or even opposition. He cautiously approached the Royal Society, recognizing that the organization might be able to help him promote his machine. His letter to the society president made conservative claims and acknowledged that Royal Society members might not believe it possible to create a machine that could handle such complex calculations without supervision. He tried to disarm potential criticism by invoking Jonathan Swift: "I am aware that the statements contained in this Letter may perhaps be viewed as something more than utopian, and that the philosophers of Laputa may be called up to dispute my claim to originality."[62]

The Laputian philosopher that Babbage wished to avoid was not the computing tailor but an inventor who lived on the flying island. This inventor claimed to have a computing machine whereby "the most ignorant person, at a reasonable charge, and with a little bodily labour, might write books in philosophy, poetry, politics, laws, mathematics, and theology, without the least assistance from genius or study."[63] The machine was a silly device, a box of shafts and gears that spun through every possible combination of words, but it was a symbol of Jonathan Swift's mockery of the Royal Society. Babbage, though he shared some of Swift's reservations about the society, desired to avoid any comparison to the mythical device. He carefully described his invention in the context of de Prony's computing organization. He explained how the computers prepared their tables and then analyzed "what portion of this labour might be dispensed with." By his count, de Prony had employed ninety-six individuals to produce seven hundred computations a day. Babbage claimed that his machine could replace all of the human computers and most of the mathematicians. Of the original staff, all that would remain would be the ten planners and one mathematician, "or at the utmost two," to direct the work.[64]

Though the Royal Society was in no hurry to pass judgment upon Babbage's proposal, the Astronomical Society rushed to give the idea uncritical praise. Based upon what they observed of the crude prototype, they offered Babbage a gold medal for his contribution to astronomy.[65] "The labour of computing equations with the pen would be immense, and liable to innumerable errors," wrote one member of the Astronomical Society, "but with the assistance of [Babbage's] machine, they are all deduced with equal facility and safety." He made a special effort to emphasize the general use of the machine, arguing that "astronomical tables of every kind are reducible to the same general mode of computation" and that the machine could even be applied to commercial tasks, such as the preparation of "Interest, Annuities, &c, &c, all of which are reducible to the same general principles."[66]

The Royal Society eventually endorsed the Difference Engine, and the English government offered to finance the construction of the machine. Eager to devote his entire energies to the project, Babbage resigned his office with the Astronomical Society in 1823. The work progressed more slowly than he would have wished, as both he and his mechanic needed to refine his design and improve their metalworking skills. The project was interrupted by the death of Babbage's wife, Georgiana, an event that disrupted his life in a way that nothing else could have. The loss "left Babbage a changed man," observed biographer Anthony Hyman. There was "an 'inner emptiness' to the man, who had only recently seen so much potential in his life." In "his public controversies, there was a new note of bitterness of which there was no trace while Georgiana was alive." Babbage left England for a Continental tour in 1827, leaving the engine unfinished.[67]

In later years, Babbage recognized that he had been naive in his attempt to build the Difference Engine. Without referring to himself directly, he confessed that a novice engineer could be "dazzled with the beauty of some, perhaps, really original contrivance," and would rush into its construction "with as little suspicion that previous instruction, that thought and painful labour, are necessary to its successful exercise."[68] Babbage worked on the Difference Engine for ten years. During this time, he was engaged in other projects, such as computing life insurance tables, forming two new scientific societies, and writing a book about manufacturing. Even accounting for these other projects, the work on the Difference Engine took longer than Babbage had anticipated, and it encountered unforeseen problems. The English government eventually grew impatient with Babbage's progress. Concluding that they would see no return on their investment, they withdrew their financial support in 1834, forcing Babbage to terminate the project.

Undeterred by his failure to complete the Difference Engine, Babbage moved to design a second, more ambitious device. He never even attempted to construct this machine, which he called the Analytical Engine. Modern writers have generally viewed the machine as an important intellectual step toward the stored-program electronic computer. One historian has gone so far as to claim that it was "a general purpose computer, very nearly in the modern sense."[69] The drawings of this machine show how Babbage anticipated the features of a modern computer, though his design used gears and levers rather than chips and circuit boards. The Analytical Engine had a means of storing numbers, a central processor, and an elementary programming mechanism. Unlike the Difference Engine, this machine was not restricted to a single mathematical method, such as the method of finite differences. The programming mechanism, which read instructions from a string of punched cards, controlled the order of operations. One observer, the daughter of the poet Lord Byron, Ada Lovelace (1815–1852), called the Analytical Engine the "material and mechanical representative of analysis," a triumph of the division of mathematical labor. Lovelace herself illustrated the nature of the machine by writing a sample program for it.[70]

Babbage would spend almost fifteen years designing the Analytical Engine. He left nearly three hundred detailed engineering drawings of his proposed machine.[71] As he worked over these drawings, he recognized that the Europe of the early nineteenth century might not be able to support large computing organizations or computing machines based upon the division of labor. The "most perfect system of the division of labour is to be observed," he wrote, "only in countries which have attained a high degree of civilization, and in articles in which there is a great competition amongst the producers."[72] As the early nineteenth century saw little competition for scientific computation, it offered little opportunity for the sophisticated division of labor espoused by Babbage and de Prony.

CHAPTER THREE

The Celestial Factory: Halley's Comet 1835

Saw the heavens fill with commerce, argosies of magic sails,
Pilots of the purple twilight, dropping down with costly bales.
Alfred, Lord Tennyson, *Locksley Hall* (1842)

IN HIS NOVEL *Hard Times*, Charles Dickens described an astronomical observatory "made without any windows" and an astronomer who "should arrange the starry universe solely by pen, ink, and paper." He used this description, which sounded more like the computing room of the *Nautical Almanac* than the staff of an observatory, as a metaphor for a factory. In this factory, the director "had no need to cast an eye upon the teeming myriads of human beings around him," just as the director of almanac computations had no need to watch the stars each night, "but could settle all their destinies on a slate, and wipe out all their tears with one dirty little bit of sponge." When Dickens used this metaphor, both the Royal Observatory at Greenwich and the British Nautical Almanac Office had adopted the basic elements of factory production, elements that included a central facility and a standard schedule of operations or, as Dickens described them, "a stern room, with a deadly statistical clock in it, which measured every second with a beat like a rap upon a coffin-lid."[1]

Both the almanac and the observatory consciously accepted such methods around the time of the 1835 return of Halley's comet. Neither organization had seen much innovation since the days of Nevil Maskelyne. Through the middle part of his career, Maskelyne had been an active leader of both organizations, acquiring new equipment for the observatory and developing new methods for the almanac. At some point in the 1780s or 1790s, he had settled into a comfortable routine and had watched innovations occur elsewhere. The major astronomical discovery of the late eighteenth century, the planet Uranus, had been accomplished by an independent observer, William Herschel, not by the observatory staff. The radical division of labor came from Gaspard de Prony. A second periodic comet, the first to be discovered after Halley's, was identified by at least four individuals, none of whom was associated with Greenwich. This comet was ultimately named for a German astronomer, Johann Franz Encke (1761–1865), who calculated the object's orbit.[2]

Following the death of Nevil Maskelyne in 1811, the British Admiralty made only a few improvements to the almanac and the observatory. The new Astronomer Royal apparently spent little time at the observatory and acquired a reputation for hiring unimaginative assistants, "indefatigable, hard-working, and, above all, obedient drudges." A critic of the observatory argued that "Men who had the spirit of 'drudges,' to whom observation was a mere 'mechanical act,' and calculation a 'dull process,' were not likely to maintain the honour of the Observatory."[3] Such assistants had not sustained the honor of the *Nautical Almanac*. In 1818 the Admiralty removed the publication from the observatory and appointed an independent supervisor.[4]

Of the two organizations, the Nautical Almanac Office was the first to undergo a serious reform and adopt factory methods. Its workers were more familiar with the demands of production, as they had labored under fixed deadlines since the founding of the almanac in 1767. The pressure to reform the almanac came not from the almanac staff or even from the Admiralty but from the Royal Astronomical Society, the group that had started its life in a London tavern as the simple Astronomical Society. "The most prominent subject of public interest," reported the society president, "was the proposing of an amended form of the Nautical Almanac."[5] He argued that the almanac devoted too many pages to the lunar distance method of navigation and not enough to tables that would assist contemporary navigators and astronomers. The society claimed, with no contradiction from the British Admiralty, that most ship officers had abandoned Maskelyne's technique for calculating the time. In its place, chronometers, the high-precision clocks, could "be found in perhaps every ship which relies upon astronomical means for her guidance."[6]

With the consent of the Admiralty, the Royal Astronomical Society formed a review board for the almanac. This committee met in the offices of the society and consisted of a broad selection of almanac users, including naval officers and astronomers, shipowners and insurance men, financiers and merchants, scientists and clergy, the Astronomer Royal and Charles Babbage.[7] The committee published its recommendations in 1830 and 1831. In spite of their objections to the lunar distance method, they refused to cut any of Maskelyne's lunar distance tables and suggested that such tables be expanded and redesigned. They also requested several substantial changes: new tables of the planets, values that would improve astronomical observations made on board ships, the "mean time of high water at London Bridge," dates of Islamic holidays ("which may be occasionally useful to officers cruising in the neighbourhood of Mohammedan states"), and the expansion of many other tables. The proposed new additions increased the size of the periodical by 50 percent, but the committee felt that the calculations could be distributed "amongst the several computers as will afford them constant employment." "With due economy," they concluded, "the whole of the additional computations may, in a short time, be obtained without much (if any) additional cost to the nation."[8]

After receiving the report of the Royal Astronomical Society, the British Admiralty appointed one of the review board members, Lieutenant William Samuel Stratford (1791–1853), to oversee the almanac and implement the needed changes. From the start, Stratford concluded that the computers would be most efficient and be most constantly employed if they worked in a central office. He gave the old staff "due notice that their services would no longer be required after the completion of the Almanac for 1833," leased rooms for an almanac office in central London, and hired new computers. Only a few of the old staff moved to London and joined the new office. Most of the old computers wished to remain with their homes and families in the country. Stratford's staff would be drawn from city dwellers and would calculate in the almanac computing room, follow a daily schedule, and be under the watchful eye of the superintendent.[9]

Stratford opened the new almanac office shortly before the 1835 return of Halley's comet. There was less anxiety over this return than over its 1758 appearance. No one, at least none of the major astronomers, questioned the basic principles of Newton's theory, no one argued that the comet was anything more than a celestial object, and no one decried the "spirit of calculation." Stratford identified five major attempts to compute the comet's orbit and reviewed all five in the first volume of the almanac that was prepared under his superintendence. He reported that all of them identified roughly the same orbit and that the "principal variation appears to be in the time of the perihelion passage."[10] He felt that the most complete calculation was done by the French astronomer Philippe Gustave Le Doulcet, Comte de Pontécoulant (1795–1874). Pontécoulant had spent five years computing the comet's path and adjusting his figures. He started with an idealized orbit, a perfect ellipse around the Sun. Step by step he added the major influences on the comet: the gravity of Jupiter, Saturn, and Uranus. He even adjusted his equation to account for the position of the Earth during the 1758 passage. His first calculations identified November 7 as the date of the perihelion. His second effort moved the date to November 13. The third retreated it to November 10. The last advanced the perihelion to the evening of November 12.[11]

Under the direction of Stratford, the almanac computers produced an adjustable ephemeris for Halley's comet. In comparison with Pontécoulant's calculation, it was a mundane activity, a practical contribution to astronomy rather than a grand test of Newton's theory. Once the comet had been spotted, this ephemeris could be used to plot the comet's

7. Halley's comet in 1835 (fifth from right) with other nineteenth-century comets

position in the night sky and to predict the date of perihelion. Stratford used the ephemeris to engage about one hundred astronomers, both professionals and amateurs, to search the sky for the comet and to record its position. He instructed the observers to send their records of the comet to the almanac office. From the data, the almanac computers revised their equations for the comet and filed their results for the next generation of astronomers to use. The Royal Astronomical Society praised this work as a "most useful and arduous task."[12] Taken as a whole, the computations for the 1835 return halved the error of the 1758 calculation. The actual date of the perihelion, November 16, fell within sixteen days of all the major predictions. One of Pontécoulant's predictions came within three days and a few hours of the true perihelion.[13]

The reform of the Greenwich Observatory began just as the comet swung past the sun and began retreating from view. In the eyes of contemporary astronomers, the observatory required many changes. It needed new equipment, a stronger staff, and revised methods of operation. Like the Nautical Almanac Office, it had become a center of production, but this production was considered a burden on the staff. The British Admiralty had given the observatory the responsibility of caring for the navy's stock of chronometers. Observatory personnel cleaned, tested, and corrected the time of every chronometer before it departed England on an ocean voyage. These tasks occupied an entire room of the observatory and the labor of several observatory personnel. The director of the observatory at Cambridge, George Airy (1801–1892), looked at the activities of the Greenwich staff and complained that chronometer work degraded the Royal Observatory "into a mere bureau of clerks" and added that "it is difficult even for the director to resist the contagion" of such tasks.[14]

Airy had made his reputation as a reformer. Under his direction, the Cambridge Observatory had been transformed from a small, uninteresting teaching facility into "one of the pace-setting scientific centers of England."[15] This transformation was the crowning achievement of his career at Cambridge. He had enrolled at the university as a sizar, a scholarship student too poor to pay the tuition, and he had graduated at the top of his class. He held the title of "Senior Wrangler," a title bestowed on the student with the top score on the Tripos exam, the honors exam for students in astronomy, physics, and mathematics. For a time, Airy had also served as the Lucasian Professor of Mathematics, the position that had once been occupied by Isaac Newton and, more recently, by Charles Babbage.[16]

In 1835, the Admiralty appointed George Airy as the Astronomer Royal. The first test of his leadership was a collection of unprocessed astronomical data that had been accumulated over twenty-five years. These numbers lay unused in observatory logs, some written with neat and refined digits, some scribbled in haste, some recorded with hands that had grown stiff and cramped from the cold night air. This backlog of data was "a lump of ore," Airy observed, which is "without value till it has been smelted."[17] The smelting process is called "reduction." When computers reduce data, they take the raw values from the telescopes and convert them into a form that astronomers can use for research. For each observation of a celestial object, an astronomer generally records four numbers: the height of the object above the horizon (altitude), the direction of the object in the sky (azimuth), the time of day, and the date of the year. These four values change as the Earth rotates. The process of reduction collapses, or reduces, these four values into a pair of fixed numbers that represent the position of the object on the celestial sphere.

The celestial sphere is an imaginary hollow globe that surrounds the Earth. Astronomers identify positions on the celestial sphere with astronomical latitude (called the right ascension) and astronomical longitude (declination). The right ascension is measured from the celestial equator, which lies above the Earth's equator. The declination is measured from the celestial prime meridian, which runs north and south through the constellation Aries, the first constellation of spring, the traditional herald of the new year. The computations for data reduction require about ten steps, depending on how precise one wishes to be, and require a firm knowledge of trigonometry.[18]

To process the Greenwich backlog, George Airy created a computing group at the observatory that resembled the staff at Stratford's almanac office. Airy's computers were mere boys, some as young as fifteen, who arrived at eight in the morning. They worked at tall desks in the original observing room of the building. This room, called the Octagon Room,

8. Original Greenwich Observatory with the Octagon Room

more closely resembled an eighteenth-century ballroom than the dome of a modern observatory. Edmund Halley had gazed through the room's tall, slender windows to map the night sky. Those same windows admitted the sunlight to illumine the desks and papers of the computing boys. At noon, the boys took a break for supper, and they resumed their places at one. In winter months, they would bring candles to compensate for the failing light as they computed through the afternoon. Only when the clock had passed through a complete cycle and reached the hour of eight in the evening would they be allowed to go home.[19]

The boys were not the hairdressers of de Prony's Bureau de Cadastre, as they came to the observatory with abilities that could be compared to those of the almanac computers of Nevil Maskelyne. They generally possessed the basic skills of mathematics, including "Arithmetic, the use of Logarithms, and Elementary Algebra."[20] Some had been educated at the Greenwich Hospital School, a school that trained boys to be seamen in the navy. Many had learned their computing skill from fathers or uncles. Among Airy's early computers were two sons of an almanac computer. The boys had learned enough mathematics to expect a career in ocean trade or as civil servants or possibly as scholars at Cambridge. They became computers only because their father had died and left the family with no fortune or income.[21]

The elder of the two boys, Edwin Dunkin (1821–1898), was seventeen when he began computing. He reported that he learned the observatory's computing procedures from a book of printed forms that had been designed by Airy. On his first day in the Octagon Room, the chief computer placed the book in front of him and indicated the work that needed to be done. "I felt a little nervous at first," he reported, "and a momentary fear crossed my mind that some time would be required to enable me to comprehend this intricate form, and to fill up the various spaces correctly from the Tables." After a little instruction from the chief computer, "I began to make my first entries with a slow and tremulous hand, doubting whether what I was doing was correct or not. But after a little quiet study of the example given in the Tables, all this nervousness soon vanished." His brother, who was two years younger, worked on simpler problems, but at the end of the day, both felt that they had mastered a small part of astronomical mathematics. "We went home tired enough to our lodgings," Dunkin recalled, "but with light hearts and the happy thought that we had earned our first day's stipends."[22]

Among the computers, Dunkin was one of the promising young men. His background gave him the possibility of a career beyond the computing room. Eventually, he would rise through the observatory ranks, become the superintendent of the computing office, and be admitted to membership in the Royal Astronomical Society.[23] Not all of his peers would recall their time in the Octagon Room with such fondness. Some compared it to the "satanic mills" of William Blake or the soot-stained buildings of Charles Dickens, where "every day was the same as yesterday and to-morrow, and every year the counterpart of the last and the next."[24] Several observatory workers described George Airy as "despotic in the extreme" and claimed that he "was the cause of not a little serious suffering to some of his staff."[25] The charges against him were similar to those in any conflict between labor and management: the pay was too low, the hours too long, the computers could be dismissed if there was insufficient work, and they were always terminated when they reached twenty-three years of age.[26] Most critics identify the twelve-hour working day as the greatest source of complaint against Airy, though they acknowledge that this schedule was due, in part, to the computers themselves. Airy eventually concluded that the length of the shift was too long and reduced it to eight hours without reducing pay.[27]

Airy's defenders have portrayed his management of the observatory as progressive and inventive. "He was a colossal-minded man," wrote one friend, "and his ideas seemed to be executed in granite."[28] Most modern scholars have recognized that Airy brought new efficiency and new strength to the Royal Observatory, but they have also argued that he embraced the very "contagion" that he hoped to avoid by organizing his computers as if they were a "bureau of clerks." "Airy was impressed by the power of contemporary industrialists and engineers to transform both society and the manufacturing arts," wrote historian Allan Chapman. Chapman went on to argue that the only thing that distinguished Airy from a contemporary industrialist "was the fact that his profit was measured in terms of public utility and scientific prestige, rather than Pound Sterling."[29] In fact, Airy and William Stratford at the Nautical Almanac Office may actually have seen a connection between prestige and "Pound Sterling." There was no market for scientific research the way there was a market for wool cloth or tin sheeting, but as Adam Smith had argued fifty years before, there was a value that could be placed on the economic efficiency of the Royal Observatory or Nautical Almanac.[30]

The economic efficiency of both the observatory and the almanac was evaluated by independent organizations. Such review organizations were one more aspect of factory operation that was adopted by scientific institutions. These review organizations played a role similar to that of the board of directors of a commercial organization by representing those with a stake in the institution. The Nautical Almanac Office had faced a thorough review in 1829–31 by a committee of the Royal Astronomical Society. Beginning in 1836, the Greenwich Observatory was evaluated annually by an independent board of visitors. This board checked everything from the quality of research to the size of the staff and the condition of the instruments.[31] George Airy may have exercised "despotic" control according to some, but he did not possess the unfettered freedom that had been allotted to Edmund Halley in the seventeenth century or to Alexis Clairaut in the eighteenth. In making any decision, Airy had to consider how his actions might be judged by his board of visitors.

As Astronomer Royal, Airy had some experience in reviewing the work of others. In 1842, the British government asked him to evaluate the Difference Engine of Charles Babbage. After inspecting the drawings of the machine and looking at the prototype parts, Airy brusquely dismissed the engine. "I can therefore state without the least hesitation," he wrote afterwards, "that I believe the machine to be useless, and that the sooner it is abandoned, the better it will be for all parties."[32] It was, perhaps, an unnecessarily harsh conclusion, and it caused Airy trouble in later years. The historian Doron Swade has observed that Babbage took his revenge by portraying Airy as an "unimaginative bureaucrat—a mediocre but influential insider."[33] Yet Airy's evaluation was more true than false. The Difference Engine could not process data and hence could not assist his computers in their efforts to reduce the Greenwich backlog. Though it might be able to prepare an ephemeris, it would be forced to do those computations in small pieces. Between each computation, the machine would have to be stopped and reset. The process of resetting the machine required substantial mathematical analyses that Babbage had not done, and hence Airy could easily believe that the Difference Engine would not be worth the effort required to operate it.

Babbage had received a gold medal from the Royal Astronomical Society for designing the Difference Engine, but the prestige of that medal did not translate into the pounds sterling needed to complete his machine. William Stratford and George Airy labored not for gold medals but for the simple approval of their actions by their oversight committees and boards of visitors. In the course of their careers, they created the premier computing laboratories for their day. The computing staff of the British *Nautical Almanac* and the computers of the Royal Greenwich Observatory would provide models for computing offices for the next eighty years. Such offices would have a central computing room, an active manager, preprinted computing forms, standard methods of calculation, and a common means of checking results. They would also have regular hours of operation, a clock on the wall beating out the hours, and an oversight board that would ensure that the work was well done.

CHAPTER FOUR

The American Prime Meridian

> The eyes of others have no other data for computing our orbit
> than our past acts, and we are loath to disappoint them. . . .
> Ralph Waldo Emerson, "Self-Reliance" (1841)

THE METHODS of computation migrated to North America with the navigators who guided the ships across the North Atlantic and the surveyors who delineated the European claims upon the continent. Throughout the eighteenth and nineteenth centuries, the Greenwich Observatory and the British *Nautical Almanac* served as sources of computational techniques for those traveling west. The observatory had employed Charles Mason before he departed with Jeremiah Dixon to survey the border between Pennsylvania and Maryland.[1] One of the almanac computers, Joshua Moore, emigrated to the United States and corresponded with President Thomas Jefferson on subjects mathematical.[2] The British *Nautical Almanac* itself was reprinted in Salem, Massachusetts, and was freely available for purchase at the ports of Boston, Nantucket, New York, and Philadelphia.

The models of organized computational labor emigrated more slowly to the United States than did the mathematical methods of astronomy, navigation, and surveying. Through the first decades of the nineteenth century, American science was the work of individuals rather than organizations, the effort of Benjamin Franklin or Thomas Jefferson rather than of an almanac office or the computing factory of an observatory. "Amongst few of the civilized nations of our time have the higher sciences made less progress than in the United States,"[3] observed the French writer Alexis de Tocqueville (1805–1859). When de Tocqueville visited the United States in 1831, he found a country that had supported only a few large scientific projects, most notably the survey expedition to the Pacific Northwest of Meriwether Lewis and William Clark. De Tocqueville drew the general lesson that "those who cultivate the sciences amongst a democratic people are always afraid of losing their way in visionary speculation," but in reaching that conclusion, he failed to understand the connection between the nature of American democracy and large scientific projects. Even in the early nineteenth century, the big scientific endeavors received government patronage. In England, the crown's government had provided financing for the *Nautical Almanac*, the Royal Observatory, Babbage's Difference Engine, and even the construction of the railroads. In the United States, the citizens deeply distrusted the power of governments and national institutions. Just before de Tocqueville visited the United States, the American Congress had rejected a proposal to create a national university and national observatory. The proposal had been drafted by President John Quincy Adams, who had a deep interest in science and learning. Adams was well read in the classic literature of political philosophy, but such knowledge did not make him a skilled leader. "It would have been difficult . . . to propose a more unpopular measure," observed historian Hunter Dupree.[4]

Before 1840, the United States government operated only one permanent scientific agency, the Coast Survey Office. As the name implied, the Coast Survey was responsible for cataloging harbors and navigational hazards of the Atlantic shore. It had been founded in 1807 but had accomplished little, as it was bedeviled by a weak leader and vacillating congressional support. Congressional interest in science changed only as the country's population began to move from farm to workshop. In 1800, less than 5 percent of the adult population was involved in manufacture. By 1840, that fraction had risen to 25 percent. The new workers brought to power the Whig Party, a group that advocated improvements to the national infrastructure, including the construction of roads, the digging of canals, the expansion of ports, and the creation of scientific institutions. The navy was the first government office to be touched by this political shift. In 1842, the naval secretary reorganized the entire command structure of the navy and created two scientific offices. The first was a small ordnance proving ground, a testing place for cannons and mines. The second was an astronomical observatory, originally given the name of National Observatory, though the title of "National" was soon changed to "Naval."[5]

The navy's actions were soon followed by a major reorganization of the Coast Survey Office. In 1843, Congress gave the survey office an expanded budget, a broader scope of operations, and a new, dynamic superintendent.[6] The next year brought the founding of the Smithsonian Institution (1844) as an office for the "increase and diffusion of useful knowledge among men."[7] These government agencies were matched by two important private scientific institutions, the Harvard Observatory (expanded in 1843) and the American Association for the Advancement of Science (1848). The last major scientific institution of the 1840s, the American Nautical Almanac Office, was created in the last year of the decade.[8]

"The peculiarity of American institutions," wrote the historian Frederick Jackson Turner, "is the fact that they have been compelled to adapt themselves to the changes of an expanding people—to the changes involved in crossing a continent, in winning a wilderness, and in developing . . . out of the primitive economic and political conditions of the frontier into the complexity of city life."[9] In many ways, the founders of the early American scientific institutions were working in on intellectual frontier, borrowing and adapting ideas from their European counterparts. The Coast Survey purchased equipment from France and Germany. The American Association for the Advancement of Science took its name and purpose from a British organization whose founders included Charles Babbage. The American *Nautical Almanac* based its operations upon the ideas of Nevil Maskelyne. Not all such ideas were successful in their transit across the Atlantic. The U.S. Navy mistakenly constructed its new observatory on a river bluff site that resembled the placement of the Royal Observatory in England. Unlike the park at Greenwich, the navy's perch over the Potomac River had little to recommend it. From the first nights of operation, astronomers complained that river mists fogged telescope lenses and that marshes bred swarms of infectious mosquitos. They might have borne such trials more bravely if the site had had the advantages of Greenwich, such as easy proximity to a navy yard or the view of ocean-bound ships. The only vessels that passed the American observatory were canal barges bound for the Ohio Valley, boats that did not need an almanac in order to find their way.

Though all of the American scientific institutions of the 1840s were touched by the "primitive economic and political conditions" of the North American continent, they were more profoundly shaped by competition with European institutions. This is especially true for the largest computing organization of the age, the American *Nautical Almanac*. The almanac was generally viewed as a practical means for improving the navigational skills of the navy and the mercantile fleet. "But for the *Nautical Almanac* of England or some other nation," claimed one supporter, "our absent ships could not find their way home nor those in our ports lift their anchors and grope to sea with any certainty of finding their way back again."[10] Yet many individuals saw the almanac as a way of demonstrating America's intellectual accomplishments. It "would be a work worthy of the nation," wrote one scientist, "and might engage our ablest astronomers and computers."[11] Such thoughts were echoed by the political leaders. An almanac "was important to the character of the country," argued one member of Congress, "important to the national pride, national honor and independence."[12]

The almanac was shaped by two individuals, Lieutenant Charles Henry Davis (1807–1877) of the U.S. Navy and Harvard College professor Benjamin Peirce (1809–1880). The two were related by marriage and lived in neighboring houses in Cambridge, Massachusetts.[13] Their front

9. Lieutenant Charles Henry Davis of the American Nautical Almanac

doors were just a few steps from Harvard Yard; their back windows looked across the fields to the distant ships at the docks of Cambridge-port. Davis's son recalled that children moved freely between the two homes and that their families "dwelt almost as one."[14] Their friendship had begun some fifteen years before, when Davis had taken a house in Cambridge following an extended voyage. He had once been a student at Harvard College, but he had left without a degree in order to take an officer's commission. During his travels, he had retained an interest in learning. One of his commanders described him as "intelligent in his profession, energetic in his character, and devoted to the improvement of his mind." While on an early voyage, he studied navigation, learned "French, Spanish and a good smattering of Italian," and read the complete works of William Shakespeare.[15] In 1840, he had returned to Cambridge for a year of intensive mathematical study with Peirce. According to his son, Davis did not always "follow the transcendent flights of Peirce's genius" but persevered in his study and ultimately acquired "a working familiarity with mathematical tools." For his efforts, he received a Harvard degree, conveniently backdated to suggest that he had graduated with his original classmates.[16]

By all appearances, Benjamin Peirce was an unlikely friend for Davis. Davis was a practical and disciplined officer from a privileged Boston family. No sign better captured his bearing than his prominent moustache, groomed with military precision. Peirce was one of the "bearded ancients" of Harvard College and sported an unruly head of hair. His lectures were rambling, difficult affairs that were often incomprehensible to students. The one quality that may have caught Davis's sympathy was Peirce's ability to maintain his intellectual bearings. In the 1840s, Cambridge was awash in the Transcendental movement, the philosophical current that saw in every aspect of nature "a symbol of some spiritual fact."[17] Peirce befriended the leader of the Transcendentalists, Ralph Waldo Emerson (1803–1882), and occasionally referred to God as the "Divine Geometer," but he never lost sight of the theories of Newton or the mathematics of calculus.[18]

When the American *Nautical Almanac* began operations in July 1849, the navy appointed Charles Henry Davis as the first superintendent. In turn, Davis selected Peirce as the chief mathematician and established the almanac office in Cambridge, Massachusetts.[19] When called to explain why he did not place the office in Washington, Davis wrote that Cambridge had "the best scientific libraries of the country—an indispensable aid in laying the permanent foundation of a work of this magnitude and importance."[20] Davis may have been rationalizing a decision that was convenient for himself and for Peirce, but he clearly wanted to create a superior publication with a staff of "first class computers." There would be no former hairdressers on his staff and no boy computers like those found at Airy's observatory. He stated that his computers "must be gentlemen of liberal education and of special attainments in the science of astronomy."[21]

At the start, Davis tried to recruit mathematicians for the almanac staff. He wrote to professors at the University of Virginia, Princeton, Rutgers, Columbia, and the Military Academy at West Point, asking each to serve on the almanac staff.[22] To this list he added the name of one foreigner, the French astronomer Urban Jean Joseph Le Verrier (1811–1877). In 1849, Le Verrier was basking in the fame of having discovered the planet Neptune, an event that dominated the public imagination of the 1840s. He was a mathematical astronomer, not an observer, and his discovery was an event of "pure calculation," "the grand triumph of celestial mechanics as founded on Newton's law of gravitation."[23] Le Verrier had estimated the location of the planet by analyzing variations in the motion of Jupiter, Saturn, and Uranus and by hypothesizing the existence of an undetected body in orbit around the Sun. He had projected the possible location of this planet and sent his estimate to the observatory at Berlin. The German observers had found the planet with a single night of searching. They waited a second night in order to confirm their observation and then wrote to Le Verrier, "Monsieur, the planet, of which you indicated the position, really exists."[24]

The discovery immediately pushed Le Verrier into public view. Many urged that the planet should be named Le Verrier, just as the comet of 1682 had been named for Edmund Halley. Observers spent their nights studying the barely perceptible motion of the planet and looking for evidence of moons. The news penetrated so deeply into public consciousness that it even reached the backwoods of Massachusetts, where it received a disparaging response from Henry David Thoreau (1817–1862) at Walden Pond. In the manuscript of *Walden*, Thoreau dismissed the mathematics that might allow an astronomer "to discover new satellites to Neptune" but "not detect the motes in his eyes, or to what vagabond he is a satellite himself."[25] Such words failed to touch Le Verrier's reputation. Had he been willing to join the American almanac staff, Le Verrier would have brought substantial fame and prestige to the periodical, but he declined Davis's request. Davis was ultimately rejected by all of the mathematics professors on his list, save for Benjamin Peirce.[26]

To build his computing staff, Davis turned to students, independent astronomers, and skilled amateurs. These individuals were not true members of the country's small scientific community, but each of them had the promise of a successful future. Most of them were either friends or former students of Benjamin Peirce. The list included a young mathematical prodigy from Vermont, Henry Safford (1836–1901); a future president of the Massachusetts Institute of Technology, John Runkle (1822–1902); and a professor, Joseph Winlock (1826–1875) of Shelby College in Kentucky. Though Davis had not been able to convince the discoverer of Neptune to join the almanac staff, he was at least able to recruit an astronomer who had achieved some fame by computing Neptune's orbit, Sears Cook Walker (1805–1853).[27]

In the fall of 1846, the fall when Le Verrier announced the existence of Neptune, Walker was an assistant astronomer at the newly opened Naval Observatory in Washington. This was not a job that he had intended to take, but he had little choice in the matter. For nearly twenty years, he had worked for the Pennsylvania Company for Insurances on Lives and Granting Annuities. By day, he was an actuary, a mathematician responsible for estimating the risk and profit of insurance policies. By night, he was an amateur astronomer, using the telescopes of the Philadelphia High School to study the moon as it passed in front of stars.[28] At the time, the high school observatory was the best equipped in the nation. Most of the instruments had been donated by Walker, who was making a substantial fortune in his actuarial practice. This life had come to an end in 1845, when a "series of unfortunate investments and commercial operations led to most disastrous results." According to a friend, this disaster left Walker "at the age of forty years utterly without means."[29]

Under the best of circumstances, Sears Cook Walker could be difficult. "I have had some differences with him," wrote Benjamin Peirce, "but they have not blinded me to his great merits."[30] At the Naval Observatory, Walker showed his great skill and his wayward nature. In the fall of 1846, he turned from his observatory duties to consider the planet Neptune. The observatory had been the recipient of small packets of data from the European astronomers, sealed with wax and addressed to the new "National Observatory." By late fall, he had a substantial collection of data that showed the planet's slow march across the celestial sphere. Walker decided that he might get a better calculation of the orbit if he could find an earlier observation of the planet, an observation that had been falsely recorded as a star. Using the data he had collected, he spent about three months computing the motion of the planet backwards from its location in 1846. He finally found his prediscovery observation in a 1795 star catalog that had been compiled by Joseph Lalande. The catalog showed a star where Neptune should have been. Using the navy's telescopes, Walker sought in vain for the star and concluded that Lalande had seen the moving planet instead of a fixed star. With Lalande's observation, Walker was able to make a refined calculation and show that the planet moved in an orbit that was nearly circular.[31] The navy treated Walker's work as a major triumph, the first important accomplishment of its observatory. "The theory of Neptune belongs, by the right of precedent, to American science," bragged the observatory director.[32]

Ironically, the most well-known of the first computers for the American *Nautical Almanac* was the sole woman, Maria Mitchell (1818–1889). When Charles Henry Davis appointed her to his staff, he presented her credentials to the secretary of the navy as "the lady who lately received from the King of Denmark a medal for the discovery of a new comet."[33] As an astronomical discovery, the comet was small and unimpressive. It was not even a periodic comet like Halley's. Once it vanished from the night sky, it was gone forever. Still, the discovery caught the public's attention as "one of the first additions to science" made in the United States.[34] Mitchell was an extraordinary scientific talent, and she lived in one of the few communities that acknowledged a public role for women in nineteenth-century America. There were no other female scientists in the United States and few women in any other field of endeavor. She had discovered her comet in 1847, the year before the 1848 Seneca Falls Conference on the Rights of Women, the event that has often been identified as the start of the American feminist movement. Behind the Seneca Falls Conference lay the Quaker Church, the only religious denomination of the time that allowed women to preach to its congregations. Four of the five conference organizers were Quakers, as were Maria Mitchell and her father, William Mitchell (1791–1869). William Mitchell supported and pushed his daughter in her career, demonstrating what the early feminist Margaret Fuller (1810–1850) called the "chance of liberality" that a father might show to his daughter but a "man of the world" might never show toward his wife.[35]

William Mitchell was a banker on Nantucket Island and an amateur astronomer. Maria Mitchell recalled that her interest in astronomy was "seconded by my sympathy with my father's love for astronomical observation."[36] He taught her how to use a telescope, how to reduce data, and how to compute time from the position of the moon. He was a friend of the scientists at Cambridge, including Benjamin Peirce, Charles Henry Davis, and the director of the Harvard Observatory. When his daughter told him that she had identified a comet, he encouraged her to notify the observatory. When she resisted the idea, William Mitchell communicated the discovery himself. "Maria discovered a telescopic comet at half past ten on the evening of the first instant," he wrote. "Pray tell me whether it is one of George's [Airy of the Greenwich Observatory]. . . . Maria's supposed it may be an old story." The observatory confirmed that the comet was new and initiated an energetic effort to ensure that Maria Mitchell was given credit for the discovery. The object of their effort was King Christian VIII of Denmark. The king awarded a medal to the discoverer of any comet. In the early winter of 1848, he was preparing to give his medal for Mitchell's comet to an Italian astronomer. The director of the Harvard Observatory did everything he could to convince the king to change his mind. He enlisted the aid of the Harvard president and the American consul in Copenhagen. Eventually, the king relented and recognized Mitchell as the first to see the comet.[37]

Charles Henry Davis recruited computers through the end of 1849 and began operations that winter. He organized the staff after the pattern that Nevil Maskelyne had established some eighty years before. Each computer took responsibility for one or two tables. Sears Cook Walker took the computations for Neptune. Benjamin Peirce handled the ephemeris of Mars and the apparent movement of the Sun. Without a trace of irony, Davis asked Mitchell to handle the computations for the planet Venus. "As it is 'Venus who brings everything that's *fair*,' " he wrote, "I therefore assign you the ephemeris of Venus, you being my only *fair* assistant."[38] John Runkle, the future president of the Massachusetts Institute of Technology, prepared part of the calculations of lunar motion, splitting the task with another computer, just as Maskelyne's computers had done.

10. Maria Mitchell

For about half of the computers, including Maria Mitchell and Sears Cook Walker, Charles Henry Davis distributed detailed computing plans through the mail. Unlike Maskelyne, Davis did not prepare hand-drawn computing forms. Though he gave the computers a rough idea of how the sheet should look, he let them organize the computations as they desired. "You may fill up the sheets as much as possible, consistently with the clearness," he told Maria Mitchell, and in so doing "can thus economize on paper."[39] He provided paper and reference books to all of the computers, sending the supplies with a private forwarding company called Adams Express. Mitchell, who had a relatively meager library, often requested books for her work. In December 1849, she asked for a copy of *Theoria Motus Corporum Coelestium in Sectionibus Conicis Solem Ambientium* (*The Theory of the Motion of the Heavenly Bodies Moving about the Sun in Conic Sections*) by the German astronomer Carl Friedrich Gauss (1777–1855). This book, one of the last major astronomical texts written in Latin, had a fairly complete discussion of the techniques of astronomical calculations. "I have directed my bookseller to endeavor to get two copies [for the almanac office]," Davis responded, "and will add a third to the list for yourself if you wish it." He concluded the letter by commenting, "I am glad you read Latin."[40]

The remaining computers, those who lived near Cambridge, worked at the almanac office. As far as we can determine, this office resembled the rooms of the British *Nautical Almanac* in London. There were worktables for the computers and a private area for the superintendent. One young computer recalled coming into the Cambridge office on a frosty January morning, taking a "seat between two well-known mathematicians, before a blazing fire." It was an informal place, where new ideas of mathematics were freely discussed between calculations, as the "discipline of the public service was less rigid in the office at that time than at any government institution I ever heard of." Each computer was expected to spend five hours a day in the office. "The hours might be selected by himself, and they generally extended from nine until two, the latter being at that time the college and family dinner hour." All that Davis required was that the work was done on time.[41]

The disjointed operation of the almanac, with some computers in Cambridge and some working from home, fit easily into the organizational structure of the navy. Davis kept track of each computer and the progress of his or her work. Four times a year, he would send pay vouchers to the computers that could be redeemed at any naval facility. The letters that accompanied these payments contain gossip, discussions of mathematics, news about astronomy, and even a few quotes from Shakespeare, who was clearly his favorite author. "Enclosed are your vouchers signed by myself," he wrote to a new computer; "you had better negotiate them with a friendly broker, rather than with one where you hold a relation of Antonio to Shylock."[42] Almanac computers earned between five hundred and eight hundred dollars per year for their work. Two of them doubled their income by checking and correcting the work of others in addition to doing their own computations.[43]

After a year of calculations, Davis could report that work for the first issue, which would cover the year 1855, was progressing nicely. He told the secretary of the navy that his "small corps of computers" was employing the theories of the "illustrious Leverrier," using the corrections of Airy, and reducing data from Maskelyne. He reminded the secretary that he had already made a special report of "a variation in the proper motion of one of the fundamental stars," which he claimed "has led to a discovery of particular interest in stellar astronomy." He summarized his progress by stating, "I have frequently expressed my wish that the *Nautical Almanac* should in every respect conform to the most advanced state of modern science and be honorable to the country and it is my determination to spare no effort by which this high object can be attained."[44]

Though Davis was confident that his computing staff would produce an American almanac that rivaled the publications of Europe, others were not so certain. The potential users of the almanac, notably the merchant navigators and the surveyors, were willing to withhold their judgment until they saw the final product. These two groups were concerned about a basic element of the almanac, the location of the prime meridian. All almanacs had to establish a meridian, a line running through the north and south poles that would serve as a reference for the positions of stars. The *Connaissance des Temps* used the Paris meridian, which passed through the Cathedral of Notre Dame. The British *Nautical Almanac* was prepared for the Greenwich meridian. Originally, this line had been drawn through the Octagon Room of the Royal Observatory, but it had since moved to the site of a new telescope mount a few yards to the west. The original proposal for the American *Ephemeris and Nautical Almanac* instructed the navy to publish an "American *Nautical Almanac*, to be calculated for the meridian of Washington city."[45]

Davis did not like the idea of a Washington meridian. He preferred to use the Greenwich line, for he felt that it would produce the most accurate longitude calculations. "Our own vessels are constantly meeting those of Great Britain on the great highway of nations," he noted, "and are in the habit of comparing with them their longitudes."[46] By adopting the Greenwich meridian, the computers of the American almanac could use without additional calculations the vast catalogs of star data that George Airy was compiling at the Royal Observatory. However, the Greenwich meridian did not satisfy the computers of the Coast Survey or the surveys of the various American states. They wanted a meridian safely on the North American continent. Their calculations would be most accurate if they could physically measure parts of the meridian and if they could work without an ocean intervening between themselves and their baseline.

From one point of view, the argument between those favoring the Greenwich meridian and those who wanted a Washington meridian was an honest scientific disagreement. Each of the two groups advocated a procedure that was best for its own needs. Each of the procedures would provide an acceptable, though not perfect, solution for the other group. Yet the argument had symbolic and economic aspects beyond the scientific construction of astronomical tables. The symbolic problem was an offshoot of Manifest Destiny and the Monroe Doctrine. The American citizenry saw themselves settling the breadth of the North American continent, from the Atlantic to the Pacific. As settlers moved west, they would use the American *Nautical Almanac* to survey the territories and set the borders of new states. Congress would insist that those borders be specified from Washington, rather than from a meridian that passed through the suburbs of a foreign capital.

Davis believed that he had a clever compromise to the meridian problem, one that would satisfy both navigators and surveyors. He proposed "to establish an arbitrary meridian at the city of New Orleans, which will be exactly six hours in time, or ninety degrees in space, from the meridian of Greenwich."[47] The idea was an elegant technical solution, as it gave the United States its own meridian and allowed navigators to compute their position relative to Greenwich with little extra effort. Davis must have believed that his idea would be accepted without complaint, but just as he was starting the computations, he reported to the navy "that a 'remonstrance' against my paper on the American Prime Meridian and against any change from the Meridian of Greenwich is circulated among the merchants and insurers of the cities of Boston and New York for signatures."[48]

The supporters of the "remonstrance" were uncomfortably close to Davis and his friends. Their leader, Ingersoll Bowditch (1806–1889), was the director of a Boston insurance firm, a friend of Benjamin Peirce, and a financial supporter of the Harvard Observatory. Their objections to a meridian at New Orleans were entirely economic. Bowditch believed that his business would be damaged by the change in meridians. He was a partner in a firm that republished the British *Nautical Almanac* in the United States. The objections of the other merchants and insurers were more speculative. They were concerned that the ports of Boston and New York would decline if New Orleans had a meridian passing through it. New Orleans already possessed a substantial advantage over the Atlantic ports, as it had a navigable waterway, the Mississippi River, with access to the vast center of the continent. With the meridian, reasoned Bowditch and his supporters, New Orleans would become a destination for ships needing to adjust their chronometers. These ships would divert trade to the south, as none of the captains would want to travel empty.

Davis attempted to counter the claims of Bowditch, but he had lost the battle almost from the beginning. He argued that his proposal was "founded on principles of science" and would contribute "towards improving the safety of navigation, and completing the geography of the seas," but neither idea swayed his opponents.[49] The debate had turned toward economic issues and way from scientific merit. In this field, Bowditch held the greater power. By the spring of 1850, Davis had abandoned his idea for the meridian and accepted a solution that divided the American almanac into two parts. The first part, the *American Ephemeris*, would be computed relative to a meridian that passed through the Naval Observatory in Washington. This meridian would be used by surveyors as they set the borders of Wyoming, Colorado, Oregon, and the other western states. These lines would fall at integral number of degrees west from the Naval Observatory in Washington, rather than from the Royal Observatory in England. The second part of the almanac, the *Nau-*

tical Almanac, would be prepared using the Greenwich meridian and be used by the nation's sailors.

The double meridian scheme put extra demands upon the almanac computers, as it required them to prepare more tables, but it had little impact upon the structure of the computing staff. Davis did not have to restructure his computers in order to make them more efficient because he had substantial support from both the navy and Congress. The navy had initially allocated $6,000 a year for the almanac, most of which was spent on the salaries of computers. By the spring of 1850, Davis had concluded that this figure was insufficient, especially with the requirement to prepare two sets of tables. He requested and received $12,000 for the second year of operations. This figure also proved to be too little. For his third year of operations, Davis asked for $18,000.[50] The navy granted this request, but the increase drew the attention of the U.S. Senate. In May of 1852, Senator John P. Hale (1806–1873) of New Hampshire rose from his desk on the Senate floor and asked the navy to justify its expenditures on the almanac. Though he expressed his concern over the size of the almanac budget, Hale was more interested in a key element of Davis's plan, the cooperation of naval officers and a civilian computing staff. He might have been less concerned if the computers had been "mere drudges" and could have been drawn from any state of the union. However, he knew that all of Davis's computers were somehow connected to Benjamin Peirce or Harvard College.[51]

Hale demanded that the secretary of the navy "inform the Senate, where, and at what Observatory, the observations and calculations for the '*Nautical Almanac*' are made." Like any skilled politician, Hale knew the answer to his question before he strode onto the Senate floor and asked to be recognized. "I think that I am not incorrect," he informed his colleagues, "when I say that all this expense has been incurred, not at the National Observatory but at the observatory of Cambridge College in Massachusetts."[52] "Cambridge College" was, of course, Harvard. He probably misidentified the school as a way of emphasizing its location. If he expected this charge to stir the Senate to action, he was disappointed. Even those who objected to government support for private institutions sat in their seats. Any discussion of the almanac was ended quickly and decisively by a senator who derided Hale as possessing "an absolute and unappeasable hostility to any connections of science with the Naval Department in any form."[53]

Though Hale's comments lasted but a few moments, they pushed Charles Henry Davis to respond. His plans would quickly fail if Congress reduced his budget or dictated the kinds of computers that he might hire. He wrote a detailed defense of his organization, addressed to the secretary of the navy but published in the *American Journal of Science and Arts* and circulated as a pamphlet. He confronted Hale on the narrow point of the attack, explaining that "no Observatory, neither that at Washington nor that at Cambridge, as has been suggested, received any portion whatever of the sum appropriated for the '*Nautical Almanac.*'" This statement was entirely true, but it did not address the point that many of the computers had a connection to Harvard. Davis tried to deflect this issue by stating that the almanac required the "most illustrious genius and the most exalted talents" and that it was not "a work of insignificant value or trifling labor." He claimed that the new almanac "is considered by American astronomers and mathematicians as a work of consummate utility and of real national importance, resembling in this respect the *Nautical Almanac and Astronomical Ephemeris* of Great Britain, the *Connaissance des Temps* of France and the *Astronomical Almanac* of Prussia."[54]

The almanac remained vulnerable to congressional displeasure, even though the crisis initiated by Senator Hale passed quickly. Through the early 1850s, Davis worked to ensure that the almanac offered no target for an easy attack. "In our work there can be no regular vacation," he advised a new almanac computer. Perhaps sensing that he had overstated his case, he conceded that computing was not a full-time activity and that there were "opportunities for occasional indulgence."[55] Without such opportunities, he would have found it difficult to retain his computing staff, as all of them had other interests. Benjamin Peirce requested a reduced computing load so that he might concentrate on mathematics. Maria Mitchell traveled to Europe and relied on her father to communicate with the almanac office.[56] The most serious threat to almanac operations was the unexpected death of Sears Cook Walker, but Davis was able to recruit Walker's brother-in-law as a replacement.[57]

The first substantive change in the almanac occurred in 1856, when Davis decided that he must put his naval career ahead of his love for astronomical calculation. In September of that year, the secretary of the navy offered him the command of a ship in the Caribbean. Davis described the order as "an alternative." He could accept the post and leave the almanac staff; "the other choice was to give up all desire for a command and to resign the active service."[58] No matter which choice he made, he had lost control of the almanac, so he accepted the command, gave his office to the senior computer, Joseph Winlock, and departed for the Caribbean.

One of the almanac computers described Joseph Winlock as "silent as General Grant with the ordinary run of men." He could be friendly and open among the computing staff, but "he had a way of putting his words into exact official form."[59] He had no interest in changing the operations of the almanac but was content to follow the computing plans of Charles Henry Davis and the advice of the ever-present Benjamin Peirce. His only

11. Scheutz difference engine used by the computers of the Nautical Almanac

step toward innovation involved the almanac in a controversy at the newly formed Dudley Observatory in Albany, New York. That observatory had acquired a difference engine, though it was not the engine that had been designed by Charles Babbage and had never been completed. This engine was a smaller machine that had been built by two Swedes, George Scheutz (1785–1873) and his son Edvard (1821–1888). The elder Scheutz had read about the Babbage difference engine in a newspaper and had built a simple model with Edvard. The younger Scheutz had created engineering drawings and found a machine shop that would build the engine. Their design drew on the technology of clocks, while Babbage had borrowed the tools and ideas of steam engines. The Scheutz engine looked like a large music box and could sit on a desk or a dining table. Recognizing that this machine was an improvement of his ideas, Babbage praised the Scheutz design as "highly deserving of a Medal."[60]

The Scheutz difference engine had reached the United States because of the efforts of Dudley Observatory director Benjamin Gould (1824–1896). Gould was yet another student of Benjamin Peirce. He had discovered the difference engine while traveling in Europe and had been impressed with its potential. He claimed that it might change the nature of observatory staffs by replacing "the toiling brain by mere muscular force."[61] He had no immediate use for the engine and was willing to let the almanac staff experiment with the device. Joseph Winlock thought that it might be able to prepare certain tables for the almanac. He sent two computers on a trip from Boston to Albany with the instructions to prepare an ephemeris for Mars. When the computers arrived at the Dudley Observatory, they found that the machine was inoperable. "The dirt, which had accumulated on the passage [across the Atlantic], and thickened oil, impeded its action greatly." A mechanic cleaned the machine and returned it to operating condition, allowing the computers to start their work.[62] The calculations proved to be more difficult than anyone had anticipated. The computers discovered that the Scheutz machine was fragile and sensitive. It could easily jam in the middle of a calculation, a problem which could require a lengthy repair effort and force the computers to restart their calculations from the beginning.[63]

After a month of work, the computers returned to Boston with two intermediate tables but without a complete ephemeris. They had discovered that the difference engine could calculate only small segments of tables. If the orbit of Mars were placed on a clock dial, the machine could compute the arc between noon and eleven o'clock before it needed to be reset. "The strictly algebraical problems for feeding the machine made quite as heavy demands upon time, and thought, and perseverance, as did the problem of regulating its mechanical action,"[64] observed Benjamin Gould. In reviewing the tables, Winlock expressed his hope "that the immense labor of astronomical calculations may be materially diminished by the aid of machinery," but he was disappointed.[65] "The result thus far has not been such as to demonstrate to my satisfaction that any considerable portion of the Almanac can be computed more economically by this machine."[66] The failure of this experiment had more impact in Albany than in Washington. Benjamin Gould had become involved in a fight with observatory trustees over several key decisions, including the purchase of the difference engine. "The responsibility of having recommended this machine I willingly accept," wrote Gould. "If the machine has not multiplied and tabulated [the donor's] fame to an amount equal to the wishes of [his] most ardent friends," he added, "it has not been my fault."[67]

Compared to the other issues facing the almanac office, the failure of the difference engine experiment was only a minor problem for Winlock. The almanac computers had fallen behind schedule. The almanac budget had been reduced by Congress. At the start of 1859, Winlock had been forced to release part of his computing staff and stop work on some of the ephemerides.[68] The trouble ended only with the return of Charles Henry Davis that summer. Davis reported "that several parts of the work were omitted or postponed; that printing had been arrested or delayed, and the efficiency of the corps of computers diminished."[69] He quickly brought the computers back to their production schedule and convinced Congress to restore the almanac budget. In less than a year, he could report that "the regular work of the office was resumed with more than common activity."[70]

Of course, the fall of 1860 was no common time. The changing political climate could only remind Davis that the scientific activity of the Nautical Almanac Office was held most firmly in the grip of forces beyond the ability of any one person to control. By December, the goal of preparing a high-quality almanac seemed less important than the need to preserve the country's social structure. Following the election of Abraham Lincoln in November, the country had reached a point of crisis on the question of slavery, and the Southern states were preparing to secede. Southern naval officers, including the director of the Naval Observatory, resigned their commissions and returned home.[71] Charles Henry Davis was summoned to Washington and discovered that he would have little time to think about the operations of the almanac. "I have found that [the director of the Coast Survey] has a plan of his own to carry out, which involves my remaining here," he explained to his family. "[The director] wishes to establish a military commission, or advisory council, to determine military proceedings and operations along the coast."[72]

Through the spring and summer of 1861, Davis tried to manage the almanac computers while simultaneously organizing surveying units for both the navy and army. He spent most of his days at the Coast Survey building. The survey office was located just south of the Capitol. Its top floors had a view of the navy yard, the Washington arsenal, and the Confederate volunteers of Virginia holding positions on the far side of the Potomac River. The sights and sounds and rumors troubled him so much that he complained, "The more I hear, the more I fear for the end." He poured out his sadness, his anxiety, and his guilt by invoking Shakespeare's murderous lord: "Like Macbeth, I'm sick at heart ('Seyton, I say!')."[73]

The words of Macbeth came freely from his pen that summer. If taken literally, Davis cast himself as the title character, the military leader who had been misled by his own ambitions, had killed his king, and had pushed his country into chaos. Increasingly, the almanac director felt that he was needed elsewhere. "My hands are not of much use in working," he told his family, "but my head might be in directing."[74] In September, he concluded that he needed to be released from the institution that he had built and led for a decade. "Today I give up the 'Nautical Almanac,'" Davis wrote. "I am very sorry to do it, but I could not retain it." He would have to trust his hopes and plans for the almanac to others. "Winlock takes my place," he acknowledged, "and he will be glad to get it."[75] Though Davis was generally not given to ironic comments, he does not seem to be blessing his successor. An era was ending, and the early computers, the "gentlemen of liberal education," would soon depart, to be replaced with more traditional computers. Benjamin Peirce had already left the almanac. The others would soon follow. Even Maria Mitchell had only a short time remaining. She would shortly be appointed the first professor of astronomy at Vassar College for Women.

CHAPTER FIVE

A Carpet for the Computing Room

"Arcturus" is his other name—
I'd rather call him "Star."
It's very mean of Science
To go and interfere!

Emily Dickinson, Poem (1859)

"I WAS SURPRISED to find how quickly one could acquire the stolidity of the soldier," wrote a computer at the Naval Observatory.[1] The observatory, from its hill on the northern bank of the Potomac River, had an unobstructed view of Confederate territory. Directly across the water sat the plantation of General Robert E. Lee, abandoned by its owner and occupied by Union troops. The navy insisted that the computers and astronomers, most of them civilians, be trained as an artillery crew. The staff dutifully reported to the ballistics range at the Washington Navy Yard, where they learned how to charge and fire a cannon. They spent a day sending shells into the opposing riverbank before they returned to their telescopes and computing desks. None of them remarked that the arc of a cannon trajectory bore a mathematical similarity to the path of a comet, and, of course, none of them was quite ready to fire a gun in anger.[2]

The war permeated Washington, D.C., filling the streets with uniformed troops and the anxious gossip of distance campaigns. Only once was the city the site of an actual battle. That skirmish was fought on the northern border of Washington, a substantial ride from the solitary cannon that had been deployed at the Naval Observatory. Taken as a whole, the Civil War had little direct impact upon the practice of organized computation, even though it reordered American society, revolutionized American government, and profoundly touched the lives of the computers themselves. For the major American computing offices, the rooms found at the Naval Observatory, the Nautical Almanac, and the Coast Survey, the war was a plague that passed over their institutions, leaving them little changed. Individual computers had to make their choice "twixt that darkness and the light," in the words of a contemporary poem, but the computing floors could operate as they did before the war.[3]

Only the computers of the Coast Survey served in battle. The survey's

12. Coast Survey Office. Computing rooms are on the top floor

computing division was actually the oldest computing office in the country. It had been formed in 1843, the year that the survey was reformed and reorganized. The division hired "gentlemen in private life," as a report described them, "to repeat the important calculations of the survey."[4] Each summer, the surveyors would disperse to the field to create triangulation surveys, a large network of triangles that connected major landmarks and navigation points along the coast. Inevitably, after the trigonometry and longitude calculations were done, the triangles would fail to align perfectly. Some sides would be too short; others would be too long. Angles would not close, and lines would cross at the wrong place. The surveyors would correct these failures by modifying their numbers, taking a small bit from one side, adding a little to an angle, rounding numbers until all the triangles matched. The Coast Survey computers in Washington duplicated the process of survey adjustment and provided a means of evaluating the judgment of the original surveyors. If the two sets of computations were sufficiently close, then the director would accept the original survey. If they did not agree, then the computers would be asked to prepare a third set of adjusted calculations.[5]

After the war began, some of the survey computers volunteered for advance survey teams. These teams prepared maps for the Union army and charts for the navy, but this work could not be called organized computation.[6] The majority of these field calculations were quite simple, a quick determination of longitude or adjustment of a triangulation point. The computers usually handled the numbers simply, writing the results in a dirty notebook or on a piece of scrap paper.[7] As the historian Hunter Dupree observed about most science of the Civil War, "experimentation was largely done on the spot, when the men were actually confronted with specific and pressing problems."[8]

For those who remained behind, the Coast Survey computing floor became a haven for scientists unwilling to take up arms. The head of the office, a German immigrant named Charles Schott (1826–1901), "did not find military life congenial" and paid a substitute to take his place in a Northern regiment.[9] The son of Benjamin Peirce, Charles Saunders Peirce (1839–1914), joined the office as a way of avoiding the draft. "I perfectly dread [military service]," he wrote the director of the survey; "I should feel that I was ended and thrown away for nothing."[10] In the computing office, he joined Schott in doing a mix of civilian and military computations. They did engineering calculations for the fortifications of Washington and analyzed the trajectories of new army cannons.[11]

In general, the Civil War cannons did not inspire much organized computation. Though the armies and navies of Europe were starting to deploy steel guns with rifled barrels, the forces of both the Northern and Southern states used less accurate, smoothbore cannons. Shells fired by these guns rarely followed the trajectories predicted by mathematical ballistics, and so most ordnance officers needed only figures that showed the approximate range of their shots. The only real mathematical ballistics of the war was done at the experimental battery in the Washington Navy Yard, the same place where the Naval Observatory computers had learned the basic gunnery drill. The battery's gun placement was a long, low building that faced the river. A flag on a small cupola warned nearby ships when the gunners were preparing to shoot. Its staff fired thirty to fifty shots a week across the water, rattling the windows of local houses and reminding the nearby members of Congress that the country was at war. From these shots, the officers estimated the trajectories of the weapons. Most of the computations were done by a single officer, John Dahlgren (1809–1870), who was more interested in estimating the stress on gun barrels than in predicting the flight of shells.[12]

The war produced three new scientific institutions within the federal government, though none of the three had any immediate influence upon the practice of organized computing. The first organization, the Naval Inventions Board, included both Dahlgren and Charles Henry Davis. It reviewed proposals for new weapons that had been submitted to the Union navy, but it conducted no research of its own and produced no computation beyond the simple figures that the board members scratched on proposals. The second institution, the National Academy of Sciences, also listed Davis and Dahlgren among its first members. This group, modeled on the Académie des Sciences in France and the Royal Society in England, was intended to organize the nation's scientists and advise the government "upon any subject of science or art."[13] The French Académie had been an active supporter of organized computation, as it published the *Connaissance des Temps*, but its American counterpart undertook no such activity during the Civil War. The early years of the institution were disrupted by controversies over the selection of the first members. The National Academy of Sciences undertook only a little research during the war, none involving large calculations.[14]

The last new scientific institution of the war was a civilian organization, the Department of Agriculture. Congress, no longer checked by Southern members fearing the expansion of government institutions, formed the new department to promote agricultural production. The department's leader, who shared the name Isaac Newton (1837–1884) with the seventeenth-century mathematician, wanted the new agency to start a new program of scientific computation, one that would gather and analyze statistics of agricultural output and consumption. "Too much cannot be said in favor of agricultural statistics," he wrote. They reveal "the great laws of supply and demand, of tillage and barter, thus enabling both [farmer and merchant] to work out a safe and healthy prosperity."[15] In spite of his enthusiasm, the department was unable to start a comprehensive statistical program during the conflict.[16]

Though the war may have produced no immediate changes in the division of computational labor, its relics slowly transformed scientific calculation. At the war's end, the landscape was littered with thousands of miles of new telegraph wires. These wires followed the armies, Northern and Southern, as they marched across the countryside. They allowed Abraham Lincoln to second-guess his generals in the field and Jefferson Davis to follow the Confederate incursions into Pennsylvania; commanders used them to coordinate the divisions of their armies. After the war's end, these lines were available for a national effort to collect and process weather statistics.

Before the construction of the telegraph, the study of the weather had been a frustrating affair. Though an individual scientist could collect weather data, such numbers were of limited use unless they could be put in the context of a region or a continent. Without a full picture, a researcher might mistake a small squall for a major front, a local wind pattern for the great movement of the atmosphere. In the 1840s, the U.S. Navy and the new Smithsonian Institution had attempted to collect weather data along the northern length of the Atlantic coast. They had recruited four hundred volunteers, including Maria Mitchell's father, given them printed data forms, and created a standard ritual for collecting data. The observers had recorded wind, temperature, and precipitation at fixed times during the day and placed the completed forms in the U.S. mail. Workers at the Smithsonian had sorted the forms and forwarded the data to Lafayette College in Pennsylvania, where a staff of computers summarized the results.[17]

The joint navy/Smithsonian meteorological project was a modest success at best. As weeks passed before the data reached Washington, the results could only be used for historical study and not for forecasting. Perhaps more troubling was a lack of discipline in the system. The volunteers were told to gather data at specific times, but in fact they recorded the numbers when it was convenient for them or perhaps did not record the data at all. The scientists in Washington could not be sure that their summaries represented an accurate picture of the weather at any specific time and day. During the 1850s, the director of the Smithsonian proposed using the telegraph as a means of collecting data more quickly and imposing more discipline on the volunteers. He wrote to several telegraph companies, requesting that they "allow [the Smithsonian], at a certain period of the day, the use of their wires for the transmission of meteorological intelligence."[18] Getting the telegraph companies to cooperate on a such a large project proved to be almost as difficult as collecting the weather data. Most of the telegraph companies owned only a single line or perhaps a small collection of lines that fanned outward from a central office. Generally, these companies were only willing to transmit the data at night, a restriction that still allowed the Smithsonian to gather and transmit their data in a day's time. By the late 1850s, the telegraph lines were sending so much data to Washington that the Smithsonian hired a staff of "expert computers" to process it all.[19]

The war disrupted the Smithsonian weather project. When the collection of weather data was resumed in 1870, it was under the control of the Army Signal Corps, which controlled many of the telegraph lines that had been built during the conflict. The Signal Corps devised a system to collect weather data three times a day; at midnight, 11:00 AM, and 4:00 PM. The observers recorded their data in a coded form and telegraphed their results to Washington.[20] Data moved toward the capital as if it were water coursing through the tributaries of a river. It flowed from farms and villages, joining other data en route and pausing at regional offices to wait for an open moment on the lines that stretched toward Washington. In general, the rising flood of data took about three hours to reach the central office of the Signal Corps.[21]

The Signal Corps created a small computing staff that processed all of the weather data in intensive two-hour shifts. Computers worked at stand-up desks, where they recorded the information on preprinted forms and blank weather maps. Each computer was responsible for only part of the data. One handled temperature, another wind, a third precipitation. They reported to their stations just as the data began to emerge from the telegraph room. Young boys carried the paper from the telegraph to a single clerk, who read each telegram aloud, translating the symbols into placenames, amounts of precipitation, and percentages of humidity. As they heard their numbers being read, the computers would add the figures to their summaries and mark their maps. After the clerk had read the last telegram, the computers required only a few minutes to complete their calculations. The "work is done as fast as the translator can dictate," reported the director of the office, "so that within an hour all the telegraphic reports are received and within the same hour all the translating and map making is done."[22]

At first appearance, the Signal Corps computers had little in common with observatory or almanac staffs. They worked in short bursts of effort, in contrast to the sustained efforts of the astronomical computing offices, and handled calculations far simpler than the process of reducing data or preparing an ephemeris. Yet the staff of the Signal Corps was building upon ideas that had been first seen in observatories. Astronomers had seen the telegraph as a means of coordinating scientific effort across large distances. The principles of electric telegraphy had been first demonstrated by the German astronomer Carl Friedrich Gauss in the 1830s. The staff of the U.S. Naval Observatory had witnessed the first test of an American telegraph in 1844.[23] The second message to travel from Washington to Baltimore in that demonstration of the telegraph, the message that followed the famous invocation from the Hebrew scriptures, "What hath God wrought!" was a question concerning astronomical research: "What is your time?"[24] A difference in time corresponded to a difference in longitude, which was often expressed in values of hours rather than in the modern unit of degrees. (In some sense, we retain this old scale of longitude when we talk about Los Angeles as three hours off New York or Iowa as an hour ahead of Colorado.) The telegraph promised the ability to make measurements of longitude that were accurate to within one hundred feet. A *Nautical Almanac* computer devised a way to make such measurements with equipment no more sophisticated than a pair of standard pendulum clocks.[25] At the Harvard Observatory, the staff attempted to extract the possible accuracy for its telegraphic measurements by installing a telegraph switch panel in its basement and placing telegraph keys on its observing chair.[26]

At the Computing Division of the Coast Survey, the end of the war brought new rigor to the adjustment calculations. For many years, the surveyors had known that the adjustments could be handled mathematically, without any appeal to the judgment of computers, by a method called least squares. Using least squares, a computer could find the minimal number of adjustments that would make all the angles meet and all the sides be the proper length. The method of least squares had been developed by the same Carl Friedrich Gauss who had demonstrated the telegraph. Though many scholars have concluded that Gauss did not invent the method, they do acknowledge that most English-speaking astronomers learned the method of least squares from Gauss's book *Theoria Motus Corporum Coelestium in Sectionibus Conicis Solem Ambientium.*[27] The fact that this book was written in Latin was a hindrance to many American computers. Increasingly, the classical languages of Latin, Hebrew, and Greek were being dropped by the nation's colleges and replaced with modern languages, notably German. During the 1850s, Davis had concluded that his computers needed an English edition of the book and had undertaken the translation himself. He had finished the manuscript just before he departed the almanac to take command of a ship in the Caribbean.[28]

Prior to the war, the method of least squares was not widely used in the survey office because the calculations were long and arduous. It was much simpler to adjust the values by instinct and trust the judgment of the computers. The calculations for least squares required two steps. First, the computer had to prepare what were called "condition equations" or "normal equations." A typical problem would adjust twenty-five values in a survey spread over two or three square miles. In all, there might be two or three hundred measurements that had been made by the surveyors that had to be reduced into twenty-five condition equations, one for each of the values being adjusted. The second step disassembles the condition equations in a messy series of calculations. The process can be compared to the job of disassembling a steel-frame building, moving it to a new site, and reconstructing it in a slightly different shape. As the building is reassembled in its new form, each girder needs to be adjusted with a cut or an extension. Like the process of moving a building, least squares computations required detailed record keeping. Any mistake in either the records or the arithmetic required the computer to begin the process again.[29] The English translation of *Theoria Motus Corporum* did much to promote the method of least squares, but it described a plan for the computations that was slow and redundant. A computer knowing only the plans described by Gauss would resist any large problem.[30] In the United States, the method became widely used only after a Coast Survey computer named Myrrick Doolittle (1830–1911) developed a more efficient means of doing the calculations.

Doolittle was yet another student of Benjamin Peirce, though he had come to Peirce later in life. He had been born in Vermont and lacked the time or the money or the inclination to attend college. Proving to be a natural teacher, he had begun his career by instructing the children of friends and neighbors. At the age of 26, he accepted a position at the New Jersey Normal College and remained there until the onset of the war.[31] Deciding that it was time for him to get a degree, he enrolled at Antioch College in Ohio. Antioch was a small religious college that embraced the abolitionist cause with all the fervor of the age. He studied at the college for two years and was awarded a bachelor's degree in 1862. That summer Doolittle volunteered for an Ohio regiment in the Union army, only to find "his physical condition preventing enlistment."[32] He remained at the college for one more year as a mathematics instructor and then left to join Peirce at Harvard.

We have no record of Doolittle's life in Cambridge. Given the nature of his mathematical interests, it seems likely that he spent some time in the Nautical Almanac Office with the computers, but he stayed less than twelve months with Peirce. In 1864, he left Massachusetts without completing a degree in order to join his wife, Lucy Doolittle, who was working in Washington, D.C. She was a volunteer with the Sanitary Commission, a precursor of the Red Cross.[33] With the help of Benjamin Peirce, Doolittle took a job at the Naval Observatory as an assistant observer. He found the observatory work "wearying," as the job required him to record the positions of stars by night and to reduce observations by day. Deciding that he wanted to live a less taxing life, he resigned his position and took a job at the Patent Office examining the applications for patents on steam boilers.[34]

During Doolittle's time at the Patent Office, Benjamin Peirce came to Washington as the director of the Coast Survey. Peirce surprised both his friends and his detractors by proving to be a competent manager. He demonstrated that he could schedule survey crews, approve new projects, draft budgets, and work with members of Congress. He also found time among the demands of his official duties to conduct mathematical research. He returned to the classical problem of mathematical astronomy, the three-body problem, and developed a new way to compute the motion of the moon. He requested that the Computing Division prepare a table of the moon's position from his equations, but this work had to be suspended "for want of a sufficient and adequate force of computers." He was able to complete the tables only when he convinced the secretary of the navy to allocate an extra $2,000 to pay the salaries of three new computers.[35]

Peirce also experimented with a new form of arithmetic which used only two symbols, 0 and 1, instead of the ten digits of decimal arithmetic. Eventually named "binary arithmetic," it would be the basis of electronic

13. Benjamin Peirce, director of Coast Survey Office

computing machines, though such machines were seventy-five years in the future and beyond the vision of Benjamin Peirce. "I have no such extravagant thought as that of a substitute for our decimal system," he wrote, though he believed that it might be interesting to compute "some of the fundamental numbers of science by a new arithmetic for the purpose of comparison and verification."[36]

In 1873, Peirce offered Doolittle an appointment as a computer. According to his family, this appointment brought Doolittle "into his life's work."[37] He quickly came to dominate the Computing Division, though he never became its formal leader. He was not a mathematician after the manner of Benjamin Peirce or even Charles Henry Davis, but he had an innate grasp of calculation and was able to produce concise, simple computing plans. In 1874, he turned his attention to the calculation of least squares. He did nothing to the method itself, but he found a simpler procedure for handling the computations. He reduced the redundancies in Gauss's procedure, dropping calculations that were handled in other ways. The result was a plan that was more orderly and more efficient. In effect, he found a way of taking down the girders and reassembling them with the fewest steps. "The results reached appear very satisfactory," reported the director of the computing floor.[38] The technique became the standard method of computation at the Coast Survey. It removed the old judgments from survey adjustment and the old "gentlemen in private life" from the computing floor.

In the two decades that followed the end of the Civil War, women began to find a place in the computing rooms. Just as their male counterparts were no longer educated gentlemen, these women were not Maria Mitchells or Nicole-Reine Lepautes, the talented daughters of loving fathers or the intelligent friends of sympathetic men. They were workers, desk laborers, who were earning their way in this world with their skill at numbers. In many ways, they were similar to the female office workers, who were increasingly common in the nation's cities. Before the war, offices were closed to women, except for those well-nurtured daughters, loyal helpmates, or resilient widows. The war had opened government offices to women, as the federal agencies needed more clerical workers than they could draw from the dwindling pool of male labor. Many of these first female clerks were, in fact, war widows, women who had married in the first exciting days of the conflict and lost their husbands on some hallowed battlefield in Tennessee or Virginia. These women earned a meager salary from the government that often had to support children and mothers in addition to the worker herself.[39]

Women began finding employment with private companies after the war ended in 1865. The nature of business had expanded with the war, had become more complex, and now required central office staffs to coordinate production. Business skills were taught at high schools, a new innovation in public education. These skills were similar to those needed by human computers, though clerks did not need to understand the calculus of Newton. Both clerks and computers were subordinate to a professional staff, both were paid a small wage, and both handled routine work. By 1875, one out of six hundred office workers was female, and within a decade, women would fill one out of fifty office jobs.[40]

In 1875, Anna Winlock (1857–1904) approached the Harvard Observatory and asked if she might be employed in calculation. Winlock was the eldest daughter of Joseph Winlock, the director of the Harvard Observatory and the former superintendent of the Nautical Almanac. Joseph Winlock had unexpectedly succumbed to a brief illness, leaving behind a widow and five children with no obvious means of support. The officers of Harvard had been kind and gracious. They had given Mrs. Winlock a decent interval to vacate the house on the observatory grounds and had helped her to find a new home in Cambridge.[41] Once they felt that the new widow was settled, they ceased all financial support of the family.[42]

It fell to Anna Winlock, the eldest child at eighteen years of age, to sustain the family. Winlock had been close to her father in much the same way that Maria Mitchell had been close to William Mitchell. She had watched him work at the almanac office and learned from his example the rudiments of mathematical astronomy. When Anna Winlock was twelve years old, she had been her father's companion on an expedition to view an eclipse of the sun. The party left Cambridge and traveled southwest to Kentucky in order to make its camp. The young girl was probably included because the trip passed through the country of her father's birth. Along the way, the senior Winlock introduced his daughter to his various cousins and sisters and aunts. Nonetheless, the expedition was a time with the astronomers, an opportunity to prepare instruments and observe an eclipse.[43]

Joseph Winlock had left the Harvard Observatory volumes of unreduced observations, a decade of numbers in a useless state. The interim director complained that he could not process the data, as "the condition of the funds is an objection to hiring anyone."[44] At this point, Anna Winlock presented herself to the observatory and offered to reduce the observations. Harvard was able to offer her twenty-five cents an hour to do the computations. Winlock found the conditions acceptable and took the position.[45] In less than a year, she was joined at the observatory by three other women. The first was Selina Bond, the daughter of her father's predecessor. Like Winlock, Bond stood in need of a steady income, as her father's fortune had been squandered through the actions of a "rascally trustee." The second, Rhoda Saunders, was the graduate of a local high school who had been recommended to the observatory by the Harvard president. The last was probably the relative of an assistant astronomer.[46]

By 1880, the Harvard Observatory employed a complete staff of female computers. The director who hired this staff, Edward C. Pickering (1846–1919), is often considered progressive and liberal for employing

14. Computing room of the Harvard Observatory

women. He was called a "true Victorian gentleman in his attitude towards women and to everyone, men and women alike," by one of his computers.[47] Pickering worked closely with a female benefactor, Mrs. Anna Palmer Draper, to finance the activities of the observatory, and he encouraged an assistant to teach mathematical astronomy at Radcliffe College, the new women's school affiliated with Harvard.[48] Yet Pickering was motivated as much by economy as by altruism. "To attain the greatest efficiency," he wrote, "a skillful observer should never be obliged to spend time on what could be done equally well by an assistant at a much lower salary."[49] The salary he offered to the women was half the prevailing rate for calculation. "[The Harvard] computers are largely women," complained the director of the Naval Observatory in Washington, "who can be got to work for next to nothing."[50]

"Pickering's Harem," as the group would occasionally be called, served as an uncomfortable example to the government computing agencies.[51] When the secretary of the navy asked why the Naval Observatory could not reduce its expenses for computation, as the Harvard Observatory had done, he was met with a defensive reply from the superintendent. He claimed that the Naval Observatory paid "its employees at exactly the same (or in some cases less) rate as in other branches of the

Government Service," deftly deflecting the issue of hiring women. Shifting to a more aggressive position, he argued, "To charge extravagance against the Observatory because its employees are paid according to a rate fixed by law for the public service at large is clearly disingenuous and tending to mislead."[52] The Naval Observatory would not hire a female computer until 1901.[53] The Coast Survey and the Nautical Almanac, who had benefited from the labors of Maria Mitchell, were slightly more progressive and hired their second female computers in 1893. However, the Nautical Almanac was self-conscious about this action and identified its new employee as a man.[54]

The Harvard Observatory has left an unusual document that suggests the daily routine of its computing staff and the challenges faced by the female computers. This document is a musical parody, entitled the *Observatory Pinafore*, based on W. S. Gilbert and Arthur Sullivan's operetta *H. M. S. Pinafore*. The parody was written by a junior astronomer, Winslow Upton (1853–1914), and so reflects the point of view of the astronomers, not the computers. It shows the women struggling with their work, confronting astronomers with problems, and working in an environment that would constrain their role or even deny that they were part of a scientific endeavor.

The *Observatory Pinafore* must be one of the first parodies of Gilbert and Sullivan's opera, a show that has been adapted and modified many times. Winslow Upton probably saw the original *H. M. S. Pinafore* during its American premier in November 1878. This production, unauthorized by Gilbert and Sullivan, debuted in the old Boston Museum of Art six months after the show opened in London. Upton acquired a copy of the vocal score and wrote his observatory version nine months later "during four rainy days of an August vacation in Vermont."[55] He replaced the opening chorus of sailors, who tell of their shipboard life, with a chorus of computers, who sing:

> We work from morn till night
>   For computing is our duty;
> We're faithful and polite.
>   And our record book's a beauty;
> With Crelle and Gauss, Chauvenet and Peirce,
>   We labor hard all day;
> We add, subtract, multiply and divide,
>   And we never have time to play.[56]

The song is in the same spirit as the original and suggests that Upton is going to follow Gilbert and Sullivan's original plot. In *H. M. S. Pinafore*, a young sailor is in love with the daughter of his captain. Prevented from marrying by the social gulf between them, the two spend much of the opera planning to elope. "Love levels all ranks," says one character, "but not so much as that." The show is filled with trios and dances, a silly song by a pompous bass singer, a touching lover's lament for the soprano, and a rousing chorus expressing English superiority. Near the end, one character announces, much to everyone's surprise, that the captain and the young sailor were switched at birth, an announcement that immediately elevates the sailor and allows him to wed his love. The other characters gleefully ignore the logical problems with this solution, notably that the young sailor is now old enough to be the father of his bride, and sing a rousing recap of the songs.

Upton could have chosen to adapt Gilbert's plot to the Harvard Observatory. He might have decided to have a young female computer, perhaps a recent graduate of Cambridge Girls' High School, fall in love with an assistant astronomer who is a member of a prominent Beacon Hill family. After two acts of rewritten dances, solos, and choruses, we might discover that there was a mix-up in the two families which, after an extended lawsuit, reverses the fortunes of the two young people and allows them to marry. However, Upton did not follow this approach, as he was not interested in exploring the relations between men and women. Instead, he chose to explore the love between a young man and his science, the love that a junior astronomer had for his position at the Harvard Observatory.

In the *Observatory Pinafore*, Upton recasts the female love interest as a young male astronomer, though he retains the female name "Josephine" from the original.[57] The male Josephine is about to lose his position at Harvard. With no other positions available, he intends to take a job at an inferior private observatory in Rhode Island. Upton expresses nothing but scorn for the Rhode Island observatory. It is the hobby of a wealthy businessman, and hence it is tainted, less virtuous than the Harvard facility. The script portrays the businessman and the director of his observatory as vain and stupid. "I'm very proud of my degree," sings the Rhode Island astronomer, "For it shows that I'm a man of extraordinary sense."[58] As the plot moves through its paces, the Josephine astronomer bemoans his departure from Harvard and eventually finds a way to demonstrate his astronomical skill by repairing a telescope. When he fixes the device, his value is recognized, and he is allowed to remain in Cambridge.

The only computer identified by name in the *Observatory Pinafore* is Rhoda Saunders. She may have been chosen because of her role in the observatory, because she was a friend of Winslow Upton, or because she possessed the best voice among the female employees. The script suggests that she was strong and confident of her skills. In a scene which has no parallel in the original Gilbert and Sullivan play, she defends her work to an agitated assistant astronomer. "Oh! Oh! Oh! Oh! Oh! It's all wrong! A fearful mistake!" exclaims the astronomer, who has just been looking at a sheet of reduced data. When the figures are identified as the work of Saunders, she quickly affirms, "I don't believe it's wrong."[59] The observer eventually calms down and agrees with her. As the script was intended to entertain the staff of the Harvard Observatory, this scene is probably an exaggerated version of a real event, though it gives us no real assessment of Saunders's character. She may have been strong and outgoing or quiet and retiring. Even if it could give us a fuller picture of Saunders, it would be of little use, for the scene is one of the last times that a computer appears in the original script. By the end of the first act, the female computers all but disappear from the show.

Though Upton changed the gender of the female romantic lead and gave most of her songs to a man, he was not such a fool as to give the best soprano song to a male voice. He assigns the lover's lament, the song "Sorry her lot," to Rhoda Saunders and uses it to express some of the frustrations of the computers. In the original opera, this song is one of the points where Arthur Sullivan's music transcends the comedy of Gilbert's words and expresses a few moments of honest human emotion. It is stately, with a delicate melody that falls from high notes to low. In *H. M. S. Pinafore,* the falling melody is a fall of resignation, a recognition that love may be eternal but that courtship and marriage are governed by the conventions of society. "Weary the heart that bows the head," she sings in the last verse, "When love is alive and hope is dead."

Upton's version of the aria nearly matches the language of the original. It is his best writing in the entire script. Saunders, the computer who had held her own against the charges of an assistant observer, sings of repetition, tedium, and self-doubt.

> Sorry her lot, who adds not well;
> Dull is the mind that checks but vainly;
> Sad are the sighs that own the spell
> Symboled by frowns that speak too plainly.[60]

To those unversed in arithmetic, computing was indeed a difficult task, but Saunders was a good computer and served at the observatory for thirteen years. She, as well as the entire staff, had known the times when fatigue made any mental labor difficult. Under such circumstances, computers could barely add a column of figures and would make error after error in the simplest of calculations. Upton underscores the problem of fatigue by contrasting the computers, who work during the day, against the observers, who work at night.

Happy the hour when sets the sun;
Sweet is the night to earth's poor daughters,
Who sweetly may sleep when labor is done,
Unlike their brother astronomers.[61]

The sweet sleep is only a brief respite from their days of toil. Upton reinforces the tedium of the work through the chorus:

Heavy the sorrow that bows the head
When fingers are tender and the ink is red.[62]

The computers used red ink to correct their mistakes. A page of red figures meant the day had been hard and frustrating.

Once the song is finished, Saunders has only a few more lines in the script about mistakes and red ink before her role is finished and she leaves the stage. Just before she departs, there is a brief exchange that suggests the anxieties created by the presence of women at the observatory. The Rhode Island astronomer asks Pickering about the size of the computing staff. Pickering replies that it is "quite large—most enough for a good dance in spare hours." After hearing this, the other astronomer exclaims, "Do you allow, sir, your assistants to dance? Physicians tell me it is promotive of inaccuracy in computation." To modern sensibilities, this response is a quaint remnant of Victorian superstition, but it actually expresses a deep anxiety of the times. No one was entirely confident that men and women who were unrelated by family ties could work together in offices without being influenced by physical attraction. Taken at face value, the script suggests that E. C. Pickering may not have thought about this question, for the observatory captain responds, "Indeed! I was not aware of that."[63] However, it seems likely that Upton is being ironic, for Pickering was comfortable with the presence of women in his observatory, while the leaders of the Naval Observatory and the Nautical Almanac were not.

In 1880, shortly after Winslow Upton had finished the *Observatory Pinafore*, the director of the real Harvard Observatory, E. C. Pickering, announced that "a handsome carpet has also been purchased for the Computing Room in the east wing."[64] It was a pleasant addition to the room, for it deadened excess noise and provided a layer of insulation against the frozen ground of the Massachusetts winter. Pickering undoubtedly purchased it with all goodwill, yet the carpet was also a symbol of divided labor, of tasks for men and women. The literature of the time talks of separate spheres for the two genders, a metaphor that borrows from the classical model of astronomy. The sphere of Mars encircles the sphere of Venus, but both spin on the same axis. Men could move freely through the observatory, even into the telescope room, "a dark and dingy place," Upton tells us, "all clattered up and smelling strong of oil."[65] The women had their computing room, with its desks and its carpet.

The Harvard Observatory marks the end of experiments with dividing the labor of computation. Subsequent computing offices draw their models, consciously or unconsciously, from the laboratories that were founded before 1880. The Harvard Observatory also marks the end of astronomy as the dominant force in scientific calculation. Indeed, the *Observatory Pinafore* makes it clear that orbits and positions are no longer the most interesting part of astronomical research. Upton's writing transforms the rousing chorus of British superiority at the end of the play into a song of praise for a photometer, an instrument that measures the brightness of stars.[66]

PART II

Mass Production and New Fields of Science 1880–1930

> The engineer, the astronomer, the mathematician, the electrician, form a mighty and always increasingly important army of male labourers.
>
> Olive Schreiner, *Woman and Labor* (1911)

CHAPTER SIX

Looking Forward, Looking Backward: Machinery 1893

> As the machine is truer than the hand, so the system, which does the work of the master's eye, turns out more accurate results.
>
> Edward Bellamy, *Looking Backward* (1888)

ON A TYPICAL WORKDAY in 1881 or 1882, Rhoda Saunders would begin her calculations in the Harvard Observatory computing room by picking up her pen, uncorking her bottles of ink, one black and one red, and opening her computing book. All of these objects would have been familiar to Edmund Halley in the seventeenth century or Nicole-Reine Lepaute in the eighteenth, but each had been subtly changed by industrialization. Saunders's pen had a preformed steel nib that easily outlasted the hand-cut point of a goose quill. Her bottles of ink were commercially produced and varied little from batch to batch. The lines in her computing book, printed by a mechanical press, were straight and true, without any of the wiggles or cramped margins found on the hand-ruled sheets of Nevil Maskelyne or Maria Mitchell.

The uniformity that could be found in Saunders's mass-produced writing implements could also be found in a new generation of computing tools. The slide rule, which became popular in the United States during the 1880s, was a mass-produced version of a 1622 invention.[1] The original slide rule had been created by the English mathematician William Oughtred (1574–1660), who used the concept of logarithms. Logarithms were a new discovery in the early seventeenth century. They are special values that converted multiplication into addition. For any number, a mathematician could find a corresponding logarithm. For example, the logarithm of 2 is .3010, while the logarithm of 3 is .4772.[2] By adding the two logarithms, we get .7782, which is the logarithm of 6, the product of 2 and 3. Oughtred created his rule by inscribing two rods with logarithmic scales that resembled the scales of an ordinary ruler, except that the space between numbers grew smaller and smaller as the values increased. By sliding these two rods, Oughtred could add two logarithms or, equivalently, multiply two numbers.

The slide rules of Oughtred and other seventeenth-century scientists were expensive, hand-crafted devices, each one designed for a specific kind of problem. Isaac Newton created a rule with three slides that could be used to solve cubic equations. England's first Astronomer Royal, John Flamsteed (1646–1720), purchased a special rule for astronomical calculation.[3] Only with the invention of precision ruling machines that could cut logarithmic scales did these rules become common. These machines started appearing in the early nineteenth century. By 1833, slide rules were so inexpensive that Gaspard Riche de Prony could report that Parisians are "commonly using 'sliding rules' at all levels, including shop keepers and artisans."[4]

In the United States, "interest in the slide rule was awakened in about 1881," according to an early historian of mathematics. Again, the driving force behind the adoption of the slide rule was a manufacturing economy that could produce them in large quantities. During the 1880s, slide rules were introduced in the American university curriculum, and they started to appear in the engineering literature. Most American scientists adopted a standard version of the slide rule known as the "Mannheim rule," named after its designer, a French army officer.[5]

Slide rules were used only sporadically by the observatories and almanacs because they could not perform calculations to the precision needed by astronomers. A good slide rule operator could get an answer to two or three digits of precision. Observers and astronomers often wanted each calculation accurate to six or eight digits. The astronomical computers were more open to the geared adding machine. Like the slide rule, the adding machine was a seventeenth-century curiosity that became an important tool only after it encountered nineteenth-century methods of mass production, especially the development of interchangeable parts and high-precision manufacturing. The adding machine can be traced to the same geared mechanism of Blaise Pascal that inspired Babbage's Difference Engine. Pascal's device consisted of eight geared wheels, each marked with the digits from 0 to 9. An operator would add two numbers by turning the wheels. Once a wheel had completed a full cycle, moving through the numbers 1 through 9 and returning to zero, it would trip a carry mechanism that would advance the neighboring wheel by one place. Pascal tried to manufacture his machine and sell it, but he was unable to build a robust device that would produce reliable results in the hands of others. As the historian Michael Williams has observed, "The mechanism, although ingenious, is rather delicate and prone to giving erroneous results when not treated with the utmost care."[6]

The first adding machine to achieve large sales, the French Arithmometer, was a contemporary of the Scheutz difference engine. It was developed by Charles Xavier Thomas de Colmar (1785–1870) during the

1840s. De Colmar, an actuary, attempted to sell his first machines to insurance firms but found it difficult to convince business owners that they could benefit from a machine that could do arithmetic. "It was not enough to see need," wrote James Cortada, historian of the adding machine industry; "the [adding machines] had to be sold."[7] De Colmar began to find a market for his machines only after he displayed them at the Paris Exhibition of 1867.[8] He drew visitors to his presentation with a giant Arithmometer, a machine approximately the size of a piano. His standard machine, which was only about two feet long, was purchased by customers from France, Britain, and even the United States.

Unlike the slide rule, the de Colmar adding machine could compute to high precision. By 1878, a British scientist observed that "English astronomers are now just beginning to use [the arithmometer] for the tedious computations continually going on in observatories."[9] Demand for de Colmar's machine began to grow in the late 1880s, largely driven by the expansion of office work. The decade produced several improvements to calculating machines, including lightweight adding mechanisms and keyboards for entering numbers. In the United States, the adding machine industry took root in the new industrial cities of St. Louis and Chicago. These cities were railroad hubs and supported substantial populations of skilled mechanics and adventurous businessmen, "the hopeful and the hopeless—those who had their fortune yet to make and those whose fortunes and affairs had reached a disastrous climax elsewhere."[10] They produced industry pioneers such as Frank Baldwin (1838–1925), William Seward Burroughs (1855–1898), and Dorr E. Felt (1862–1930). Baldwin created a lightweight machine that resembled a hand-cranked coffee grinder that had been laid on its side. Burroughs designed a more substantial machine for banks. His device, also named the Arithmometer, stood four feet tall on long legs and was covered by ranks of buttons that were labeled with the numbers 0 to 9. Felt called his device the Comptometer.[11] It was about the size of a cigar box and weighed no more than a college dictionary. It was the first machine purchased by one of the major computing laboratories. In 1890, the Coast Survey computers acquired one for Myrrick Doolittle, who used it to adjust triangulations with his method of least squares. The next year, the Coast Survey Comptometer was joined by a Burroughs Arithmometer.[12]

The last of the new computing machines, the punched card tabulator, was developed at the same time that Baldwin, Burroughs, and Felt were building their first adding machines. Like the slide rule and the adding machine, it combined ideas from older technologies with the new methods of mass production. Punched cards had been developed to control looms in the early nineteenth century. The tabulator had a geared counting mechanism, like the mechanism of Pascal. The influence of mass production came not from machine tools or the concept of interchangeable parts but from the mass processing of data at the United States Census Office. The Census Office was assembled every ten years to fulfill the constitutional requirement of enumerating the American population. Over the ten censuses that had been performed since 1790, the office had developed an elaborate system of clerks and forms to summarize the population counts. Through the 1880 census, the office employed no real machine to assist the work beyond a frame that held a role of paper.

One of the workers at the 1880 census was Herman Hollerith (1860–1929), a recent graduate of the Columbia University engineering school. Hollerith would later recall that his interest in mechanical tabulation began on a day when he was inspecting the office operation with the census director. Watching the clerks repeat similar operations again and again, the director remarked that "there ought to be a machine for doing the purely mechanical work of tabulating population and similar statistics."[13] It took three years for Hollerith to conceive the basic principles of a mechanical tabulating device and refine these ideas into a workable system. In later years, he would claim that he found his inspiration by watching a train conductor punch holes in a ticket. By 1885, he was recording data as holes in cards, where each card held information for a single person. Was the subject female? There was a hole to punch. Was she a resident of Massachusetts? There was another hole to punch. Did she own property? There was a third hole to punch. His tabulator read the data from each card. The card was placed on a little frame, and an array of wires was lowered upon it. If the wire encountered a hole, it would complete a circuit and advance a counter. If it encountered cardboard, it would do nothing.[14]

Hollerith tested his tabulating system in 1886 by computing vital statistics for Baltimore, the numbers of births and deaths in the city. The test gave him an opportunity to see how his machines would fare in a production situation and correct any design flaws. The Census Office followed the progress of the test and the accomplishments of the machines. In 1889, they gave the machine a formal trial for the 1890 census and compared it with two other methods that had been proposed to tabulate the records. Of the three, Hollerith's method was the only one that involved a new computing machine. The director of the census judged that the three "methods have certain features in common," as each recorded the census information on cards and then sorted the cards to tabulate the data. The other two methods sorted the cards by hand and then used racks or preprinted paper forms to make the final counts. When the three systems were compared in a trial, Hollerith's system proved to be twice as fast as the other two. "The total probable saving," wrote the director of the census, "from the use of punched slips and the electrical counting-

15. Hollerith punched card tabulator and operator

machine would amount to $579,165," enough money to pay the salaries of five hundred census clerks.[15]

In early 1890, when the tabulating machines were installed in the Census Office, they created a sense of mystery and reverence among some of the staff. "One who has not had personal experience in handling cards cannot conceive the stimulating effect which they have upon the imagination of the statistical computer," wrote one worker. He described the cards as images of living beings "whose life experience is written upon their face in hieroglyphic symbols." The tabulating machines, which not only counted the cards but also sorted them into little bins, seemed to presage a final judgment. One human computer claimed that the sounds of the tabulating machine were "the voice of the archangel, which, it is said, will call the dead to life and summon every human soul to face his final doom."[16]

The tabulators were not motor driven and hence did not produce the roar of a machine shop, but they made a loud click each time the wires descended upon the card and rang a bell to signify that the data had been counted. A visitor to the Census Office recorded that the facility was far noisier than an ordinary office. "Upon first entering the [computing] room," he wrote, "the whizzing of the electric fans and the ringing of bells, which are attached to each machine and respond to every touch upon the keys, are confusing." The Census Office paid a weekly lease on the machines and was eager to complete the tabulation as quickly as possible. Clerks were required to punch 700 records a day. Tabulator operators processed 50,000 records on each shift. The visitor observed that the census director demanded so much of the staff that many workers lasted only a few days and that, finally, "none but the industrious are left."[17]

The Hollerith tabulators received their first public display at the World's Columbian Exposition, the Chicago World's Fair of 1893. Hollerith was a reluctant participant, perhaps concluding that his invention would be overlooked at an event that offered 65,000 individual exhibits spread over a 633-acre campus.[18] "He doesn't want to go one bit," his mother-in-law recorded, "and will only exhibit his machines in the Census Office exhibit." His machines were rewarded with a bronze medal as a "novel electrical tabulating system," but scientific computing and computing machinery were only a small part of the exhibit.[19] The Hollerith machines occupied one corner of the government display. Slide rules from Germany and America could be found between bolts of cloth and gleaming steam engines. The adding machines invented by Dorr Felt and William Burroughs were grouped with accounting ledgers and wooden desks. The U.S. Navy gave a small case to the *American Ephemeris and Nautical Almanac* as part of a grand display of naval power. The Coast and Geodetic Survey proudly showed its maps and field instruments to the visitors, but only a discerning observer would have appreciated that the agency had a large computing office.[20]

"The Exposition itself defied philosophy," wrote historian Henry Adams. "Since Noah's Ark, no such Babel of loose and ill joined, such vague and ill-defined and unrelated thoughts and half-thoughts and experimental outcries as the Exposition, had ever ruffled the surface of the

16. Exhibit hall of the World's Columbian Exposition

Lakes."[21] For all of its importance to Chicago and to American culture, the exposition was only a trade show, a grander version of the American county fair. The direct connection to the rural agricultural shows could be found at the south end of the grounds, where farmers displayed well-scrubbed hogs and carefully nurtured ears of corn. Perhaps the only common theme to be found in the exposition were the statistics of economic growth that accompanied almost every product. Exhibitors used numbers to describe the cost of their products, the value of labor they would save, the lives they would improve, and the profits they would make. Henry Adams confessed that he stood before one exhibit and was "obliged to waste a pencil and several sheets of paper trying to calculate exactly when, according to the given increase of power, tonnage, and speed, the growth of the ocean steamer would reach its limits."[22] An extreme use of statistics could be found in the German pavilion, where Krupp Industries displayed a new cannon that weighed 42 tons, had a 17-inch bore, and could throw a 2,300-pound projectile 16 miles at a cost of $1,259 per shot. In jest, Krupp engineers had suggested that the exposition committee might fire the gun at the conclusion of the fair, as they had calculated that the "concussion would undoubtably knock down all the great buildings in Jackson Park and thus save a lot of labor in their removal."[23]

Taken by themselves, these commercial statistics offered little to human computers, as they were not considered to be part of scientific calculation. When placed in the larger context of the fair, the presence of social statistics suggested that many individuals and organizations were attempting to bring the precision of scientific calculation to the phenomena of social, economic, and personal life. The use of statistics was clearly seen at the intellectual part of the exposition, a series of meetings called the World's Congress Auxiliary. The Congresses, as they were called, were held at the newly finished Art Institute building in center-city Chicago rather than at the fairgrounds. In contrast to the fair directors, who were most interested in material products and inventions, the Congress organizers took as their motto "Not things but men; not matter but mind." Between May and October, there were 1,200 separate meetings that covered almost every aspect of human endeavor. Some of these congresses were conferences of existing professional organizations, while the rest were special events that covered topics such as women's rights, religion, art, philosophy, engineering, history, literature, and science.[24]

The Congress most directly related to computation, the meeting on mathematics and astronomy, revealed signs of a split between the two disciplines, the decline of an old alliance that had nurtured human computers. Representatives of the astronomers and the mathematicians had agreed to hold a single meeting, but there was no Benjamin Peirce, no central figure to pull the groups together. The mathematicians, desirous of showing their independence and sophistication, were discussing the theorems and proofs of German mathematics. It was a time when German scholarship, particularly German scientific research, "became the focus of extravagant excitement and admiration."[25] German mathematicians emphasized abstraction, generality, rigor, and formal proofs. Of the forty-six mathematical speakers at the Congress, half came from Germany. Congress organizers were so interested in German methods that they arranged for a special train to take them to the fairgrounds so that they might study a display prepared by the German universities of Göttingen, Bonn, and Berlin.[26]

At the Mathematics and Astronomy Congress, the established computing offices could offer nothing that could compare to the German speakers. The first American to address the group was the librarian of the Coast and Geodetic Survey Office, Artemas Martin (1835–1918). Martin was the kind of mathematician that had once been common in the United States. He had been raised on a farm in western Pennsylvania and had taught himself the basic elements of mathematical practice. While selling produce at a local market, he would fill the margins of his account books with mathematical problems and their solutions.[27] Though he was well known for his "rare and happy faculty of presenting his solutions in the simplest mathematical language," his contributions seemed overwhelmed by those of the German contributors.[28]

The astronomy talks were also divided between the old and the new. The senior astronomers were more interested in the new methods of astrophysics than in the classical calculations of positional astronomy. Edward Pickering, whose computing floor at Harvard remained one of the largest astronomical computing groups, talked about his analysis of the light reflected from the moon, an analysis that identified the chemical composition of the lunar surface. The discussion of calculation was left to the junior astronomers: a French computer, Dorothea Klumpke, from the Paris Observatory; Maria Mitchell's replacement at the Nautical Almanac Office, David Todd; and Harvard computer Wilhemina Fleming.[29]

The Congress on Electrical Engineering, which was held the same week as the Mathematics and Astronomy Congress, suggested that organized computing was starting to move into the commercial and manufacturing applications of science. An employee of the new General Electric Company, the engineer Charles Steinmetz (1865–1923), gave a talk on the mathematical model of alternating current. Near the end of his presentation, he described his company's computing division. For those familiar with the computing offices of observatories and almanacs, the name was misleading, for the division had no large staff of human computers. The computing division consisted of a small staff of engineers who reviewed the designs of electrical circuits, motors, controllers, and other devices. Using the techniques that were being developed by Steinmetz and others, these engineers verified that the electrical devices would behave as they were intended to behave. This work required a lot of calculation, but for the moment, all of the arithmetic was handled by the engineers themselves. A decade would pass before they would divide the work with a staff of human computers.[30]

The Congress on History provided a starting point for the use of numbers in the social sciences and the need to process large amounts of statistical data. The congress included a paper by University of Wisconsin professor Frederick Jackson Turner, who began his talk by referring to the reports of the 1890 census. Using numbers that had been tabulated by Herman Hollerith's machines, the reports stated that the United States no longer had a large area that could be considered an unpopulated frontier. "This brief official statement marks the closing of a great historic movement," observed Turner to his audience. "Up to our own day, American history has been in a large degree the history of the colonization of the

Great West."[31] Turner's conclusion may have surprised his audience, but it built upon a traditional relationship between statistics and the discipline of history. Through the end of the nineteenth century, the study of statistics was related more closely to historical research than to mathematical study. The term "statistics" was taken to mean the numbers of the state, the numbers that described the strength, wealth, and health of a country.[32] Most of the early American statisticians were either physicians or historians. The physicians were using numbers to measure problems of public health, while the historians were interested in social stability.[33] The Statistical Congress at the fair spent little time on mathematical issues and debated how numbers could be better used for governance and management.[34]

By 1893, statistical methods had begun to spread to other fields of research, notably economics, agricultural research, and the field that would ultimately be named "Sociology." The Congress on Social Progress caught the first discussions of this new discipline. One of the key speakers, the Chicago social worker Jane Addams (1860–1935), based her ideas on the practical needs of the city dwellers, but she reached for a deeper understanding of society that could only come through numbers. She not only spoke of individual cases that appeared at Hull House, the institution that she had founded, but also tried to give a fuller picture of social needs in the city of Chicago. Her ideas were echoed in other discussions that touched upon social issues, notably the Congress on Women's Progress and the Congress on Labor.[35]

Henry Adams, who spoke at the History Congress, clearly saw the rising importance of statistics and numbers in the study of social life but was uncomfortable with such tools. "At best [I] could never have been a mathematician," he wrote, "but [I] needed to read mathematics, like any other universal language, and [I] never reached the alphabet."[36] Numbers tended to suggest a scientific certainty, fundamental laws, ultimate goals. To him the fair and congresses suggested that Americans seemed to be "driving or drifting unconsciously to some point in thought, as their solar system was said to be drifting towards some point in space,"[37] but he could not identify that point. Within the field of computation, it is hard to find a single idea at the fair that summarized the position of human computers in 1893. One can find the influence of the traditional computational fields: astronomy, calculus, surveying, and navigation. Equally prevalent were the new ideas of German mathematics, social science, mathematical statistics, and computing machinery. Tying these themes together were the familiar strands of mass production and the division of labor. By 1893, most observers could see that the industrial economy had both benefits and drawbacks. Companies rewarded their workers unequally. Factory methods eliminated some of the skills that workers had passed from generation to generation. The industrial economy had only a few places for women, even though colleges were educating women in record numbers. Industrial leaders, including scientists, could develop products and ideas that were not always beneficial to society as a whole. The innovations in scientific calculation that came with mass-produced calculating machines were not as easy or as obvious as the lessons in divided labor. If they were headed toward a single point in space, that point encouraged the expansion of scientific methodology to problems beyond astronomy, the demand to use resources efficiently in research, and the requirement to have accurate results.

CHAPTER SEVEN

Darwin's Cousins

I was quite certain that . . . the contemporary woman would
find her faculties clear and acute from the study of science. . . .
Jane Addams, *Twenty Years at Hull House* (1910)

IN 1894, when the playwright George Bernard Shaw (1856–1950)
needed to invent a character that captured the challenges faced by the
young women of his age, he made her a mathematician. Vivian Warren,
the central character of the play *Mrs. Warren's Profession*, is a graduate
of Newnham College, a women's school at Cambridge. Such colleges
were still new in the 1890s and were trying to find their way amidst the
older and wealthier men's schools. One measure of success for the
women's schools was the scores of their students on the Tripos, the Cambridge mathematical honors exam. In 1890, a Newnham student had
drawn national attention by besting all of her male peers and achieving
the top score on the Tripos, an achievement that would have made her
First Wrangler but for her gender.[1] In *Mrs. Warren's Profession*, Shaw has
the fictional Vivian Warren achieve the third-highest score on the exam,
a detail that was probably added in consideration of Shaw's friend, the
mathematician Karl Pearson (1857–1936). When Pearson was a student
at Cambridge, he had been the Third Wrangler in the Tripos.[2] As a
friendly jab at Pearson, who was somewhat sensitive about the fact that
he did not get the top score on the exam, Shaw has Warren confess that
she took the Tripos exam only because her mother agreed to pay her fifty
pounds "to try for fourth wrangler or thereabouts." Even though she
bested her goal, Warren concludes that the Tripos "doesn't pay. I
wouldn't do it again for the same money."[3]

In *Mrs. Warren's Profession*, Vivian Warren is identified as an actuary,
but she does the work of a human computer. She describes her work as
"calculations for engineers, electricians, insurance companies, and so
on." In one speech, she talks about how much she enjoys working in an
actuarial office in the city of London. Her days are spent in calculations.
"In the evenings we smoked and talked, and never dreamt of going out
except for exercise. And I never enjoyed myself more in my life."[4]

The play opens with the trappings of a domestic comedy: a young
woman, a young man, a country house, a wise friend, and a mother, the

Mrs. Warren of the title. As the plot unfolds, we learn that Mrs. Warren is not simply a wealthy landowner but also a former prostitute and the manager of a brothel. As Vivian Warren learns her mother's story, she systematically rejects the other characters of the play and retreats into mathematics, as if it is the only thing that is pure and untainted. Her suitor is the easiest thing to reject, as he proves to be her half brother. She also declines a marriage with the brothel's financier, rejects the conventional advice of the wise friend, and firmly expels her mother from her life.[5]

Like Vivian Warren, the new computers of the 1890s were college graduates, though none left a record quite so dramatic as the one described in Shaw's play. Many were graduates of the new women's colleges: Newnham and Girton at Cambridge, Bedford in London, Radcliffe and Bryn Mawr in the United States. Most of these colleges had been formed in the late 1870s or early 1880s. Though only a small fraction of their students studied science, the numbers were growing, as were the expectations that the graduates would find useful work. "If it had been wasteful in the 1870s for women to sit idly home," wrote the historian Margaret Rossiter, "it was much more intolerable for college graduates to lack useful and respectable work." Rossiter notes that women moved quickly into laboratories but that they were "introduced in ways that divided the ever-expanding labor but withheld most of the ever precious recognition."[6] For the women of the 1890s, the social and biological sciences offered new opportunities for employment. These fields were incorporating new methods of statistical analysis, methods that required not only the traditional measuring of samples and tabulating of data, but also the more sophisticated calculations of the new mathematical statistics.

The advance of statistical analysis was closely tied to Charles Darwin's theory of evolution in much the same way that astronomical calculations were linked to Newton's fundamental laws of motion. Darwin's theory suggested that biological organizations were shaped by the force of natural selection, that natural selection was still operating in the nineteenth century, and that the effects of natural selection might be measured in both animals and people. If it could be measured, it might provide an explanation for a host of biological and social phenomena, just as Newton's theory of gravitation provided an explanation for Halley's comet. Darwin claimed that evolution could explain the size and shape of animals. His followers speculated that evolution might explain differences in intelligence, behavior, and even social standing. "Those whom we called brutes," quipped George Bernard Shaw, "had their revenge when Darwin shewed us that they are our cousins."[7]

One of Darwin's human cousins, Francis Galton (1822–1911), worked to find a mathematical way of verifying the presence of natural selection.

Galton has been portrayed as "a romantic figure in the history of statistics, perhaps the last of the gentlemen scientists,"[8] a characterization that describes his family background and captures the unorganized nature of the science he pursued. His father was a wealthy Birmingham banker, and his mother was the daughter of a wealthy physician and the aunt of Charles Darwin. He had enrolled in Cambridge with the intent of taking the Tripos and pursuing a career in mathematics. The strain of study broke his health and forced him to temporarily withdraw from school. "It would have been madness to continue the kind of studious life that I had been leading,"[9] he concluded. After a year of rest, he returned to Cambridge and completed an ordinary degree, without taking the Tripos and without honors.[10]

Without an honors degree, it would have been difficult for Galton to find an academic appointment. Like Charles Babbage, Galton had inherited a substantial fortune, and again like Babbage, he used his funds to finance his interest in science. He spent several years traveling through the Middle East and recording his observations of the land and its inhabitants. At times, his travels seemed to be more a rite of passage for a wealthy young man than a genuine scientific expedition. The historian Daniel Kelves reported that Galton sailed down the Nile River "lazing the days away half dressed and barefoot."[11] The trip was not entirely an adventure, for Galton brought a modicum of rigor to his work. Writing his brother from East Africa, he reported, "I have been working hard to make a good map of the country and am quite pleased with my success. I can now calculate upon getting the latitude of any place, on a clear night to three hundred yards."[12] He did not suggest that he had mastered the more difficult calculation of longitude.

In his records of the trip, Galton shows that his ideas on quantification were crude and often uncertain. In one episode, often retold, his work could have been lifted directly from Jonathan Swift's description of Laputa. In East Africa, Galton reported to his brother that he had found a community in which the women "are really endowed with that shape which European milliners so vainly attempt to imitate," adding that they had "figures that would drive the females of our native land desperate— figures that afford to scoff at Crinoline." To quantify the shape of these women, Galton had measured the dimensions of their bodies as the Laputan tailor had measured Gulliver. "I sat at a distance with my sextant, and as the ladies turned themselves about, as women always do to be admired, I surveyed them in every way." Once he had recorded the angles, he "subsequently measured the distance of the spot where they stood— worked out and tabulated the results at my leisure."[13] If Francis Galton had moved in literary circles and had been as familiar with Charles Dickens as he was with Charles Darwin, this letter might be considered a joke, a satire on scientific practice, a sly way of telling his brother that he had spent the day studying half-naked women with a telescope.[14]

Upon his return to England, Galton found a position at the Kew Observatory, a government-funded weather research station. He spent most of his time testing new meteorological instruments, but he found some time to consider problems that were suggested by the different sizes and shapes of the Africans.[15] He tried to put his investigations in the context of Darwin's theories and tried to derive mathematical methods that would verify the action of natural selection. At first, he attempted to find a way of measuring economic and social success across the generations of a single family. "As a statistical investigation, it was naive and flawed," wrote historian Steven Stigler, "and Galton seems to have realized this."[16] In his second approach to this subject, he considered physical traits, such as those he had measured with his sextant in East Africa. The standard methods of statistics were largely confined to the tabulation of data and gave him no obvious way to approach the problem.

Galton was more comfortable with graphical techniques than with computations or formulas. In one problem, he used a graph to find a mathematical relationship between the heights of parents and the heights of their fully grown sons. His set of data included measurements on 928 people, 205 pairs of parents and 518 sons. His first step was to reduce the heights of the two parents to a single value, a value that he called the "mid-parent." The mid-parent was an average of the values with a slight adjustment to place the mother's height on the same scale as the father's. Once he had computed the mid-parent value, he paired this value with the height of the son and created a graph. "I began with a sheet of paper, ruled crossways, with a scale across the top to refer to the statures of the sons," he explained.[17] The scale down the side referred to the mid-parents. For each pair of data, he drew a small pencil mark on the grid.

The final picture looked like an oval. Tall parents tended to have tall sons, and short parents seemed to produce short sons. He summarized that relationship by drawing a line from one of the narrow ends of the oval to the other, a line that split the oval in half. The slope of that line, when adjusted for scale, would be known as the correlation coefficient.[18] A correlation value close to 1 indicated that the quantities would be highly related. A value close to zero indicated that they had no relation. Unsure of the underlying mathematics, he turned to a Cambridge mathematician, who confirmed the "various and laborious statistical conclusions with far more minuteness than I had dared to hope."[19] In confirming the work, the Cambridge mathematician could produce no simple formula for the correlation coefficient. The only way that Galton could compute a correlation was to draw the picture with its ovals and lines. That restriction did not seem to bother him, as Galton at first believed that he had solved a special problem with limited application. It took him about five years to appreciate that he had created a general method for studying any statistical data that shared the same mathematical properties as his height data. "Few intellectual pleasures are more keen," he wrote, "than those enjoyed by a person who . . . suddenly perceives . . . that his results hold good in previously-unsuspected directions." Still, he was embarrassed that he had not recognized the importance of his discovery and confessed fear that "I should be justly reproached for having overlooked it."[20]

Galton's influence on organized computation began in December 1893, when he established the "Committee for Conducting Statistical Inquiries into the Measurable Characteristics of Plants and Animals." This committee, which reported to the Royal Society, was a test of organized scientific research. It was a time of "trial and experiment," wrote one observer. "The statistical calculus itself was not then even partially completed," and "biometric computations were not reduced to routine methods." The first work of the committee was to support the research of the biologist W. F. Raphael Weldon (1860–1906). Weldon had discovered the methods of Galton in the late 1880s and applied them to the study of shrimp and crabs. He "was on the look-out for a numerical measure of species," wrote one biographer, and sought in his measurements evidence that one type of animal was evolving into two species. He was an energetic researcher and pushed the committee beyond its ability to support his work. None of the members could provide the mathematical advice he needed, though they did "ask for a grant of money to obtain materials and assistance in measurement and computation."[21]

Through most of his early research, Weldon's chief computer was his wife, Florence Tebb Weldon (1858–1936). Florence Weldon was one of the first college-educated human computers. She had graduated from Girton College at Cambridge, a companion to Newnham. By working for her husband, Florence Weldon received little recognition but probably found a substantial scope in her scientific work. She did the same tasks that her husband handled. The two of them spent their summers traveling around England and visiting Italy. Typically, they would collect about a thousand specimens, clean the animals, and measure them. In an early study, they took twenty-three measurements on each specimen. Wife and husband shared the labor of research, tabulated the results, and calculated averages, "doing all in duplicate." They "were strenuous years in calculating," recorded a friend. "The Brunsviga [calculator] was yet unknown to the youthful biometric school." The Brunsviga, a favorite of English statisticians, was similar in design to the machine invented by Frank Baldwin in St. Louis. It was small and light and used sliding levers, rather than keys, to record data. Having no calculating machine of any kind, the Weldons "trusted for multiplication to logarithms and [the tables of] Crelle."[22]

Florence Weldon proved to be a greater help to her husband than Galton's committee. "The committee did not possess a mathematician to put on the break," claimed Shaw's friend Karl Pearson, "and Weldon attempted too much in too short a time." As W. F. Raphael Weldon began to publish his results, he was met by the same kind of hostility that had greeted the calculations of Halley and Clairaut. W. F. Raphael Weldon's work was far from perfect, and his mathematical formulas did not always demonstrate the properties he had hoped to illustrate; but the response to his work was not in proportion to its flaws. "The very notion that the Darwinian theory might, after all, be capable of statistical demonstration seemed to excite all sorts and conditions of men to hostility," observed Pearson. W. F. Raphael Weldon worked with the committee through the mid-1890s and then took a position at Oxford University.[23]

The methods of organized statistical calculation, especially the calculation of correlation values, developed in the laboratory of Karl Pearson at the University of London. Pearson was a professor of mathematics, but he had broad interests that ranged from history and politics to religious faith and the relations between the sexes.[24] Born Carl Pearson, he was the son of a London attorney and the product of a traditional Cambridge mathematics education, including the third-place finish on the Tripos. As a young man, he had a crisis of faith that caused him to abandon conventional Christianity and embrace socialism. After two years of study in Germany he adopted the German spelling of his first name, *Karl*.[25] Pearson was a radical but not a bomb thrower. George Bernard Shaw described such people as Pearson as "unconventional in a conventional way." "[He] was in many ways poorly socialized," observed his biographer, Ted Porter, "a thoroughly original character who, while drawing deeply and repeatedly from the cultural resources of his time, rejected many of the conventions of his class and profession."[26] Before turning his attention to statistical theory, he wrote books on the philosophy of science and organized a selection of his friends into a Men and Women's Club. According to a historian of the club, "discussions ranged from sexual relations in Periclean Athens to the position of Buddhist nuns, to more contemporary discussions of the organization and regulation of sexuality, particularly in relation to marriage, prostitution, and friendship." Pearson's presentations were highly intellectual, often laced with Darwinian ideas, and were occasionally beyond the grasp of other club members.[27]

Pearson's influence over the practical issues of computation began in 1895, when the Royal Society added Pearson to the Committee for Conducting Statistical Inquiries into the Measurable Characteristics of Plants and Animals. Pearson was not particularly impressed by the organization of the group or its method of operation. He later recalled that "Weldon's work was hampered by the committee" and suggested that the members had neither the inclination nor the ability to help him.[28] Pearson's contribution to the group was a mathematical formula for correlation, a formula that turned correlation analyses from a lengthy graphical procedure requiring a certain judgment to a straightforward equation. This formula required the computers to summarize the data in five quantities. In Galton's example, two of the quantities were computed from the children's heights, two more came from the mid-parent heights, and the last was calculated from products of the two sets of data. A final computation of four multiplications, three subtractions, one square root extraction, and a long division produced the correlation value.[29]

The formula for correlation became one of the first mathematical tools for a small computing group that Pearson formed at the University of London in an organization he would call the "Biometrics Laboratory." "All the work of computing undertaken in my Department," Pearson explained, "[was] entirely done by volunteer workers," a group of computers that included students, friends, relatives, and his wife.[30] Though Pearson dominated the group, he liked to think that they were all collaborating as equals. The first large project of this group began in the summer of 1899. At "Hampden Farm House in the Chilterns," he reported, "we had at our disposal a considerable strip of garden covered with Shirley poppies." The poppies, with their distinctive seed pods, provided the basic material for a study of inheritance. Pearson recruited fifteen friends to help with the research, a collection of eight men and seven women that included the Weldons. Several of this group had been members of the Men and Women's Club. Pearson treated this project as a socialistic endeavor, an effort in which all contributed equally. Though he was directing the work, he took his turn with the more mundane tasks, such as tending the plants, measuring specimens, and harvesting the seed.[31]

Pearson's socialism did not prevent the summer from having an informal elegance, the air of a comfortable English country house during the last summer of Queen Victoria's reign. The group would meet on Fridays, at what Pearson called the "biometric teas," a term that invoked men with high collars, women in long dresses, and an attentive servant pouring tea and offering cucumber sandwiches. In the leisure of a long summer afternoon, they would review their progress, discuss new ideas, and make plans for the weekend. Pearson reported that "Saturday and Sunday . . . were given to calculation and reducing weekly work."[32] In this case, the term "reduction" meant not the processing of astronomical data but the calculation of the five basic terms for correlation analyses. The re-

17. Karl Pearson with Brunsviga calculator

sults of this experiment were published in a paper without an author, though a footnote acknowledged that Pearson had drafted the report. The paper listed the contributions of all sixteen workers, including the women. It also invited interested parties to join the research. "To any of our readers willing to assist in further observations on 'first flower,' the Editors will most gladly send seed of pedigree poppies with suggestions for further work."[33]

The cooperative spirit of Hampden Farm would quickly dissipate, but at least a few of the workers would have a long relationship with Karl Pearson and the Biometrics Lab. One of the summer workers, Alice Lee

(1859–1939), became a student of Pearson's. She held a bachelor's degree from Bedford College, a women's college at the University of London. She supported herself by teaching mathematics and physics at the college, occasionally covering Greek and Latin courses whenever there was a need. To supplement her income, she lived in the college rooms and oversaw the students. This assignment gave her little time to herself, but it paid for her room and board.[34]

Initially, Lee worked as a volunteer in the Biometrics Lab. Her relationship to Pearson was awkward at best. At times she was his student and at times a staff member. Occasionally, he thought of her as a peer but never as an equal. Twice she declined to be listed as the coauthor of a paper with Pearson. She wrote Pearson, "I have done nothing but the Arithmetic, and I suppose a machine could do most of that."[35] She may have been modest about her accomplishments or scrupulously honest in her dealings with Pearson, but she may also have suspected that indiscriminate credit might hurt her reputation as a scholar. If her name appeared on a paper for simply doing arithmetic, then other scientists could conclude that she was nothing more than a computer. She was attempting to establish a reputation in the field of craniometry, the measurement of skulls. The task is hard enough if the subject is dead but considerably more difficult if the object of measurement remains alive. She developed a statistical model that estimated the cranial volume of living subjects from external skull measurements. When she presented the results of her study, some faculty claimed that the work was a simple elaboration of Pearson's ideas. Pearson intervened in the evaluation of her work, defended the merits of her approach before his colleagues, and argued that he had no claim to her results. Based upon his presentation, the faculty reconsidered the case and awarded the degree.[36]

Pearson's treatment of Alice Lee and his other computers vacillated between radical ideas and common stereotypes. He could defend the contributions of women to Francis Galton, writing that "their work is equal at the very least to that of the men. They are women who in many cases have taken higher academic honours than the men and are intellectually their peers."[37] Yet in private he could express doubts about the quality of their work and was not always able to treat women as equals. He once complained of Lee, "I am here, as on other occasions, apt to be vexed by her want of power of expression." Rather than blame her education or even her upbringing, he found the cause in her gender. "On the whole I think it is characteristic of most women's work."[38]

Through 1903, Pearson had no regular source of funds to support the Biometrics Laboratory. "I live on students fees practically," he wrote to a colleague, "and any step which leaves my department under incomplete supervision tends to impair its efficiency [and] affect my income."[39] That year, he received a grant of £500 from a civic organization that had been established by the businessmen that manufactured and traded cloth, an organization known as the Worshipful Company of Drapers.[40] The funds brought some measure of order to the laboratory and a salary to Alice Lee. She received £90 a year, about half the salary given to an assistant professor. For this money, she worked three days a week, arriving at 9:30 AM and leaving at 5:00 PM. Pearson allowed her half an hour for lunch, which was presumably taken at her desk or, when the weather allowed, in the courtyard of the school. Her duties included reducing data, computing correlation coefficients, creating bar charts—charts that Pearson was now calling "histograms"—and calculating a new kind of statistic, which Pearson had denoted χ^2 (usually pronounced *chi-squared*). The χ^2 statistic promised to be just as important as the correlation statistic. It allowed researchers to test scientific theories in a formal, mathematical way.[41]

In addition to her computational work, Lee did "all the hundred and one things that need doing here." She acted "more or less" as the laboratory secretary, excerpted books in the library, formed indexes, and organized catalogs of data. By the standards of office work, the job was not well paid, a fact that slightly embarrassed Pearson. A young female typist or stenographer was usually able to feed a mother or a child on an office worker's salary. He confessed, "It is a post well suited to a woman living with her family in London and keen on scientific work."[42] Even with all of this clerical work, Lee continued to pursue her own research projects, which led to four papers published in her own name and contributions to twenty-six others.

The Draper grant allowed Pearson to hire the sisters Cave-Browne-Cave as part-time computers. Beatrice and Frances Cave-Browne-Cave were the daughters of a senior civil servant and graduates of Girton College. They met both Pearson and Francis Galton at a "reading party," a meeting that was probably a public discussion of natural inheritance or statistical methods.[43] The elder, Beatrice Cave-Browne-Cave (1874–1947), taught mathematics in a girls' school located in south London. She worked as a part-time employee of the Biometrics Lab, doing calculations in her home. Before the Draper grant gave her a small stipend, she did Pearson's calculations without compensation. The only reward she received was to be listed as a joint author on two papers, one published jointly with Pearson, the other as a collaborative effort of five authors.[44]

The younger sister, Frances Cave-Browne-Cave (1876–1965), remained at Girton College as a teacher. She had been the top student in her class and had tied the Fifth Wrangler in the Cambridge Tripos exam.[45] Like Alice Lee, she had a research program of her own. She was performing a correlation analysis of weather data that had been collected from the east and west sides of the Atlantic Ocean. Pearson guided the work and outlined the basic mathematics but let Cave-Browne-Cave work at her own pace. She was a popular teacher at Girton and enjoyed socializing with the young women at the college. She devoted many of her evenings to her students, talking with them and helping them with their studies.[46] Only on those rare occasions when she was left alone in her room did she find time for her research. "The magnitude of the computations," she recorded, "almost precluded the idea that any individual worker or workers can hope to complete such a task within a reasonable period."[47] The project required her to calculate hundreds of correlations with Pearson's formula. Even though she worked on her own, she was able to complete two substantial projects that demonstrated patterns in the weather as it moved across the ocean.[48]

Even with the Draper grant, Pearson's largest projects retained the collaborative aspects of the Hampden Farm experiment. In 1903, he oversaw a large study of child development, "a cooperative investigation extending over a number of years, and depending upon a body of collaborators." The project collected physical measurements and character assessments from 4,000 children and their parents in order to establish evidence of the inheritance of what Pearson called "moral qualities," attributes that we would now identify as aspects of intelligence or personality. Both sisters Cave-Browne-Cave were among the six collaborators who worked on the project. Unlike the Hampden Farm experiment, this project seemed firmly in the control of Pearson. He was the one setting the scale of the project and posing the questions to answer. His collaborators gathered the data by measuring and observing the children. Beatrice Cave-Browne-Cave collected data from her high school students. Only a few of the assistants, including both Frances and Beatrice Cave-Browne-Cave, processed the data, created tables, and computed the correlations.[49]

In 1904, the Worshipful Company of Drapers pledged an annual grant to the Biometrics Laboratory, giving Pearson a measure of financial security.[50] With this money, the laboratory slowly lost its collaborative feel and acquired the more conventional feel of a university office. The lab workers split into three distinct strata: professors, students, and staff. Most of the staff worked as computers, and most, though not all, of the computers were women. These women occupied an isolated corner of academic life. They tended to interact either with Pearson or with other women. Lee taught women at Bedford; Frances Cave-Browne-Cave was a professor to women at Girton; her sister Beatrice worked at a girls' high school. Their experience was not that different from the life of other female computers. At Harvard, the observatory computing staff had been augmented by women who measured photographs and conducted research of their own. Most of these women were aware that their positions were quite different from those of the men. "I have to see to all the changes of household linen, etc. and gather together the family wash," wrote a member of the Harvard Observatory staff. "Alas! How matter of fact and different from the Sunday morning duties of other officers of the University."[51]

The increasing stratification of labor in the 1890s made the computing laboratories vulnerable to the labor troubles of the age, though none of them faced anything as severe as the strike at the Homestead steel mill in Pennsylvania or the riots in Chicago's Haymarket Square. The most dramatic incident occurred when the Greenwich Observatory was bombed by a French anarchist. The explosion shook the building, though it left the structure undamaged and the computing staff untouched.[52] Investigators looked for a motivation for the explosion among the observatory employees and found only minor complaints. Among the human computers, the most pressing concern was the policy, established by George Airy, of dismissing any computer once he or she reached the age of twenty-three. "The mathematics and kindred subjects which have been acquired [at the observatory] become absolutely useless," complained one former computer, "as men are not wanted in the market unless having had a good business experience. So they swell the already overcrowded unskilled labour multitude. This rotten system seems to be maintained by the Government solely to save money."[53] Investigators found it difficult to connect such sentiments to the group that prepared the bomb. They eventually concluded that the explosion was detonated accidentally and that the observatory was probably not the target. The writer Joseph Conrad fictionalized the incident in his novel *The Secret Agent*. He suggested that the bomber was targeting the institutions of capitalism and the British government, not the observatory itself.[54] In the story, the bomb was thrown "into pure mathematics," and it had "all the shocking senselessness of gratuitous blasphemy."[55]

Most of the labor problems among computers were the ordinary conflicts between labor and management and hence were more irritating than dramatic. At the American *Nautical Almanac*, the "gentlemen of liberal education" had been replaced long ago by computers drawn from the civil service rolls. Computers had to pass a special exam to gain a place on the almanac staff. This exam was difficult, and it regularly failed to identify enough qualified applicants to fill the vacancies in the computing room.[56] The director of the almanac office was Simon Newcomb (1835–1909), who had begun his career as a computer under the direction of Charles Henry Davis. Through his work at the almanac, he had achieved a certain measure of fame and was probably the best-known American scientist of his day. He had a conservative nature and had little sympathy for the complaints of workers, labor movements, and strikers. When he was asked his opinion on the growing use of machines in manufacture and their impact upon workers, he replied that "the laboring man is earning higher wages now than he did . . . before the introduction of labor saving machinery." He watched the labor unrest of the 1890s with a sense of alarm and concern. He placed the blame for such discontent directly on the workers. "Too many men will not do more than they are compelled to do," he said; "they do not work cheerfully and become malcontents ready to destroy."[57]

For the most part, Newcomb's computers were German immigrants from Foggy Bottom, the working-class neighborhood of Washington that served as a home to the almanac office. His most promising computer was a Swiss-German named John Meier. Meier "was the most perfect example of a mathematical machine that I ever had at my command," Newcomb reported. Meier was hardworking and skilled at arithmetic. Newcomb also observed that, "Happily for his peace of mind, he was totally devoid of worldly ambition." Meier lived the turbulent life of urban working classes and regularly needed Newcomb's assistance "as an arbitrator of family dissensions."[58] Meier suffered from an illness that he called "nervosity." Newcomb gave no name to the disease, though he clearly believed it to be alcoholism. Meier, who had been able to limit the problems caused by his "nervosity," began to lose control of his life when his wife left him. His children, a boy and a girl, proved more than he could handle alone. His son, testy and combative, showed that he was more than ready to pick fights with his father. The daughter, seventeen years old, had no one who could discipline her and was often found "in company with young men."[59] Newcomb avoided intervening in the failing marriage, but he advised Meier's son, counseled the daughter, and sought support from the family's pastor, the minister of the neighborhood German church. After several months, Newcomb tired of the demands upon his time and concluded that Meier simply was not capable of working for the almanac. He relieved Meier from service and requested the return of all books belonging to the almanac office. Using the popular notions of inheritance to justify his actions, he wrote that Meier "illustrates the maxim that 'blood will tell'" and then added, "which I fear is as true in scientific work as in any other field of human activity."[60]

Newcomb faced a second labor conflict that was resolved only after a hearing by the secretary of the navy. In this incident, Newcomb accused a senior staff member of being "incapacitated for effective work," a phrase that probably implied that the worker arrived at the office drunk, and of taking "one week to do what a skilled computer should do in one or two days." The staff member defended himself by claiming that Newcomb had showed favoritism to incompetent computers, that he was only

18. Simon Newcomb

concerned with "advancing his personal reputation," and that he had "diverted practically three-fourths of the appropriation made for the support of the Almanac Office for years past to a purpose for which it was not intended."[61] The hearing was reported in embarrassing detail by the local papers, but the issue was ultimately resolved in Newcomb's favor.[62] However, the departure of the troublesome employee provoked the secretary of the navy to impose a little more control over almanac operations. "To avoid further trouble," he wrote, he would "remove the almanac office from the Navy Department Building to the Naval Observatory, where it naturally belongs."[63]

Simon Newcomb was an astronomer, but like Pearson, he had a wide range of interests. He met Pearson during a trip to Europe in 1899.[64] There was probably no way the two could have been friends, as their political values seemed to have little in common. Pearson, who was just starting the Hampden Farm experiments, seems to have treated Newcomb with respect, but there is no evidence that the two men corresponded after Newcomb's visit. The two came into contact again only after four years had passed and Newcomb had become the director of the Congresses that were planned for the 1904 World's Fair in St. Louis.

Newcomb wrote to Pearson and asked if he would come to the United States to discuss the methods of statistics at one of the Congresses. Pearson had no interest in such an event and brusquely declined, stating that "I see no possibility of my being able to afford a visit to America from either standpoint of time or money."[65]

By then, Newcomb had retired from the Nautical Almanac Office and had turned to promoting the use of mathematics in "other branches of science than astronomy," especially in the "examinations and discussions of social phenomena."[66] As this concept seemed to be related to the goals of Pearson, Newcomb wrote to the English statistician and asked him to help found an "Institute for the Exact Sciences." "The nineteenth century has been industriously piling up a vast mass of astronomical, meteorological, magnetical, and sociological observations and data," he explained to Pearson. "This accumulation is going on without end, and at great expense, in every civilized country." His proposed institute would collect and process this data. One division of the organization would concentrate on data from experiments. A second group would assemble data that had been collected by observing social phenomena. The third division of the institute would be a large computing laboratory. The computers would process the data gathered by the other two divisions and would develop new mathematical methods that could be applied "to the great mass of existing observations."[67]

The new institute would be expensive to organize and to operate, but Newcomb believed that he could find funds at the Carnegie Institution of Washington, a philanthropy founded by U.S. Steel president Andrew Carnegie (1835–1919). As it operated in 1904, the Carnegie Institution was a granting agency that provided small amounts of money to researchers scattered across the country. Newcomb believed that this strategy was misguided. "We find that centralization is the rule of the day in every department of human activity," he argued. "Two men anywhere will do more when working together than they will when working singly."[68] He argued that a transformed Carnegie Institution, one that followed his model for an institute of exact sciences, would make better use of Carnegie's money and would be "in the true spirit and intent of its founder."[69]

Pearson showed no enthusiasm for Newcomb's plan. Unlike Newcomb, who had spent all of his career working for a military agency, Pearson knew what it was like to ask for research funds with cap in hand and suspected that it would be difficult to extract money from the Carnegie Institution and nearly impossible to transform the organization, as Newcomb envisioned. He also may have felt threatened by the proposed organization, as the proposed Institute for the Exact Sciences would do work similar to that done at the Biometrics Laboratory. Newcomb was not easily dissuaded by Pearson's objections, and he pushed the statistician to support the idea.[70] It was a simple plan, he told Pearson, and it was important to avoid "thinking that I have in view something more comprehensive than I really have."[71] However, Pearson would not be moved and replied through a secretary that "Professor Karl Pearson is very much obliged for your letter re: Carnegie Institution proposals. He still considers the matter extremely difficult of execution."[72]

By 1906, Pearson's Biometrics Laboratory could handle most of the tasks that Newcomb outlined for his Institute for the Exact Sciences. To be sure, it was smaller than Newcomb's proposed institute, and its mathematical methods had taken a circuitous route from the observatory and almanac before they reached the problems of evolution and human behavior. With each passing year, the computing staff was gaining skill and experience with different forms of calculation. By 1906, Pearson could report that the group had mastered the art of mathematical table making. He had set his staff to work evaluating the functions that described the average behavior of random quantities. A typical function was the bell curve, sometimes called the normal curve. This curve described how certain quantities, such as the heights of people or the width of a crab's body, clustered around a central average value. Statisticians need to know the area underneath the bell curve, a value that is tedious and time-consuming to compute. Pearson had his computers tabulate these values as a service to the general scientific community. "It is needless to say that no anticipation of profit was ever made," wrote Pearson; the computers "worked for the sake of science, and the aim was to provide what was possible at the lowest rate we could." When he published a book of these tables, he apologized for having to set a price on the work but claimed "That to pay its way with our existing public, double or treble the present price would not have availed."[73]

The statistical tables were only a small part of the computations at the Biometrics Laboratory. The bulk of the computations summarized large sets of data and were difficult to undertake without an adding machine or other calculating device. Indeed, Pearson often referred to the work of calculation as "cranking a Brunsviga," a phrase that understated the role of computation at the Biometrics Laboratory. Through the first decades of the twentieth century, every member of the laboratory undertook at least a little calculation each day. In 1908, one visitor complained "that preoccupation with mastery of details of calculation and technique obscured, to some extent, the full meaning and scope of the new science."[74]

This new science, the science of mathematical statistics, offered a new way of studying a vast range of human problems, including those found in medicine, anthropology, economics, sociology, and even psychology, a field that was not quite separated from the discipline of philosophy.

However, in the first decade of the twentieth century, Pearson's new science was still linked, at least partially, to the study of human inheritance, a field that had acquired the name of eugenics. Francis Galton had been an early proponent of eugenics and had established a laboratory that collected family trees and looked for patterns of human inheritance. In his eighty-eighth year, Galton proposed to donate his laboratory and his fortune to University College London. The money would be used to support a professor of eugenics, a position that was given to Karl Pearson. Pearson's interest in eugenics is well documented and has been the subject of several scholarly studies. "When it came to biometry, eugenics, and statistics," wrote historian Daniel Kevles, "[Pearson] was the besieged defender of an emotionally charged faith."[75] Grateful for the financial support, Pearson accepted the position, which put him in charge of two laboratories. "There is undoubtedly work enough for two professors," he wrote, "but it is an ideal of a distant future."[76]

CHAPTER EIGHT

Breaking from the Ellipse: Halley's Comet 1910

> Once more the west was retreating, once again the orderly stars
> were dotting the eastern sky. There is certainly no rest for us on
> the earth.
>
> E. M. Forster, *Howards End* (1910)

"BEFORE THE 1835 RETURN [of Halley's Comet] there were at least five independent computations of the orbit," complained the English astronomer Andrew Claude de la Cherois Crommelin (1865–1939), "and it is difficult to understand why an equal amount of interest is not shown in the approaching return."[1] As Crommelin well knew, astronomers had no pressing questions that would be answered by calculating the comet's orbit. Newton's analysis of the solar system had been accepted by astronomers as the laws of celestial motions. The contradictions to these laws, which were being explored by Albert Einstein, offered no idea that might be tested by the return of a comet. In 1909, less than a year before the expected return, there could be found only two calculations of the date of perihelion. The first had been done by Pontécoulant, the computer of the 1835 return. Pontécoulant had cycled his 1835 equations through one more orbit, though he had added data to his analysis and included new terms to account for the gravity of Neptune. As no new planets had been discovered since 1846, many scientists felt that there was nothing to add to the calculation, but Andrew Crommelin disagreed. "Doubtless [Pontécoulant] regarded it as certain that there would be numerous investigations when the time drew nearer," he argued. "This is borne out by the fact that there are certainly some slips or misprints" in the computations.[2]

During the late nineteenth century, some astronomers had experimented with nontraditional ways of computing cometary orbits. The Swedish astronomer Anders Jonas Ångström (1814–1874) had adopted a statistical approach. Making no effort to address the physics of the orbit, he confined his attention to the dates of the comet's returns. From his analysis, he discovered that the average time between perihelions was 76.93 years and found that this period varied in a cyclical manner. With this information, he constructed a simple equation to predict the date of the perihelion. Applying this equation to all of the known sightings of the comet, he computed the date of every return. In most cases, the equation missed the actual date by only a few months. Though these values did not approach the accuracy of Pontécoulant's 1835 calculation or even Alexis Clairaut's 1758 work, they were far more accurate than the equation's prediction of the 1910 return. Ångström's equation suggested that the first return of the twentieth century would not occur until 1913, three years after the generally accepted prediction. Crommelin was not impressed with this work. "We have here a curve which admirably fits 25 successive passages," he wrote, "and yet the first time it is used to predict a return it breaks down utterly, the error being almost 3 years or three times the largest previous error."[3]

In spite of his French name, Crommelin was a British subject and an assistant astronomer at the Greenwich Observatory. He reported to George Airy's successor and oversaw the observatory's program of data reduction. The computing staff included ten full-time adult computers and twenty-six occasional workers, who reduced observations or prepared ephemerides when the research demanded it. They occupied a new facility behind the original observatory building which was equipped with calculating machines, slide rules, mathematical tables, and other computing aids.[4] This office structure, which combined the central computing room of William Stratford with the part-time workers of Nevil Maskelyne, could also be found in business offices of the time. Increasingly, office managers were forming a single pool of secretaries and stenographers to handle normal correspondence and were relying on outside workers for the extra demand of peak seasons. The United States Navy had also adopted this model for their computers, combining the calculating staffs of the Naval Observatory and the American *Nautical Almanac* into a single staff of eight, and relying on the contributions of twenty part-time computers at key times of the year, such as during the final preparation of the almanac.[5]

With the Greenwich computing staff at his disposal, Crommelin could have been tempted to improve Pontécoulant's calculations simply by devoting all of his resources to the task. He could have divided the calculation into smaller steps and assigned all of his computing staff to the problem. Instead, he returned to the original analysis of the comet's motion and removed a key element that had been the basis for all cometary calculations, the elliptical orbit. Beginning with Edmund Halley's work in 1695, astronomers had assumed that the comet followed the idealized orbit of an ellipse and computed how the planets stretched and altered this path. As a calculation stepped the comet around the sun, it would eventually place the object so far from its original orbit that astronomers would have to recalculate the ellipse. This adjustment was a time-consuming task, but it was essential in order to maintain the accuracy of the work. It "is a frequently troublesome process and does away with much of the advantage that there is in assuming that the comet is moving in an ellipse," wrote Crommelin.[6]

Rather than attempting to improve Ångström's work or correcting the equations of Pontécoulant, Crommelin argued that he would get the best results by "discarding the elliptical hypothesis altogether and proceeding by mechanical quadratures."[7] Mechanical quadratures is a method for solving differential equations, the basic mathematical expressions that model the motion of bodies in space. Differential equations describe relationships between the position of an object, its velocity, and the forces acting upon it. Newton solved a differential equation when he analyzed the way in which one body moves around another under the influence of gravity. He had used the techniques of calculus to solve this equation and discovered that one of the possible results was an ellipse. Mechanical quadratures, a technique now called "numerical integration," was an alternative to Newton's calculus. It solves a differential equation solely by numerical methods, with no reference to the original ellipse or any other curve.

Crommelin, with the assistance of an observatory colleague, Phil Crowell (1879–1949), identified the basic differential equations that described the path of the comet and created a computing plan for the Greenwich Observatory computing staff. The computing plan had certain similarities to the plan that Clairaut had used in 1757. It located all the key objects in space and described the forces acting between them. At each step of the calculation, the computers advanced the comet, Saturn, Jupiter, and the other planets forward by a small distance. They did not worry about elliptical orbits but instead followed the direction of the forces. Once they had moved the objects, they had to recalculate all the forces. It was a slow and methodical process, one that required much grinding of Brunsvigas and other calculating machines. Without adding machines, only the most dedicated computers would attempt to use mechanical quadratures to compute the orbit of Halley's comet. Even though Crommelin and Crowell had the assistance of the Greenwich Observatory computing staff, they found that the work took longer than they had anticipated. "Owing to the pressure of time it has not been possible to do the whole of the work in duplication,"[8] they confessed. They were able only to confirm that there were no gross errors in the result, which suggested that the perihelion would occur on April 17.

When Halley's comet appeared as a faint speck in September 1909, the two astronomers returned to their calculations in order to get an exact date for the perihelion. Rather than recalculate the entire orbit using mechanical quadratures, they simply employed the traditional formula for an elliptical orbit. "This assumption is amply sufficient to give the date of

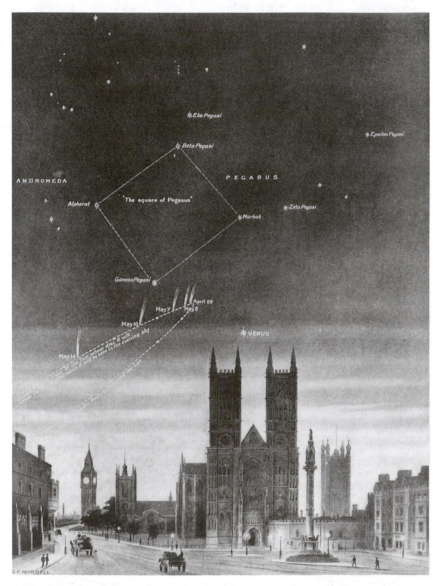

19. Path predicted for Halley's comet in 1910 as seen from London

perihelion," they wrote.[9] This calculation gave a date of April 20, and though this result was of some interest to astronomers, it proved to be of limited importance to the larger public in both Europe and the United States. Those outside of the scientific community looked for some larger meaning in the comet, something beyond the periodic cycling around the sun of a material lump. The author Mark Twain saw it as the grand end of his era. "I came in with Halley's Comet in 1835," he wrote. "It will be the greatest disappointment of my life if I don't go out with Halley's Comet."[10]

"We had the sky up there, all speckled with stars," Twain had written in his novel *The Adventures of Huckleberry Finn*, "and we used to lay on our backs and look up at them, and discuss about whether they was made or only just happened." Twain's protagonist could not believe that anyone or anything could have created the stars because "it would have took too long to make so many," but he conceded that they might have been laid, like the eggs of a frog. In the years that had passed since Twain wrote this novel, astronomers had developed some tools that could gather information on the constitution and origin of the comets. By dividing the reflected light of the comet into its individual colors, they were able to identify some of the substances to be found in its head and tail. Such information was of interest to an age that was anxious about the fact that the Earth would pass through the tail of the comet. A French astronomer was quoted in the *New York Times* as stating that the comet "would impregnate the atmosphere [of the Earth] and possibly snuff out all life on the planet."[11] Of all the predictions, both good and ill, that circulated in the spring and winter of 1910 concerning the comet, only Twain's prediction of his own demise proved to be accurate. Twain died on April 20, 1910, just as the comet was passing through its perihelion and was starting on its outbound course.

As others had before him, Crommelin and his staff returned to their calculations after the comet passed and assessed the accuracy of their results. Taken as a group, the calculations of 1835 fell within sixteen days of the true date of perihelion. The single 1910 computation was within two days, sixteen hours, and forty-eight minutes of the correct value, an improvement by a factor of about five. Crommelin claimed that the method of mechanical quadratures, "if pressed to the extreme accuracy of which it is capable, will give results of higher degrees of accuracy than any previously published method of dealing with this comet." He attempted to demonstrate the validity of this claim by making a more refined calculation of the comet's orbit. After expending much effort on the computation, he found that his revised work was no better than his original estimate. He did not withdraw his claims for mechanical quadratures but instead speculated that something other than mathematical accuracy was causing the discrepancy. He suggested that some additional force, an interstellar drag, was slowing the progress of Halley's comet.[12]

After the comet had again been lost to view in the outer reaches of the solar system, just a few years after Crommelin had complained that no was one engaged in computing the orbit of Halley's comet, the English mathematician Edmund Whittaker (1873–1957) wrote that "there has been a great awakening of interest in [computation]; and it is now included in the syllabus for the [British Civil Service] Examination." The calculation of a comet's orbit might no longer pose an opportunity for new discoveries, but there were other fields that looked to scientific calculation as a way of providing detailed, precise answers. Whittaker observed that a knowledge of calculation was "required by workers in many different fields—astronomers, meteorologists, physicists, engineers, naval architects, actuaries, biometricians, and statisticians."[13] In 1913, he was a professor at the University of Edinburgh, in Scotland, and was in the process of forming a laboratory devoted to the topic of computation. Most of his efforts were devoted to cataloging the methods of calculation and to creating new techniques that might be applied to broad classes of problems, just as the method of mechanical quadratures might be applied to problems other than those of comet orbits.

Whittaker had been trained as an astronomer and, for a time, had served as the Astronomer Royal of Ireland. Like other astronomers before him, Whittaker had been drawn to the mathematics of insurance and, according to his biographer, had "been influenced by his friendship with the great actuaries of the period."[14] His laboratory borrowed from actuarial practice, a practice that demanded that computers follow calculating plans to the letter and that they record their intermediate values in ways that would allow others to verify their work. Whittaker gave detailed recommendations for every aspect of computing. He told computers to write numbers in pairs, for it is "found conducive toward accuracy and speed," and argued that "every computation should be performed with ink in preference to pencil; this not only ensures a much more lasting record of the work but also prevents eyestrain and fatigue." He felt that all computation should be done on specially prepared paper that was "divided by a faint ruling into 1/4″ squares, each of which is capable of holding two digits." His recommendations even included the desks that should furnish the computing room. He stated that the desks "used in the mathematical laboratory of the University of Edinburgh are 3′ 0″ wide, 1′ 9″ from front to back and 2′ 6 ½″ high. They contain a locker, in which computing paper can be kept without being folded, and a cupboard for books, and are fitted with a strong adjustable book-rest."[15]

Beginning with his first writings on the subject, Whittaker was a strong advocate for direct numerical calculation. Many scientists still relied on graphical means for calculation. They would find a way of constructing some two-dimensional shape that had an area equal to the desired result. Once the shape was finished, the correct answer could be determined by measuring the area of the shape. Edmund Halley had used such a method when he first looked at the orbits of comets. Just as Halley had abandoned this technique, Whittaker advised others to do the same. He wrote that at his Edinburgh laboratory "graphical methods have almost all been abandoned, as their inferiority has become evident, and at the present time the work of the Laboratory is almost exclusively arithmetical."[16]

The staff of the Edinburgh Mathematics Laboratory worked on a variety of problems, but they devoted most of their efforts to the construction of mathematical tables, to computing values of complicated mathematical functions that appeared in many different forms of scientific or engineering endeavor. Simple examples of mathematical tables include the familiar pages of logarithms, sines, and tangents and the probability functions of Karl Pearson and his Biometrics Laboratory. The Edinburgh computers tried their hands at such tables, but they were most interested in tabulating the Bessel function. The Bessel function was named for Friedrich Wilhelm Bessel (1784–1846), a German scientist who had been known as a mathematician and as the director of an astronomical observatory in Königsberg. "Many mathematicians, usually working in celestial mechanics, arrived independently at the Bessel function," wrote the historian Morris Kline.[17] Bessel wrote the first systematic treatment of this function in 1824 as an outgrowth of his study of planetary motion. The task of tabulating this function was substantial, as it appeared in two different forms and behaved in a wide variety of ways. In the late nineteenth century, it emerged as one of the great problems for scientific computers. The orbit of Halley's comet was computed once every 75 years and then packed away for another generation. The Bessel function found applications in ever-broadening fields of science. It was useful to anyone studying problems of vibration, including vibrating drums, vibrating air, and vibrating electrical signals. It also could be applied in the study of heat and the diffraction of light.[18] As scientific computing moved from Greenwich to Edinburgh, from Halley's comet to the Bessel function, it slipped, at least slightly, from the grasp of astronomers and was picked up by scientists who were studying phenomena that were much closer to the earth.

CHAPTER NINE

Captains of Academe

> War rolled swiftly up the beach and washed the sands where
> Princeton played. Every night the gymnasium echoed as platoon
> after platoon swept over the floor and shuffled out the basket-
> ball markings.
>
> F. Scott Fitzgerald, *This Side of Paradise* (1920)

DURING THE LAST DAYS of July 1914, in the final hours of peace, the European powers positioned themselves for the impending conflict. Germany prepared to march its army through the supposedly neutral country of Belgium. The French hurried to throw their military might between
the advancing troops and Paris. The English, perceiving that they had interests on the Continent, organized an expeditionary force to send into
the fray. Karl Pearson, the great admirer of German culture, found himself caught on the European side of the English Channel. He hurried
home to London on the first day of the conflict and declared that the
needs of his country were more important than his personal ambition or
his love of science. "On August 3, 1914," he wrote, "I at once put the
whole Laboratory staff at the service of any [British] Government department that was in need of computing or statistical aid."[1]

In 1914, the Biometrics Laboratory employed a staff of ten computers,
four men and six women. The military could have utilized all ten. They
would have enlisted the men qualified for service and made them navigators and surveyors. The women, and those men who were unfit for military duty, would have been given jobs as clerks or engineering assistants.
Pearson argued that the group should remain intact and under his control. "The Laboratory can do far better work nationally as a whole than
scattered, as it is trained to work together."[2]

At first, he was willing to accept relatively menial assignments for his
computers. Beginning that fall, the Biometrics Laboratory "provided
weekly some 500 or more graphs showing the state of unemployment of
insured and of uninsured trade men and women." The work required no
advanced mathematics and was relatively straightforward, even though it
kept the computers on a strict production schedule. "Six and sometimes
eight series of these graphs were kept running and carried to date each
week," Pearson reported. The staff worked "without a break through all

the vacations up to July [1915]," balancing the requirements of the statistical reports with the demands of Pearson's biometrical research.[3]

When the laboratory staff returned from the summer vacation of 1915, Pearson learned that half of his workers were not satisfied with the role that they were playing in the war effort. The conflict still retained its heroic potential, its romantic promise of bold actions and daring deeds. On those days when the wind blew from the southeast, the more alert residents of London could hear the report of cannon and the muffled thud of exploding shells. Three of the male computers had tried to enlist in the army with the hope of serving on the front. All three were rejected by the recruiting office, but each had found a position in industries that were supplying the military. An equal number of female computers had also left the lab, one to serve in a hospital, one to teach, and one to rejoin her family.[4] Pearson, conscious that he would have to train and recruit a new staff, decided to temporarily withdraw from war work.

On September 9, the war reached the Biometrics Laboratory. "We are all congratulating ourselves that we have seen a Zeppelin at last," wrote one of the female computers. The Zeppelin had crossed the sea from Germany and followed railroad tracks into London. "I was coming home in a tram just before 11 PM," she added, "when the driver called out that there was a Zep." She got out of the tram and began walking home, keeping an eye on the big, hulking shape in the night sky. "Nobody obeyed the Instructions to seek shelter. We could see the flashes from the anti-aircraft guns but they all went very wide of the mark." The airship was looking for the Charing Cross railroad station, which was situated on the river Thames, but it dropped a bomb near the university "and nearly every window is smashed and numerous shops were destroyed." The computer reported that the population seemed to regard the event as an adventure rather than as a threat, claiming that the "whole of London is in a state of subdued excitement."[5]

Pearson claimed that he was able to face the bombings with resolution. "I just went about my usual tasks," he wrote after the war: "I made belief that it was nothing."[6] By December, he felt that his computers were ready to resume war work. This time they created shipping reports, "preparing graphs dealing with the tonnage required for various imports for the use of committees controlling these matters."[7] He reported that the work occupied all of the computers' time, but at least one of his computers was still doing biometric computations. Pearson continued with his statistical research but spent increasing amounts of his time doing analyses for aircraft factories. The aviation industry was only a decade old, and its engineers had much to learn about structure, motion, lift, and drag. Most of the problems undertaken by Pearson involved the flexing of structural parts: propellers, wing struts, airframes. The mathematics of this analysis was somewhat tricky, but Pearson had studied the subject before he had become interested in mathematical statistics. The work brought him into direct contact with the military and introduced him to the big mathematical problem of the war: the calculation of bomb and shell trajectories, a subject known as mathematical ballistics.[8]

Mathematical ballistics lay behind the most brutal weapons of the war. It allowed artillery crews to aim their guns at distant targets, mortar crews to lob gas-filled shells from behind the protection of a hill, moving Zeppelins to bomb stationary structures, and defense gunners to destroy invading aircraft. It had a history as old and venerated as the history of mathematical astronomy, and it had a problem every bit as difficult as the three-body problem of Halley's comet: the problem of modeling the atmospheric drag on the flying shell. This problem had been first encountered in the fifteenth-century cannonball experiments of Galileo Galilei (1564–1642). Galileo argued that the air had no effect upon the motion of the balls and concluded that trajectories were graceful parabolas through the sky.[9] Newton placed Galileo's analysis on a more formal foundation, but like his predecessor, he assumed that the cannonball was moving *in vacuo*. By the early eighteenth century, scientists had discovered that air resistance had a substantial impact upon ballistics trajectories, and they had also found it to be a complicated phenomenon. As a projectile neared the speed of sound, it created shock waves that dissipated energy and greatly increased the drag. For extremely high velocities, such as those achieved by projectiles as they left the barrel of a gun, the nature of the drag changed again. These variations thwarted any attempt to make a simple calculation of a trajectory. Scientists of the mid-nineteenth century could compute a trajectory only by using methods similar to those that had been used by Alexis Clairaut for Halley's comet. They would track a projectile along an idealized path and adjust its position to account for the drag.[10]

A clearer understanding of air resistance began to emerge in the 1860s, as military engineers began to amass a large collection of data. The first of this data came from pendulum tests. Gunnery crews would fire a shell at a large pendulum and measure the displacement of the bob. From these measurements, the engineers could determine the velocity of the shells and, ultimately, the atmospheric drag. By 1870, the armies of Prussia, Great Britain, France, Russia, and Italy had all estimated the nature of drag. These estimates took the form of graphs rather than mathematical expressions. No simple expression described the relationship between the velocity of a shell and the drag. The French estimate, usually called the Gâvre function, after the French proving ground, was generally considered to be the most reliable of the time. It was used by an Italian professor, Francesco Siacci (1839–1907), to create a simple means of computing trajectories.

Siacci was a teacher at the Turin Military Academy and had served briefly as an artillery commander with the army during a brief conflict with Austria.[11] Rather than calculating the entire flight of a shell, he concentrated on four factors: the range of the trajectory, the time of flight, the maximum height of the shell, and the velocity at impact. Each of these quantities could be used by an artillery officer to plan and direct artillery. The time of flight was used for setting fuses so that the shells would explode over the heads of enemy troops rather than in the ground. The maximum height was used for mortar fire over hills and tall buildings. The terminal velocity gave the amount of energy in a shell and suggested the extent of damage it could produce. Rather than compute these values for all guns and all shells, Siacci chose a strategy that resembled Nevil Maskelyne's approach to the lunar distance method of navigation. He prepared a set of mathematical tables that described idealized trajectories and then gave rules for transforming values from these tables into the motion of a real shell. The final results were only approximations of the real trajectory, but they were sufficiently accurate for the cannon of the day, including the old smoothbore cannon that had been used in the American Civil War and the new rifled steel barrels that were being produced by Krupp Industries in Prussia.[12] Siacci's ballistics tables were quickly adopted by the armies of the industrialized countries. American officers translated Siacci's original paper into English less than a year after it appeared in Italian.[13]

Had the First World War been only a duel of large guns across the fields of Flanders, then Siacci's method would have sufficed for most of the ballistics calculations. This method had its flaws, but most of them could be fixed with only a little effort. The biggest corrections would have included a refinement of the atmosphere drag function and the introduction of more detail into the mathematical equations. Ballistics officers were now able to measure the velocity of shells with electrical instruments. Their experiments had shown the need to incorporate new factors into the equations, including the density of the atmosphere and the direction of high-level winds. These changes could be handled without a large, permanent computing staff.[14] Computers were needed to help with artillery problems that came from the growing use of aircraft. Since Siacci's tables presented only a few points of the trajectory, they could not be used for antiaircraft artillery, for bombing, or for aerial combat. Antiaircraft defense, a problem that became increasingly important during the war, was like duck hunting. The gunnery crew would fire an explosive shell into the air, hoping that it would explode near an oncoming plane.

To place that shot close to the plane, the gunners needed to know how fast a shell travels along the upward slope of its trajectory, information that was not easily gleaned from Siacci's theory.[15]

The British army first turned to a Cambridge professor, John Littlewood (1885–1977), for help with ballistics analyses. They gave him a lieutenant's commission and assigned him to work at the Woolwich Arsenal, a military facility located to the east of Greenwich.[16] Littlewood was an expert with differential equations and could quickly produce rough approximations of "vertical fire," the term used by the army for antiaircraft trajectories. When he needed more detailed calculations, he asked for assistance from the faculty and students at Cambridge. Often, he turned to Karl Pearson's former computer Frances Cave-Browne-Cave at Girton College. Professor Cave-Browne-Cave gladly offered to do calculations for Littlewood, but she had many responsibilities and occasionally needed help herself. "We had to ask my Girton sister to come home before she had finished her work on guns," reported her sister Beatrice, "so I have been checking some of the most urgent of her work for her."[17]

The calculations for antiaircraft and anti-Zeppelin fire were substantially more complicated than those of the old Siacci theory. Computers needed to produce a complete trajectory, not just the endpoint and the time of flight. They calculated these trajectories using the method of mechanical quadratures, the same method that Andrew Crommelin had used to prepare the ephemeris of Halley's comet. They called their version of this technique "the method of small arcs," for it divided the full trajectory of a shell into a series of tiny, curved steps. At each step of the calculation, they would advance the shell, estimate how much it had slowed in that interval, and recompute the drag. The process was simpler than computing the orbit of a comet, as the shell was influenced only by air resistance and earth's gravity rather than by the conflicting pulls of the planets. Even with this relative simplicity, the computation of a single trajectory by the method of small arcs still required at least a full day of effort.

By the spring of 1916, the computers of the Biometrics Laboratory, including Beatrice Cave-Browne-Cave, were accepting requests for trajectory calculations directly from the Ministry of Munitions. At first, they gave low priority to the ballistics work. "These [trajectories] you told me to leave till the last as you might not have them done," Beatrice Cave-Browne-Cave wrote to Pearson; "shall I go on to this now?"[18] Pearson approved this request, but he was still trying to devote as much time as possible to his statistical research. He asked the computers to clean and measure a collection of skulls that spring, a task that uncovered an infestation of insects.[19] He also kept at least one of the computers away from the ballistics calculations. This computer, a Norwegian student, did most of her work outside of the Biometrics Laboratory.[20]

As spring moved to summer, the Ministry of Munitions expanded its requests for ballistics calculations from Pearson and his staff. Though the computers faced an increasingly rigid production schedule, Pearson attempted to sustain the same egalitarian air that he had shown during the days of experiments at Hampden Farm. When he received a packet of printed materials from the Ministry of Munitions, he found that he, rather than his staff, was being credited with producing the calculations. "Please do not place my initials on the charts and tables," he replied to the ministry. "It would have the appearance of arrogating to myself work due to a number of people of whom I am only one." He had come to refer to his combined laboratories as the Galton Laboratory, and he requested, "If a mark of this kind is needful will you please place GL upon them, which will be quite as distinctive as KP and cover the whole staff of the Galton Laboratory."[21]

Though the war demanded self-sacrifice, it also offered new opportunities to the computers and encouraged them to look beyond the walls of University College, London. In August 1916, Beatrice Cave-Browne-Cave announced that she would leave the laboratory to take a better-paying position with the Ministry of Munitions. Pearson had not anticipated this news and was not pleased to be losing so experienced a computer. "I may be quite wrong," he wrote Cave-Browne-Cave, "but frankly I do not consider you have 'played the game.' " In part, he was distressed because Cave-Browne-Cave was abandoning a recently signed contract with the college, but he also felt that the computers should be motivated by something beyond money or position. "I should never attempt to hold an assistant, who wishes to leave," he explained, "because it is evidence to me that their heart is not in their work and that they have not full loyalty to the ideas of our Founder, [Francis Galton]." He ended their collaboration by stating that "under the circumstances I should prefer, as it would save friction which is not compatible with the pressure of present work, if you did not return at all to the Laboratory."[22]

By the end of the year, ballistics computations dominated all the work of the laboratory. In one request, an assistant at the Ministry of Munitions apologized for the demands it was making upon the Biometrics Laboratory staff. "Your people, being trained to work together, could probably get results out very much more quickly than we, being amateurs at computation." To emphasize the demands placed upon the ministry, he added, "we have large amounts of work we could take on, if only there was a possibility of getting them finished."[23] By that time, Pearson was finding it difficult to replace the men who had gone to war and the women who had found more profitable employment in the war industries. Increasingly, Pearson recruited young boys to be his computers. Unlike the boy computers of George Airy's observatory, these computers were students at prestigious public schools and were serving for three or six months before they were called into the army.[24]

As the conflict moved into its third year, the pressures of the war began to weary Pearson. The demands for ballistics calculation had only increased. The Ministry of Munitions needed more tables, more detailed computations, and more extensive analyses. They had increased the upper limits for antiaircraft trajectories, which had originally been set at an altitude of 15,000 feet, to 17,000 feet, then to 20,000 feet, next to 25,000 feet, and finally to 30,000 feet.[25] Attacks on London only increased the demands upon the laboratory. Following a damaging attack in September 1917, a distressed ministry official wrote to Pearson, "From the experience of last night it is evident that the work of the Anti-aircraft Experimental section will have to be tackled with even greater energy than at present if we are to make any impression on the Huns."[26] However, Pearson had little energy left. He was tired of recruiting and training new computers, saw no end to the ballistics calculations, and was frustrated that his research had stagnated. Just one week later, when he received a review of recent calculations, he gave full vent to his frustration and fatigue. The review was a gentle note in a neutral language with no obvious evidence of personal rancor. It stated there must be "some mistake in the sign of correcting terms—small things added instead of subtracted." The writer, an official with the Munitions Ministry, stated that there was "nothing to be done except to call your attention to the matter" and ended with the "hope it will prove to be possible to set matters right without imposing new calculations."[27]

The comment angered Pearson, in a way that few things ever did. He had faced scientific criticism before and had accepted it as part of the scientific process. To him, this event seemed to be an unwarranted interference, a challenge to his authority. He took it as a personal affront and demanded an apology. The official, surprised by Pearson's reaction, stated that he had intended no offense, but this was not enough to placate the statistician.[28] Surprised by the extent of the controversy, a friend of Pearson tried to intervene, but he, too, was rebuffed. "I cannot understand why you should not believe my statement," he told Pearson, "that I, at any rate, have neither heard nor seen nor made any statement other than one of respect for your work."[29] It took nearly six weeks for the ministry officials to restore a working relationship with Pearson. The chief of ordnance research offered a full apology and acknowledged "how much we were asking of you in wishing you to accept dogmatic rules without explanations or reasons."[30] For his part, Pearson agreed that there were errors in the calculations, which had been created when one of his computers misread an intermediate value.[31]

Pearson returned to work in December 1917, but he no longer seemed to be fully engaged in war calculations. When he wrote the ministry that "I don't think our people are likely to stick it out so long as [the Germans] will,"[32] he was speaking as much for himself as for the British people. Three months later, just as the Germans launched a threatening offensive, Pearson announced that he wished to withdraw from war work and return to his statistical research. This time, no one in the Ministry of Munitions attempted to change his mind. They transferred most of the Biometrics Laboratory computers to a government building and assigned a military officer to oversee them.[33] Pearson was sanguine about the change. "I have promised to disappear," he wrote to one computer, in order "to gain room for spring cleaning during the first week of April." If the weather was good, he might go hiking; "otherwise I suppose to go on steadily with the work at some place in the neighborhood."[34]

The American computers were not in the war long enough to follow the path set by Karl Pearson, but all had to weigh the claims of patriotism, personal glory, and scientific accomplishment. Unlike the staff of the Biometrics Laboratory, most of the American computers were promising graduate students or young mathematics faculty. As a whole, the computers were part of a generation that ardently supported American intervention in the European war. Long before President Woodrow Wilson declared war on Germany, college students formed campus battalions, practiced formations in gymnasiums, and spent the summer at special army training camps. "The muster rolls at [the camps]," observed one historian, "sounded like *Who's Who* and *The Social Register* combined."[35] The most adventurous of the college men joined Canadian regiments or, like the young Ernest Hemingway, volunteered to drive ambulances for the Red Cross. For Hemingway, the Red Cross was not a humanitarian service but the chance "to die in all the happy period of undisillusioned youth, to go out in a blaze of light."[36] Harvard, Columbia, and Yale sent large proportions of their student body to the war. Princeton University, where President Wilson had once taught political science, was home to the most active recruiting station in the nation.[37]

In their part of the preparation for combat, scientists formed a civilian research organization within the National Academy of Sciences. The organization, called the National Research Council, was intended "to bring into cooperation governmental, educational, industrial and other research organizations." The group was endorsed by President Wilson but was never fully embraced by either the army or the navy. On the eve of the war, the army assumed control over the National Research Council, placed it under the authority of the Signal Corps, and gave its members commissions in the army reserve. Most of war's scientific research was done under the close supervision of military officers.[38]

Ballistics research, including trajectory computation, was overseen by the Army Ordnance Department. The leader of the computing activity was the Princeton mathematician Oswald Veblen (1880–1960), nephew of the economist Thorstein Veblen. While the elder Veblen is usually associated with the liberal strains of American thought, Oswald Veblen was conservative and an advocate of American intervention in the European conflict. He began looking for a position in the military before Woodrow Wilson asked for a declaration of war. After an initial rebuff, the Department of Ordnance offered him a captain's commission and placed him in charge of experimental ballistics.[39] He spent the first summer of American involvement in the war, the summer of 1917, waiting to be called for duty. During the intervening months, he read ballistics treatises and corresponded with mathematicians interested in working for the war.[40] He was inducted on August 30, given basic training on November 20, and ordered to report to the army's new Aberdeen, Maryland, Proving Ground on January 18, 1918.[41]

The Aberdeen Proving Ground was the largest military project of the war, the Manhattan Project of its age. The army spent $73 million to build the facility.[42] They acquired 35,000 acres of Chesapeake Bay shoreline and evicted 11,000 residents, including the owners of thirty farms and the entire population of a substantial country town.[43] The Aberdeen Proving Ground replaced an older testing facility at Sandy Hook, New Jersey, a spit of land that stuck into New York Harbor. The army chose the Aberdeen site because it lay on the rail line between Baltimore and Philadelphia. It would be the last stop in a manufacturing process that would begin at factories in Pennsylvania, Maryland, New York, and Ohio. The factories would ship guns, shells, and charges to Aberdeen, where ordnance officers would test, or "proof," these devices before deploying them to army arsenals.

Waiting for Veblen at the Aberdeen post office was a letter from a professor who had taught him at the University of Chicago. It praised the young mathematician as being "High in Academic and in Military Life!!"[44] They were flattering sentiments, but they must have seemed far removed from the physical reality of the proving ground. The military base was little more than a construction site. There were a few wooden buildings, lots of tents, and miles of dirt roads. The winter was desperately cold, the worst on record. Water froze on the Chesapeake, and the wind raced unhindered through the army tents. Veblen was not able to start an experimental program until early February. He described his first

20. Captain Oswald Veblen before departing for the Aberdeen Proving Ground

efforts as "more picturesque than satisfactory." The test ranges were not finished, so Veblen gathered data by firing a cannon across an open field. After a series of rounds, he would "go down the range on horseback while the firing was suspended for meals, identify the shell holes and place distinctive marks in them to enable them to be identified by the surveyors." The bitter weather slowed the work, though Veblen noted that the winter storms deposited fresh sheets of snow, which erased the craters from earlier trials and provided an unblemished surface for a new round of shots.[45]

His computational problems began after he completed the range work. His computing staff consisted of three army officers who worked at Sandy Hook. They remained at the old facility in order to teach trigonometry and surveying to new recruits. On top of their teaching duties, they were expected to analyze Veblen's data and compute the trajectories for a range table, a table that artillery officers would use to prepare their campaigns. Their first job was to reduce the data, a task similar to reducing astronomical data. They had to adjust the data so that all the values indicated how the gun would perform under what artillery officers called "standard conditions." Under standard conditions, the air is still, with a temperature of fifteen degrees centigrade and a density of .075126 pounds per cubic foot.[46] Veblen had given them formulas to do the reductions, including ways of correcting for crosswinds and "jumps" or motion of the gun barrel during firing.

From the start, the computations went badly. The "difficulties in long range correspondence about the many technical details were very great," Veblen observed.[47] The computers had not observed the firings and did not understand all of the data. Some of the shots seemed anomalous and the results contradictory. The ballistics formulas were confusing, and Veblen's instructions were incomplete. In early March, the computers asked Veblen to come to Sandy Hook and help them with the calculations. After he returned to Aberdeen, they requested a second visit. Ten days later, with the tables still incomplete, they asked for one more session. On this last trip, Veblen worked with the computers until the calculations were done. The four of them finished checking the results a few hours before dawn on the morning of March 26. Veblen's diary for the day contained the single line "3:00 A.M. Finished tables."[48]

This first project was an opportunity for Veblen to experience every step that was needed to gather ballistics data and create range tables, an opportunity to learn the problems of ballistics calculation. He seemed to have learned his lessons well, for his new computing staff at Aberdeen never had this kind of frustrating experience. He transferred the three Sandy Hook computers to the new proving ground in order to form the core of the new group. Their efforts were augmented by three Princeton graduate students. To oversee this group, he recruited Joseph Ritt (1893–1951) and gave him the title "Master Computer." Ritt was a professor of mathematics at Columbia University and Veblen's link to the old, traditional computing labs.[49] In 1911, Ritt had spent a year as a member of the Coast and Geodetic Survey computing floor. The group had a new name, the Department of Longitude, but its methods and procedures were little changed. Myrrick Doolittle, who had become the grand old man of scientific calculation, arrived in his office every day to adjust triangulation surveys, as he had thirty years before.[50] From the Coast Survey, Ritt moved to the combined Computing Division of the Naval Observatory. At the observatory, he spent one year reducing data for the astronomers and a second year preparing ephemerides for the Nautical Almanac Office.

Though he chose a Master Computer who had been trained at conventional computing offices, Veblen did not follow the traditional model for computing floors. He placed computers on the firing ranges of Aberdeen and put them "under the direct observation of the firing officers."[51] The first facility to include a computing staff was called the "water range" because the shells overflew an island and landed in the Chesapeake Bay. The range was not completely finished when it began testing guns in April 1918. Veblen remembered: "It was necessary to haul ammunition and guns over roads which were often two feet deep in mud. The only conveyance which was able to get through was a six-mule team. Even the Ford was powerless."[52] One computer would hike or ride a horse to the gun mount. Two or three others would take a small boat to observation towers on the far side of the bay. These computers had a telephone connection to the range officers, who notified them of each firing so that they could time the length of flight.[53] From measurements of the splash made by the shell, they would compute the length of the shot and its deviation to the left or right.

Following the shots, the observing computers would phone these numbers to the computer at the gun mount. The computer reduced the data to standard conditions using the air temperature, gun temperature, humidity, jump of the barrel, direction of the wind, and other factors. Under this system, Veblen reported, "it is possible to know the results of any firing for range [distance] within a few minutes after the last shot was fired." At the end of the day, after the boat picked up the observers, the range computer would take the day's calculations back to the central computing lab, where the staff would complete the work. Veblen confessed that the lab was really nothing more than a "small shack" that was "ordered for this section [the computing group] on Friday morning and the section moved in on Saturday afternoon."[54]

Veblen implemented this plan on all of the Aberdeen ranges, though some of the facilities lacked proper equipment. On one of the antiaircraft fire ranges, before the computers acquired range finders, they resorted to techniques that might have been borrowed from Francis Galton or a scientist of Laputa. They would fire the shells vertically into the night sky and photograph the bursts against the background of the fixed stars. After the photographs were developed, the computers would measure the distance between the shell burst and the stars. With this data, they would use the tables of the nautical almanac to compute the altitude of the burst.[55]

Aberdeen had no residences for women and hence had no female computers. The only women who worked at the base were a pair of secretaries, who were forbidden to spend the night at the proving ground.[56] The army began hiring female computers only when it opened an office of experimental ballistics in Washington, D.C. This office was directed by Major Forest Ray Moulton (1872–1952), who was a professor of astronomy at the University of Chicago before he accepted a reserve commission. Moulton took the job of preparing ballistics materials for ordnance officers. In the spring of 1918, he was given an office in a temporary building on the Washington Mall and told to hire a staff. As other officers had already discovered, Moulton found it difficult to hire enough men, so he offered positions to women.

The young women of 1918 could not attend the special training summer camps, volunteer for a Canadian regiment, or even fire a 12-inch gun at the Aberdeen Proving Ground. Though they were generally enthusiastic about the war, their feelings were checked by the roles that they were offered in the conflict. The first year of the war was also the year of women's suffrage, the year that a corps of committed women moved to Washington, D.C., in order to win the right to vote. While the men were preparing to fight in France, women were picketing the White House, lobbying the members of Congress, and marching up and down Pennsylvania Avenue. One congressional aide recalled seeing "cultured, intellectual women arrested and dragged off to prison because of their method of giving publicity to what they believed to be the truth."[57]

The suffrage movement surrounded the world of Elizabeth Webb Wilson (1896–1980). Wilson was the daughter of a Washington physician, a flaxen-haired, dimple-cheeked member of the capital's wealthy classes. She studied mathematics at George Washington University, a school just a few blocks west of the White House. Her daily trolley ride took her past the lobbyists, the pickets, and the marches. Like herself, many of the suffrage leaders were the daughters of physicians. As far as we know, she took no part in the effort that ultimately caused President Wilson to endorse the suffrage amendment, Congress to approve the measure, and the states, one by one, to give their consent. Yet, in the spring of 1918, she took her own small stand for women's equality. When she heard that the federal government would employ women in war offices, she applied for a job that would "release a man for the front." It was the "patriotic thing to do," she recalled. But when she was offered a position, she refused it on the ground that it was "insufficiently mathematical." She had been the top mathematics student in her graduating class, the first woman to win the school's mathematics prize. Though her college peers had judged her

21. Elizabeth Webb Wilson in the spring of 1917

a quiet and timid woman, she stood her ground and stated that she would only take a war job where her mathematical talents "could be utilized to the fullest."[58] The personnel office offered her a second position, which she also declined, and then a third. In all, she rejected nine jobs before accepting an appointment as Moulton's chief computer.

Elizabeth Wilson's persistence was a small victory for feminism. Washington had already established itself as a city of opportunity for single women. The novelist John Dos Passos described contemporary stenography offices in the capital, where "the typewriters would trill and jingle and all the girls' fingers would go like mad typing briefs, manuscripts of undelivered speeches by lobbyists, occasionally overflow from a newspaperman or a scientist."[59] A woman could earn seventeen or twenty dollars a week, enough to pay the rent on her own apartment, send some money to her family, and have enough left to occasionally purchase clothes from one of the finer department stores. The Washington office of experimental ballistics employed both men and women and put them in situations where they had to work together. The computing staff had sixteen workers and was split equally between female and male. Like Wilson, the other seven women were the offspring of prosperous homes and graduates of coeducational schools with strong mathematics programs. Two of the women had studied at the University of Chicago, two at Brown University, and one each at Cornell University, Northwestern University, and Columbia University.[60]

The staff of the Washington ballistics office stood midway between the mathematicians and engineers of Aberdeen and the officers in the army's artillery corps. Using data from the Aberdeen ranges, they prepared tables and documents for those who would actually command the guns in battle. The work was often sensitive or political in nature, requiring Moulton to negotiate among different branches of the military to establish standard operating procedures. "As a consequence of the various interests involved," he wrote about one problem, the work "had to be taken up somewhat formally."[61] In this environment, the mathematicians worked closely with the computers to prepare material that was both accurate and appropriate for the situation. During Wilson's first weeks in the office, she and Moulton had to prepare a range table for a French gun firing American ammunition. Wilson demonstrated "both personal mastery of the technical operations involved," wrote one of the mathematicians in the office, "and skill in supervising and checking the work of others."[62]

Reflecting on the experience, Moulton saw an unusual camaraderie between the mathematicians and the computers. "The unfailing courtesy and the evidence of mutual helpfulness which were manifested in numerous ways," he wrote, "were inspired not alone by military customs and the proprieties of the situation, but much more by sincere mutual respect and personal regard." Wilson, the only member of the office staff from Washington, acquired the role of chief computer and social leader. She hosted a dinner for the office staff at her parents' home and a party at her father's club.[63]

When the ballistics computers recalled their service in the First World War, they would remember the summer of 1918. "It would be difficult to gather in any way an equal number of individuals who would have more in common and whose relations would have been more harmonious," wrote Moulton.[64] The number of firings at Aberdeen increased each day, and the data flowed from the gun mounts to the proving ground's computing building and from there to the experimental ballistics office in Washington. Veblen spent much of the summer traveling in search of new computers, but whenever possible, he would catch an early train back to Aberdeen so that he could join the artillery crews on the range. He would fire the guns until the dimming light made further observations impossible.[65] He recruited mathematical talent from universities, from industry, and even from the offices of the *Encyclopedia Americana*. "The demand was immediate," remembered one computer, so "[I] terminated my [job]. I took the next train to New York, where I changed for Aberdeen."[66] In less than two months, Veblen added twenty-three graduate students or new PhDs to his staff, bringing the total number of computers at Aberdeen to thirty. Twice that summer, the army engineers had to double the size of the computing building.[67]

During those months, the computers gathered range data for naval guns mounted on railroad cars, tested new designs of streamlined shells, and uncovered a major design flaw in existing shells, a problem that Moulton judged to be "so great that the guns were of little value."[68] Yet the problem that most interested them was a major revision of Siacci's ballistics theory. "Upon entering the army," wrote Moulton, "a hasty examination of the classical ballistic methods showed . . . that they were wholly inadequate for current demands." He argued that Siacci's analysis "contained defects of reasoning, some quite erroneous conclusions, and the results were arrived at by singularly awkward methods."[69] His criticisms were unduly harsh, as the First World War armies were using the theory for problems that Siacci had not foreseen. Siacci had not planned for the problems of high-altitude fire, antiaircraft guns, and long-range artillery, nor had he anticipated that an army might have at its disposal a staff of forty-six human computers.

By all accounts, Siacci's theory continued to work fairly well for short- and intermediate-range artillery, but events in the first years of the war had shown that it failed dramatically in long-range artillery and high-angle mortar attacks, in addition to its shortcomings for antiaircraft fire, bombing, and fire from aircraft. While the staff would be able to correct some of these deficiencies, Moulton concluded that he should develop a new, comprehensive ballistics theory that could be used to analyze any circumstance. The work engaged the computing staffs in both Aberdeen and Washington and challenged them to use the most sophisticated mathematical concepts at their disposal. Moulton created the central outline of the theory. Other mathematicians handled specific problems with the theory, such as adjustments for altitude, the spinning of the projectile, or the rotation of the earth. Elizabeth Webb Wilson, who helped prepare the tables for the theory, recalled discovering that "the Germans had the advantage because the earth turned toward the east, therefore as they were shooting toward the west, their bullets carried further into the allies' lines."[70]

The computers were mathematicians, not ordnance engineers, and were most interested in the mathematics of Moulton's ballistics theory. One computer, ignoring the advantages to the gunnery crew, recalled how the work "made a brilliant use of the new theory of functionals,"[71] a concept that was then on the frontier of mathematical research. Isolated from friends and family, separated by an ocean from the dangers of war, the computing staff at Aberdeen lived the life of extended adolescence. They hiked through the countryside and conducted unauthorized, and probably unscientific, experiments with TNT and smokeless powder. From the collected scraps and rubble of ordnance experiments, they invented surreal versions of checkers and chess. At night they would gather in the computing shack, play cards, and do the things that young men do when they are at war, though perhaps they were unique in calculating their winnings and loses on adding machines which by day computed the trajectories of shells. After the cards were dealt, the conversations would wane as the players computed the probabilities that their opponents held winning hands. When they spoke, they talked not of lost opportunities or of distant family members, but of the mathematics that they loved and the theorems that they would prove.[72] One computer wrote that the experience "furnished a certain equivalent to that cloistered but enthusiastic intellectual life which I had previously experienced at the English Cambridge."[73]

The trenches in France had their own intellectual life, at least for the English troops. "The efficiency of the postal service made books as common at the front as parcels from Fortnum and Mason's," wrote critic Paul Fussell, "and the prevailing boredom of the static tactical situation . . . assured that they were read as in no other war."[74] For the scientists at the front, there were opportunities to observe and speculate. The English meteorologist Lewis Fry Richardson (1881–1953) found that many of his days were uninterrupted by combat or even by rumors of combat. The battles most often occurred at sunrise or sunset, when one side or the other was shielded by the glare from the low-lying sun. He served as an ambulance driver and had little responsibility beyond the work of caring for his vehicle. During the quiet hours, he worked at his science, theorizing about the movement of the winds, the distribution of humidity, the impact of the light from the sun. "My office was a heap of hay in a cold rest billet," he recalled. His approach to the problem was not a statistical method, like that used by the U.S. Weather Service in the 1870s, but a differential equation model that had much in common with the mathematics of astronomy and ballistics.[75]

Richardson described his analysis of the weather as "a scheme of weather prediction, which resembles the process by which the *Nautical Almanac* is produced."[76] He derived a series of differential equations that described how the weather changed moment by moment. These equations tracked seven basic properties of the atmosphere: its movement (in three dimensions), density, pressure, humidity, and temperature. The computing plan for these equations divided the globe into "a special pattern like that of a chessboard," a grid of longitude and latitude lines that marked 2,000 points where the weather would be computed in increments of three hours. "It took me the best part of six weeks to draw up the computing forms," he recorded, and when he attempted the calculations, he discovered that he required an equal amount of time to calculate a single advance of the weather at one of the points. His duties at the front had prevented him from fully concentrating on the arithmetic, and at one point, he had misplaced his manuscript. "During the battle of Champagne in April 1917," he recalled, "the working copy was sent to the rear, where it became lost, to be re-discovered some months later under a heap of coal."[77] When he finished the calculations, he concluded that "with practice, the work of an average computer might go, perhaps, ten times faster." From this one exercise, his little respite from the war, he concluded that it would require 32 computers to keep pace with the weather at one grid point, 32 computers to complete a single three-hour prediction in exactly three hours. As there were 2,000 points in his scheme, he would require total of 64,000 human computers to track the weather for the entire globe.[78]

"After so much hard reasoning," Richardson asked, "may one play with a fantasy?" He had not come to France in search of glory. He was a Quaker and a conscientious objector to war. Rather than serve in the military, he had chosen to drive an ambulance. His fantasy was a world where the soldiers massed on the Western Front were put to work reproducing the earth's weather in numbers. He would take 64,000 soldiers from the front lines and make them computers in a giant spherical computing room, a room that would have been larger than any sports stadium of Richardson's day. The internal "walls of this chamber are painted to form a map of the globe," he wrote. The North Pole would be on the ceiling, while Antarctica would be marked on the floor. England would be found three-quarters of the way toward the top, its shape nearly hidden by one of the many balconies that ringed the inside of the room. Upon these balconies, Richardson imagined, "a myriad computers are at work upon the weather of the part of the map where each sits." He suggested that the computers might work on large sheets of paper and then display their results on "numerous little night signs" so that others could read them.[79]

The calculations would be directed by a senior computer, a scientist who had worked through every step of the mathematics. This computer would stand on the top of a tall and slender column that rose from the floor like the column for Admiral Nelson in Trafalgar Square or a tall skyscraper in a city of lesser office buildings. "Surrounded by several assistants and messengers," the senior computer would "maintain a uniform speed of progress in all parts of the globe." Richardson suggested that this computer "is like the conductor of an orchestra, in which the instruments are slide-rules and calculating machines." In his scheme, the conductor's baton was replaced by a pair of colored spotlights. The computer would turn "a beam of rosy light upon any region that is running ahead of the rest, and a beam of blue light upon those who are behindhand." Outside of the computing room, the computer oversaw a scientific compound, which included a radio station to transmit the weather predictions, a secure storeroom to hold old computing sheets, and a building that held "all the usual financial, correspondence and administrative offices." Next to the sphere was a training room and a research lab, as "there is much experimenting on a small scale before any change is made in the complex routine of the computing theatre." Richardson's weather compound ended at the border of Arcadia, the land where numbers stood for things of nature, rather than the flight of artillery shells, the output of a factory, or the commerce of men. "Outside are playing fields, houses, mountains and lakes," wrote Richardson at the end of his fantasy, "for it was thought that those who compute the weather should breathe of it freely."[80]

The fantasy of a giant computing laboratory was, perhaps, not so unrealistic to those who served on the battlefields or carried the wounded to safety. During his three years with the ambulance corps, Richardson saw hundreds of miles of trenches that were filled with thousands and thousands of soldiers. All he had imagined was the simple act of winding those trenches into a ball and using their occupants for the peaceful end of computing the weather. However, in 1918, weather did not represent the same kind of threat as the German army, so that no government had any interest in building a laboratory for 64,000 computers, 6,400 computers, or even 640 computers. Only one facility approached Richardson's vision of a computing compound with houses and lakes and trees. It employed but forty-two computers and was located in a Maryland town named Aberdeen.

CHAPTER TEN

War Production

> This was a strange and mysterious war zone but I suppose that
> it was quite well run and grim compared to other wars. . . .
> Ernest Hemingway, *A Farewell to Arms* (1929)

BY THE SUMMER OF 1918, the whole European war seemed to be packed into the city of Washington, D.C., an area barely ten miles on a side. The army tested ordnance in the city's northwest quadrant while the navy built guns in the southeast. Residents rented any available room to the new war workers who arrived each day by train. Downtown, entire government bureaus were squeezed into offices that once had held three employees, or two, or even just one. Outside, construction crews labored to create new buildings on the one large, unoccupied piece of land in the center of the city: the parklands of the Washington mall. Watching all of this activity from his office, the director of the army's Department of Ballistics concluded that the government needed nothing so much as time, "time to build manufacturing capacity on a grand scale without the hampering necessity for immediate production; time to secure the best in design; time to attain quality in enormous output to come later as opposed to early quantity of indifferent design."[1] For those who worked in Washington, the war was a problem of production more than it was a problem of military strategy. The offices in the city had to clothe, equip, and feed the members of the American Expeditionary Force. They had to transport every soldier to the front, provide the forces with weapons and medical care, and return each veteran to the United States. The tools of their war were typewriters, forms in triplicate, adding machines, accounting sheets, and statistics. For many a government manager, the most powerful machine was the punched card tabulator. "Calculators . . . went to war," wrote historian James Cortada, "but the dramatic examples of data processing at work were punched card gear."[2]

The tabulator of 1918 was substantially more sophisticated than the simple counting machine of 1890. Following the success of his first census, Herman Hollerith had created a tabulator that could add numbers and accumulate sums. He devised a method for representing numbers as holes in a card. A number was punched into the card one digit at a time, with the location of each hole representing the value of the digit. According to historian Emerson Pugh, Hollerith created this tabulator in order to sell his machines to the railroad companies. "His primary target was the New York Central, which had become the second largest railroad system in the United States." He promoted the tabulators as a tool for processing waybills, the paperwork that followed the movement of freight.[3] He also found a ready market for the new adding tabulators within the United States Census Bureau. In 1900 and 1910, the Census Office had used the tabulators to gather agricultural statistics. Such a process was beyond the ability of the original tabulators, as it required the machines to sum the number of cultivated acres in each township, the total cattle in each county, and the bushels of grain produced in each state.[4]

The 1910 census had been the last for Herman Hollerith. In 1911, Hollerith sold his company to a syndicate of investors, who merged the company with two other firms to create the Computing, Tabulating and Recording Company, or CTR. The president of this new company was Thomas J. Watson (1874–1956), who had made his reputation as a salesman for the National Cash Register Company. Watson saw a great future in punched cards and worked to find new markets for the equipment in manufacturing, banking, retail, and government.[5] When the United States went to war in 1917, he was ready to provide tabulating gear to any government division that needed to prepare statistics, to direct military operations, or to manage war production.[6] With CTR tabulators, the United States government created nearly a dozen new computing offices during the war, each office with a production capacity that rivaled the abilities of the Census Bureau. All of these offices were associated with temporary agencies that managed the wartime economy, and all of them processed substantial amounts of statistical data. The largest of these offices was part of the War Industries Board, the agency that oversaw American manufacturing. This board, under the direction of financier Bernard Baruch, worked to provide the American Expeditionary Force with all the equipment it needed while, at the same time, ensuring that the civilian economy had an adequate supply of goods and services.

The agriculture production board, known as the Food Administration, was directed by Herbert Hoover (1874–1964). Hoover was then known as a successful mining engineer and the chair of a committee that had successfully brought food relief to Belgium. Hoover faced one of the most complex managerial tasks of the war. The country had hundreds of thousands of food producers, ranging from the kitchen gardens of housewives to the massive plants of the meatpackers. As Hoover started assembling his staff in one of the new buildings on the mall, he recognized that the United States government had only a vague understanding of how producers and processors delivered food to the American market and what resources might be redirected to the war effort without causing undue hardship at home. The government's principal research office on food distribution, the Bureau of Markets, had been created only four years before by the United States Department of Agriculture. Hoover moved quickly to form a comprehensive statistics office and appointed as its director Raymond Pearl (1879–1940).

Raymond Pearl came to Washington from the Agricultural Experiment Station for the state of Maine. He was one of the many scientists, like Oswald Veblen and Forest Ray Moulton, who were eager to bring their expertise to the conflict. He had spent nearly ten years as the chief scientist of the experiment station, which was a common name for an experimental farm. At the station, he had overseen tests of new farming methods, fertilizers, equipment, plant stocks, and animal breeds. Pearl had arrived at the station with a traditional degree in zoology and a year of study with Karl Pearson at the Biometrics Laboratory. He had traveled to London in the fall of 1905 and had learned the statistical methods that Pearson and his colleagues were using to study the problems of natural inheritance. His first publication after leaving Pearson looked like nothing more than a freshwater version of the research done by W. F. Weldon and Florence Tebb Weldon in the 1890s. Instead of looking at crabs, Pearl had studied variations in crayfish. At the Maine Agricultural Experiment Station, he took the statistical methods that he had learned from Pearson and put them to work on the concrete problems of improving farm output.

Arriving at the Food Administration in July 1917, Pearl had to put experimentation aside in order to assemble a staff capable of processing a large flow of data. In many ways, his computing office resembled the Smithsonian project that had summarized weather statistics in the nineteenth century. At the center of his organization was a punched card processing room with a staff of several dozen machine operators. Surrounding this facility were several smaller offices, each charged with preparing statistical reports on a specific aspect of the agricultural economy. One office handled sugar production, another food storage, and a third retail prices. These smaller divisions collected data from a staff of volunteers, prepared cards for the tabulating machines, and edited the final reports. The division that dealt with retail prices was typical of these offices. It began operation in September 1917 with "a few clerks and stenographers," according to Pearl. It "grew and expanded until . . . [it] consisted of about thirty five people."[7] Each week, it received data from its volunteer staff, "a selected number of reliable people in every town and city of 3,000 or more inhabitants." The volunteers, who were almost exclusively women, in contrast to the male computing staff, were given a package of forms and instructions to collect the prices of specific items. "In the majority of cases, the prices reported were those taken by the reporter from her own sales slips," wrote Pearl. The completed forms were sent to Washington once a week. Seven hundred of these forms arrived in the first week of October, the first week of tabulating data. This number grew to 1,000 by the first of December. By the end of the winter, the retail price division was processing 2,000 forms a week.[8]

In general, the numbers produced by Pearl's tabulators, clerks, and computers were used by Hoover and the senior leaders of the Food Administration to describe the state of the agricultural economy. They were rarely used to explore the operation of agricultural markets and suggest policies that might increase production. If the leadership of the Food Administration had been astronomers in the days of Edmund Halley, they would have been content to review the known observations of the comet of 1682 and accept the notion that it returned to view every 75 years, more or less. They would have been unmoved to compute the comet's orbit or use it as a test of Isaac Newton's theory of gravitation. The notion that the sprawling agricultural economy could be described with differential equations or probed with statistical calculations was not widely accepted in 1917–18. Long after the war was over, the *Washington Post* ran an editorial that mocked attempts to study the economy mathematically as "Hog astronomy" that required computations with "Hogarithms."[9] While some of this mockery came from the general resistance to scientific analysis, a resistance that paralleled the objections made by Jonathan Swift to the calculations of comet orbits, part was political. Many involved in business did not want any intervention in the market that might restrict their ability to make a profit. Hoover preferred to appeal to the "voluntary effort of the people" in order to achieve his goals and placed his trust in "the spirit of self-denial and self-sacrifice in this country."[10] The records of the Food Administration are filled with posters and advertisements that encouraged the recycling of fat, the preparation of meatless meals, the production of more pork, and other ways of making the best use of the nation's food.[11]

Hoover's approach to agricultural management often conflicted with ideas proposed by specific parts of his administration. Each of the major food products had a board or commission that reviewed the reports from Pearl's statistical office and recommended policies. There were commissions for all of the major agricultural products, including beef, sugar, milk, and wheat. These commissions were more familiar with the production of their specific commodities and were often more willing than Hoover to intervene in the markets. It was within these commissions that the power of punched card tabulation met the mathematics of scientific calculation.

The Swine Commission, which was described as seven "expert swine men from all over the United States," has left us the clearest record of its experiments with advanced mathematics.[12] Well before the start of the war, the production of pork was already handled in a highly systematic way. Most of the country's pork came from an area known as the "Corn Belt." Roughly this area was centered on Chicago and stretched from Iowa on the west to Indiana on the east. The northern border of the Corn Belt could be found in Minnesota and Wisconsin, while the southern edge was in Missouri. "Corn-belt agriculture was integrated into national markets almost from the time the Midwest was settled," wrote agricultural historian Mark Friedberger.[13] Pork production began with corn, which Midwest farmers grew as feed for their hogs. When the hogs were grown and fattened, they were loaded onto trains and shipped to slaughterhouses in Chicago. Rail lines led directly into the buildings where the hogs were killed and butchered on long, moving "disassembly lines." Once it was dressed and packed, the meat was loaded onto refrigerated rail cars and shipped to the urban centers of the East. The process of turning corn into bacon and hams and chops was described by Upton Sinclair, the great novelist of meatpacking, as "porkmaking by machinery, porkmaking by applied mathematics."[14]

At the Swine Commission, the applied mathematics of pork-making was developed by a young newspaper editor from Iowa named Henry A. Wallace (1888–1965). Wallace served the commission as an unofficial staff member. He had come to Washington as an assistant to his father, Harry C. Wallace (1866–1924), one of the seven "expert swine men." The Wallace family paper, named *Wallace's Practical Farmer,* had a strong circulation in the central Corn Belt. It offered readers snippets of gossip, reports of agricultural news, market prices, advertisements, Bible lessons, recipes, advice, and general guidance for the farmer, all under the motto "Good Farming, Clear Thinking, Right Living."[15] Through the paper, the elder Wallace spoke with a respected voice on agricultural, social, and political issues. He was a member of the Republican Party, though he was part of the party's progressive, pro-agriculture wing.[16]

The younger Wallace, Henry A., had a great faith in numbers, in the ability of data to reveal the problems of society and suggest a governmental policy that would improve the life of ordinary people. It was a faith that was scientific and yet was deeply rooted in his family's religious heritage. Throughout his life, he would return to a biblical story of famine. "Behold, there come seven years of great plenty throughout all the land of Egypt: And there shall arise after them seven years of famine; and all the plenty shall be forgotten in the land of Egypt; and the famine shall consume the land."[17] The hero of this story was a young man named Joseph, one of the sons of Israel. Through his spiritual insight, Joseph foresaw the famine and showed the Egyptians how to prepare for the years of dearth. When the famine arrived, Egypt had food enough and to spare, "corn as the sand of the sea, very much, until he left numbering; for it was without number."[18]

At the start of the war, America had no want of food but the lot of the farmer was precarious. In general, farm income had been declining since the end of the Civil War with only a few periods of increase. "Who will be like Joseph?" Henry A. Wallace asked in an article he wrote for *Wallace's Practical Farmer*.[19] Though he never rejected spiritual solutions, he looked to numbers and statistics to provide the guide for the nation's farmers. After completing a degree in agriculture at Iowa State College, he had started a systematic study of mathematics and statistics. He had learned calculus with the assistance of a professor at Drake University in Des Moines. When he was ready to learn the basic concepts of mathematical statistics, he acquired a book that had been written by a member of Karl Pearson's Galton Laboratory and studied the work himself.[20]

At the Swine Commission, Henry A. Wallace found the opportunity to apply his mathematical knowledge in a computation that described the relation between the corn and hog markets. His father had argued that "hogs are simply condensed corn" and that farmers would produce more hogs only if they could make more profit by selling pork than by selling corn.[21] If this were true, then the government could increase hog production by guaranteeing a higher price for hog flesh than for the equivalent amount of corn. The younger Wallace invented a value he called the "corn-hog ratio," which was the price of a bushel of corn divided by the price of one hundred pounds of hog flesh, and began the calculations to justify his father's theory. The work was not as difficult as the 1758 computations of Halley's comet, but they would have required a substantial computing staff if Wallace had not been able to use the machine-tabulated reports of Raymond Pearl's statistical office. He computed corn-hog ratios by hand from these reports and concluded that hog production would increase if the guaranteed price for one hundred pounds of hog flesh exceeded the price of thirteen bushels of corn.[22]

When presented with the calculations of Henry A. Wallace, Herbert Hoover was unable to repeat the blessing that the biblical pharaoh had bestowed upon Joseph: "Forasmuch as God hath shewed thee all this, there is none so discreet and wise as thou art: Thou shalt be over my house, and according unto thy word shall all my people be ruled."[23] Hoover viewed the proposal as an unacceptable form of government intervention in the marketplace and had no intention of implementing it. If anything, he wanted to remind the Wallaces of the last line of the pharaoh's blessing: "in the throne will I be greater than thou."[24] This disagreement marred the last two months of the war for the Swine Commission. Hoover remained ever obdurate, and Harry C. Wallace attempted to force the issue by bringing public pressure to bear on the man who was credited with saving the people of Belgium from starvation. Wallace wrote articles in his family farm paper and recruited the assistance of other papers in the Corn Belt. At times, the language of this campaign was abusive, suggesting that, perhaps, the corn-hog ratio did not have the same heavenly source as Joseph's plan for Egypt or that the biblical pharaoh was more accepting of new ideas than Woodrow Wilson's food administrator. As the Allied troops began their final campaign against Germany, Harry C. Wallace accused Hoover of acting with "such chicanery and deceit as an experienced businessman knows how to use in case of emergency."[25] Unmoved, Hoover held his position.[26]

Had there been more time, had the United States been engaged in the war for a longer period, the Food Administration might easily have been forced to undertake more mathematical analyses of the agricultural economy and find policies that were acceptable to all, but as the country entered the last month of the conflict, discipline at all of the major computing organizations began to falter. At Aberdeen and at the Experimental Ballistics Office in Washington, the computers started growing restless. The final campaign of the war coincided with the start of the academic year, and a few computers left the proving ground in order to return to their universities. Among those that remained, a gentle discontent began to grow. Forest Ray Moulton, who had become slightly bothered by the military discipline, quipped, "It has sometimes been a little irritating for men of national and international reputations as scientists to be compelled to show their photographs before they were permitted to stand in line at a cheap cafeteria."[27] Others tired of the military discipline that invaded their mathematical idyll. One computer wrote that Aberdeen was a "queer sort of environment, where office rank, army rank and academic rank all played a role, and a lieutenant might address a private under him as 'Doctor' or take orders from a sergeant."[28]

Part of the restlessness came from the realization that only a small part of the Aberdeen computations were going to reach Europe. Early in the war, American ordnance officers had decided to delay the deployment of American guns in order to release space on the transit ships for troops and other supplies.[29] The gunners of the American Expeditionary Force used French and British weapons. Though a few range tables for these weapons were prepared in the United States, the majority of the calculations had been done at the Galton Laboratory in England or at Gâvre in France. The only prominent weapons to make the Atlantic crossing were large naval guns that were mounted on railroad cars. The guns were deployed in September and quickly moved toward the retreating German army. In an action that could be interpreted as a fit of pique or as a show of Allied dominance, one railroad gun battery used range tables prepared at Aberdeen to place the last shot of the war. The battery crew fired the shot at 10:57 AM on November 11. It flew for nearly two and a half minutes, striking its target seconds before the armistice time of 11:00 AM.[30]

The armistice released a tremendous mathematical energy on both sides of the Atlantic. Less than twenty-four hours after the firing stopped, the American army began to terminate ballistics experiments and demobilize the computers at Aberdeen. "On November 12, the telegraph wires fairly hummed with cancellation orders emanating from Washington," wrote historian David Kennedy. "Within a month, the [ordnance] department had unburdened itself of $2.5 billion of [weapons contracts]."[31] The Aberdeen computers felt the impact of these cancellations in less than three weeks. The number of test firings peaked on November 26, the busiest day of the war, and quickly began to decline.[32]

The proving ground released the first computers in early December as it began to conclude ballistics experiments and range table production. When the computers looked for jobs as civilians, they looked for positions as mathematicians rather than in fields related to ballistics or computation. "For many years after the First World War," recalled mathematician Norbert Wiener (1894–1964), "the overwhelming majority of significant American mathematicians was to be found among those who had gone through the discipline of the Proving Ground."[33] Wiener's observation probably exaggerated the influence of the forty-two mathematicians who served with Veblen and Moulton, but it suggests the reputation that this group had acquired in the weeks and months after the end of the fighting. Most of the proving ground computers, including Wiener, were able to use their war experience to advance their careers. Shortly after leaving Aberdeen, Wiener was offered a position at the Massachusetts Institute of Technology.[34]

The female computers, the women of the Experimental Ballistics Office in Washington, had fewer opportunities than their male counterparts. None of them held advanced degrees in advanced mathematics, and hence they were not qualified for the few positions that were open to women at the nation's universities and colleges. Some hoped to find positions as computers, but they soon found that computing jobs declined in times of peace, and again, these jobs went to men first. Elizabeth Webb Wilson, perhaps the most ambitious of the group, tried to find a computing job in Washington, D.C. One of the ballistics officers described her as looking for "employment in which her somewhat exceptional preparation can be made useful in the national service."[35] She was no less aggressive in attempting to use her war record than the men, but she could not insist upon a job that made full use of her mathematical talents, as she had in March 1918. After a year of unemployment, she became a high school mathematics teacher in Washington.[36]

22. Final picture of army ballistics computers in Washington, D.C.

The task of closing the army computing offices fell to Oswald Veblen. At the exact moment of the armistice, he was with the American Expeditionary Force in France.[37] Anticipating the end of the war, the army had sent him to inspect the ballistics facilities of Britain, Italy, and France before they were disbanded. He packed his bags with the latest calculations from the proving ground to give to the artillery command in France. He also took a new tuxedo, in case he was invited to any formal parties when abroad.[38] During his trip, he took every opportunity to meet with European mathematicians. He visited Cambridge in England, the École Polytechnique in France, and the University of Rome in Italy.[39] He returned to the United States in March and relieved Forest Ray Moulton at the Washington Ballistics Office. For the next six weeks, he prepared reports, summarized experiments, secured office records, and demobilized the few remaining computers.[40] With the experimental program coming to a close, he had time to attend the opera, take long walks through the city, and make plans for the future of American mathematics. "Range tables are not being worked on to any extent nowadays," was the final word from Aberdeen.[41]

The armistice allowed Karl Pearson to reclaim the leadership of organized computation. He had withdrawn from ballistics computations six months before the end of the war and a full year before Oswald Veblen resigned his commission. He spent the spring and summer retrenching, a

metaphor borrowed from the front lines in France. He hired new computers, evaluated the state of his laboratory, and started on a new plan of
research. In many ways, the war clung to him longer than it touched the
lives of the computers at Aberdeen or at Washington. On a visit to his
country house, he wrote a long, elegiac memoir of his time during the
conflict. He confessed to having become sensitive to the sound of thunder
and associating the smell of pumpernickel bread with the odor of explosives. "I want instinctively to whinny like the dogs, if there be a sudden
clap of thunder, and will-power has still to be exercised to avoid it."[42]

Armistice Day found Pearson sitting in a hospital recovery room next
to the bed of Leslie John Comrie (1893–1950). Between the two of them
was a Brunsviga calculator. Pearson was explaining the finer points of
machine calculation, while Comrie was asking how certain problems
might be handled by the device. L. J. Comrie, as he preferred to be called,
had been a late recruit for the war. He was part of a New Zealand regiment that had been assembled to replace troops from the home island.
He had studied chemistry at the University of Aukland before joining the
army, but he had a deep love of astronomy and a special affection for the
classical problems of positional astronomy. As his troop ship steamed
across the Indian Ocean, he had occupied himself by tracking the ship's
course. In ordinary circumstances, it would have been a harmless diver-
sion, but in time of war, when troop movements were secret, it defied military discipline and could have earned him a court-martial. He arrived in
France, either undiscovered or forgiven, only to meet with one of the
many meaningless events of the war. A munitions accident badly wounded
him and forced the army surgeons to amputate one of his legs. While he
convalesced in London, volunteer nurses visited him and asked if he
would like to be trained for some trade or occupation that might be
suited for the handicapped. Comrie replied that he would much prefer to
continue his university education and become an astronomer. This conversation made its way to Pearson, who was always looking for potential
computers. Brunsviga in hand, he found his way to Comrie's hospital
ward, where the two began a friendship over computation.[43]

Pearson and Comrie had little in common beyond their mutual ambitions and their love of numbers. Pearson was an imperious man, a scientist who could speak from the mountaintop of his grand visions and his
mathematical methods of proof. His biographer wrote of Pearson's
"fierce intellectuality and disposition to theorize about everything from
religious faith to sexual love."[44] Comrie was a scrapper, always impatient
to show that he was no one's inferior. Once his health had recovered sufficiently, he started working in the Galton Laboratory, now the formal
name for the office that Pearson had started as the Biometrics Laboratory. His heart was not in the study of mathematical statistics, and he cer-

23. L. J. Comrie at calculator

tainly did not share Pearson's infatuation with the Brunsviga calculator, but the laboratory gave a focus to his life while he prepared for the future. In all, he spent nine months with Pearson before a scholarship for New Zealand veterans allowed him to depart for Cambridge University and the study of astronomy.

During Comrie's term at the Galton Laboratory, Pearson brought his computing staff back to full strength and began a new round of statistical research. Either through Comrie's influence or from his observations of the scientific world, Pearson realized that he had become one of the world's experts on scientific computation. As he labored to train new workers, Pearson was "struck by the absence of any simple text-book for the use of computers and still more by the absence of obviously necessary auxiliary tables."[45] Before the First World War computing had been a craft skill, a loosely organized body of techniques that were passed from generation to generation like the skills of a carpenter or the knowledge of a butcher. One generation of computers had learned their techniques from Nevil Maskelyne. Another from Benjamin Peirce. A third from Myrrick Doolittle. At that juncture, Pearson realized that a new generation was learning their methods from him.

Pearson proposed to codify the methods of computation in a series of pamphlets, entitled *Tracts for Computers,* which would provide solutions for most "practical difficulties of the computer." The name may have been inspired by the Edinburgh Mathematics Laboratory, which had published a series of tracts on the theory of numerical methods. If Pearson borrowed the title, he did not borrow the goal of the Edinburgh series. He intended that his tracts would present practical lessons, such lessons "as we have met with [in] our own experience."[46] With these lessons, a computer could develop a computing plan for any kind of numerical problem. Of all the computers of the First World War, the staff of the Galton Laboratory had handled the largest variety of problems. They had reduced data and computed ephemerides for the University of London astronomical laboratory. They had tabulated census data for the government and handled statistical correlations for Pearson. For the Munitions Ministry, they had computed trajectories and adjusted surveys. This expertise had been scattered during the last months of the war, but Pearson remained in contact with many of the computers who had served with him and could recruit a substantial pool of talent to prepare the *Tracts.*

In all, the friends and staff of the Galton Laboratory completed twenty-six pamphlets. L. J. Comrie wrote one of them, and Pearson prepared two of the *Tracts.* Pearson's contributions dealt with the techniques of interpolation, the process of filling in the points between two existing values. Pearson had hoped that most of the tracts might deal with similar methods, but he was only able to publish four booklets on such subjects, including the two that he contributed. The other two methodological pamphlets dealt with mechanical quadratures, or the method of small arcs, and the technique of smoothing, the mathematical means for drawing a simple curve through clouds of data.[47]

In one pamphlet Pearson tried to catalog the available literature of computation. Tables and notes on computation could be found in the books and journals of at least a half dozen fields, far more than an ordinary computer could follow. He asked a colleague to prepare a bibliography of logarithm tables by reviewing the literature of physics, astronomy, optics, surveying, and engineering. Pearson claimed that scientists regularly asked him for a bibliography of tables, but he did not seem fully committed to this kind of research. In the preface to the bibliography he asked, "Has [the author] adequately supplied an admitted want?" His reply was not especially confident. "I hope it may be so," he wrote, "but only the critics, present and future, can provide a satisfactory reply."[48]

Comrie's tract was a table of tangents and logarithms. In all, twenty-one of the twenty-six pamphlets were mathematical tables, far more than Pearson probably intended. He claimed that these tables had "special value to the practical computer,"[49] but they were an odd collection of special functions, sampling numbers, and probabilities. Many of these were originally computed during the war "because the required tables [had] not yet been published to the necessary numbers of figures, or because we did not know, or still do not know, if such tables were ever computed."[50] The Galton Laboratory computers prepared these tables for publication by checking the original text for errors, proofreading the typeset table, and preparing an introduction. The introduction often proved to be the most valuable part of the tract, for it described the mathematics behind the table and showed how the values might be employed.

The largest table in the series filled eight volumes. It possessed the grand title *Logarithmetica Britannica* and embodied the nationalism that had contributed to the start of the Great War. "When it came to my knowledge that the French proposed to issue a fourteen figure table and the Germans a fifteen figure table," Pearson wrote, "it seemed to me that it was fitting that the land wherein logarithms were cradled should rise to the occasion and issue a standard table . . . to twenty figures." Through most of the nineteenth century, computers had used logarithm tables to simplify calculations by turning multiplication into addition. When the *Observatory Pinafore* computers sang of using the tables of Crelle, they were referring to the use of logarithms for astronomical calculation. By 1919, logarithms had only a limited role in scientific calculation, as they had been replaced by calculating machines. Pearson claimed that people who used ordinary logarithm tables for calculation "are either ignorant of the existence of slide-rules and mechanical calculators or else unfortunately cannot afford them." The one use he saw for logarithms was in high-precision calculations, and it was for that reason that he agreed to publish the twenty place values of the *Logarithmetica Britannica*. It was not, he said, "an enterprise of profit."[51]

Pearson hoped to publish at least one tract describing the features of calculating machines and the techniques of machine operation, but he never found the time to write such a pamphlet or identified anyone else to do the job. The first part of this work, the description of the machines, was already covered in a German book, *Die Rechenmaschinen* (*The Calculating Machines*).[52] The second part, far more difficult to write, required contributions from many individuals, as no one could claim to be an expert operator of all calculating machines.[53]

The *Tracts for Computers* probably achieved the goals that Pearson set for them. Judging from the worn condition of most library copies, we can conclude that at least some of those computers who came of age between 1920 and 1939 learned their lessons from the wartime staff of the Galton Laboratory. At the same time, these little booklets received no critical response from the scientific community. Few scientific journals printed notices of their publication, and only one or two offered reviews of the pamphlets. Even the mathematicians most qualified to pass judgment, such as those who had served at Aberdeen, expressed no opinion on the series.[54] They were simply part of the war production, part of the contribution that computers had offered to the conflict.

CHAPTER ELEVEN

Fruits of the Conflict: Machinery 1922

> Can a man sit at a desk in a skyscraper in Chicago and be a
> harnessmaker in a corn town in Iowa . . . . ?
>> Carl Sandburg, "Accomplished Facts,"
>> *Smoke and Steel* (1922)

THE ARMISTICE LEFT the United States with a vast pool of equipment, energy, and vision. Beginning in the winter of 1919, train after train arrived at the Aberdeen Proving Ground with field artillery pieces that had been built for a final offensive into Germany. The proving ground staff unloaded the weapons, one by one, and towed them to the large fields where Oswald Veblen had conducted his first range tests. They placed the guns in long, straight lines to await the next war to end all wars. As the army was starting a period of slow decline, they sat in winter snows and summer heat, leaving the base only when some veterans' lodge requested a gun to use as a lawn decoration or to serve as a memorial to fallen comrades.

The trains carried surplus punched card tabulators and sorters and punches north from the government office buildings to the warehouses of the Computing, Tabulating and Recording Company in New York. "Rising inventories became a problem in 1919 and 1920," wrote historian James Cortada, "before commercial enterprises could sift production back to civilian levels."[1] Unlike the field artillery in storage at Aberdeen, some of this equipment would form the basis for a new class of scientific computing laboratory, a type of laboratory that would combine the tabulators with the expertise of human computers. Two veterans of the Food Administration would shape this new kind of computing facility, Henry A. Wallace and Howard Tolley (1889–1958).

Howard Tolley was one of the last links to Myrrick Doolittle and, through him, to Benjamin Peirce. He had come to Washington in 1910 and become a computer for the Coast and Geodetic Survey office. "[The job] consisted of sitting in an office up . . . on Capitol Hill and running a computing machine," he later wrote, "computing such things as the latitude and longitude of particular triangulation stations in different parts of the United States and computing the altitude of different hilltops and mountaintops."[2] Doolittle was only a few years from retirement, but he was still a guiding force in the office and taught the new computers his

24. Howard Tolley (back row right) and Henry A. Wallace (front row center) at U.S. Department of Agriculture

method of computing least squares adjustments to surveys. Tolley was initially intrigued with this work, thinking that it was "the only part that required any knowledge of real mathematics," but before long he recognized that the calculations "follow[ed] a regular fixed routine, requiring no judgement."[3] After a few months of adjusting surveys, Tolley tired of the work. "What is there to this?" he complained. "I [would] come over to the office every morning at nine o'clock and I [would] work on computing these things, adding, multiplying, running these computing machines, deciphering what's in the books of these surveyors." The job paid $100 a month, a sum that had been unchanged for nearly twenty-five years. "In effect it's all spent before I draw it, and I [had] a pretty hard time keeping good clothes on my back."[4]

Tolley's frustration was compounded by the knowledge that the major surveys of the North American continent were complete and that he was only handling refinements and detailed adjustments. "Just being a computer in the Coast and Geodetic Survey was completely futile," he later remembered; "it wasn't helping the world any." He considered returning to college for graduate study or even joining a survey team for the Alaska railroad, but he concluded that graduate study was expensive and that the surveyors "didn't want a desk mathematician."[5] One morning, when Tolley was chatting with his supervisor in the Coast and Geodetic Survey offices, he learned that the Department of Agriculture was seeking a general-purpose mathematician to work on some "problems that were related to genetics—Mendelianism,—and on some that were related to the capacity of farm silos."[6] When he went to interview for the job, Tolley discovered that the Department of Agriculture was most interested in what he had learned from Doolittle, the "knowledge of least squares and the adjustment of observations."[7] He accepted a position with the department and worked on a number of issues that were fundamentally economic in nature. During the war, he assisted Raymond Pearl with the statistical work of the Food Administration and, in 1921, became one of the first members of the department's new Bureau of Agricultural Economics.[8]

In January 1921, the United States inaugurated a new president, Warren G. Harding, and the Department of Agriculture welcomed a new secretary, Harry C. Wallace. One of Wallace's first acts was to create a new research office called the Bureau of Agricultural Economics. This office collected together all the employees of the department, including Howard Tolley, who were engaged in studying problems of production, markets, and financing. Within this group, Tolley promoted the method of least squares as a means of analyzing agricultural data. This application of least squares was substantially different from its use in survey adjustment, even though the method of calculation was unchanged. In agricultural studies, least squares did not adjust data. Instead, it took data apart in order to identify underlying causes or forces. It could identify the effect of fertilizer on crops or the feed that increased the weight of farm animals. It was sometimes called "regression analysis" or "the analysis of variance."

In the fall of 1922, Tolley gave a series of lectures in order to introduce the staff of the Bureau of Agricultural Economics to the ideas of statistical least squares.[9] Tolley illustrated the theory with an example that concerned the damage done to cotton crops by the boll weevil. His data, gathered in the southern states by agents of the department, contained the extent of damage on each field, the typical size of the cotton plants, the time of year, the amount of rain recorded in the area, and the daytime temperature. Using least squares analysis, Tolley showed how it was possible to determine which of these factors were present in the most heavily damaged fields. According to his results, the plants were most vulnerable at a certain stage of their development.[10]

Though the method of least squares promised much to the agricultural researchers, it was of little use unless the department could provide a computing office. The calculations were every bit as daunting as the least squares calculations of survey adjustment. A survey calculation would be done only once. A least squares analysis of a specific problem might have to be repeated every season. Tolley believed that at least some of the calculations might be handled with punched card equipment. In agricultural statistics, as in survey adjustment, least squares computations had two distinctly different parts. The first part reduced the data to a series of normal equations. By itself, this activity was especially demanding for agricultural researchers, as they were often dealing with large collections of data that spanned counties or states or regions. The punched card equipment of 1922 could summarize data for many applications, but it needed to be used in a special way for least squares calculations. The normal equations required multiplications, and punched card tabulators could only compute sums. There was a simple way to force the machine to perform multiplications, but it required a skilled and attentive operator. The method, called progressive digiting, reduced multiplication to primitive additions. A simple product, such as 24×127, would require six cards. Four cards would have the number 127 punched on them. The other two would have the number 1270. To compute the product, the tabulator would sum the six cards.

When applied to real problems, progressive digiting seemed to be an awkward process, a complicated operation that should have been straightforward. It was something akin to counting the number of sheep in a field by summing the number of legs, adding the number of ears, and dividing the result by six. To handle real problems, operators had to punch multiple cards, sort them, sum them in a tabulator, shift the values, and sum again. It was difficult work, but it was faster and more accurate than the alternative of doing the computations by hand. For large collections of data, those that had been gathered from one thousand or two thousand farms, the punched card equipment provided the only practical way of preparing the normal equations.

Punched card technology offered no help with the second step of least squares analysis, the step of processing the normal equations. The only way to do it was to give the numbers from the tabulating equipment to a staff of human computers and let them complete the work. They would use a mechanical calculating machine and the mathematical method invented by Myrrick Doolittle some forty years before.[11] Even with Doolittle's method, this part could be time-consuming. Tolley advised researchers to minimize the labor by doing all calculations with only two digits after the decimal point. He defended this procedure by noting that agricultural research was imprecise and that "astronomical accuracy is really not necessary."[12]

After teaching his class on least squares analysis, Howard Tolley

25. Computing room at the U.S. Department of Agriculture

moved to create a central computing laboratory for the Bureau of Agricultural Economics, an office that had punched card equipment and the
expertise to perform least squares calculations.[13] In 1922, three separate
offices of the bureau had punched card equipment, but none of them was
fully utilizing its equipment. "The installation of the Cost of Marketing
Division is busy most of the time," he reported, but "that in the Division
of Land Economics about half the time, and that in the Division of Statistical and Historical Research something like one-third of the time."
His research suggested that none of these computing offices really understood how to prepare a problem for machine tabulation and that at least
one office had started problems that it had been unable to complete.[14]

In the winter of 1923, Tolley received the approval of the department's
senior statisticians to create his centralized office. Even with this approval, the task was challenging, as none of three offices was prepared to
surrender its punched card equipment. The possession of an expensive
piece of machinery was a sign of power and importance, even if the machine was not well used. Tolley worked with each of the offices, arguing
that a centralized tabulating facility could "rearrange schedules and adjust them to cards in such a way that both the cost of tabulation and the
elapsed time between the collection of data and the completion of tabulation would be materially reduced." By April, he had convinced all three
offices that they would benefit from a central computing division.[15]

For the new computing laboratory, Tolley had to choose between com-
peting brands of equipment. Two of the offices leased their equipment
from the CTR company, which was in the process of changing its name

to International Business Machines. The third office used the tabulators of the Powers Accounting Machine Company. Powers had been founded by a former census employee. His machines could tabulate the International Business Machines cards, but they operated in a slightly different manner.[16] Tolley chose to keep a unified shop and leased only International Business Machines equipment. He completed the work by the first of May and announced that the "tabulating and computing services will be available to all" engaged in economics research.[17]

When the office began operations, it was able to handle only the first part of the least squares computations, even though it employed thirty workers. The staff of this office were called "operatives," not computers. For most calculations, they punched cards and ran tabulators, following instructions that had been prepared by researchers in the Department of Agriculture.[18] The office was led by a statistician who provided advice to the various divisions of the department. He showed others how to organize their data and prepare it for punching, but he did nothing with advanced mathematics. The second step of least squares calculations was left to the bureau researchers and their assistants.[19]

A second computing laboratory was created with the assistance of the younger Wallace, Henry A., at Iowa State College in Ames. Henry A. Wallace did not move to Washington with his father but remained in Iowa to edit the family newspaper. From this position, he continued his statistical research. With the help of friends and subscribers, he would gather data from all over the state of Iowa, punch it onto cards, and summarize it with the tabulators of a Des Moines insurance company.[20] He used the results to recommend farming strategies to his readers. During the early 1920s, he urged Iowa farmers to change their mix of crops, a campaign he summarized as "More Clover, Less Corn, More Money." Though he generally used only simple numbers to support his ideas, Wallace could not resist the temptation to demonstrate his mathematical sophistication. By 1922, he was conducting least squares analyses of agricultural data and publishing the results in *Wallace's Practical Farmer*. At the end of one article, he proclaimed, "For the benefit of our statistical friends, we may say that our predicting formula has a multiple correlation coefficient of .91, which indicates a very real degree of accuracy."[21] Of course, few of his statistical friends read Iowa farming papers, and few of the hog farmers understood the mathematics of multiple correlation.

With his father running the Department of Agriculture, Henry A. Wallace had the opportunity to meet Howard Tolley and followed the development of the new computing facility at the department.[22] In the winter of 1923, Wallace followed Tolley's example and taught his own class on the subject of least squares. His classroom was at his alma mater, Iowa State College. The students were college researchers, senior faculty, and a graduate student or two. Some already knew the basics of least squares. One had served on the Food Administration as one of the "expert swine men."[23] In ten Saturday sessions, Wallace developed the fundamental theory of least squares, demonstrated how to prepare the basic equations, and showed how Doolittle's method could be used to compute the final answers. For most of the classes, he had a desk calculator on hand, a "cheap key-driven machine," he later recalled. For the last class, he decided to show how to compute correlations with a punched card tabulator. He borrowed a punched card tabulator in Des Moines, loaded it into the back of a truck from the family farm, and drove north to the Iowa State campus. He spent the morning explaining how the machine worked and demonstrating statistical calculations with it.[24]

Assisting Wallace in the classroom was an Iowa State College mathematics professor named George Snedecor (1881–1974). Snedecor held a master's degree in physics and had been hired in 1913 as part of an effort to improve the university's science and engineering departments. After his arrival at the campus, he had quickly recognized that the school had a greater need for a statistician. He had no connection to Karl Pearson, who was recognized as the founder of mathematical statistics, but he had studied some statistical methodology in college and was able to teach courses in elementary methods. He spent much of his time compiling the methods of statistical practice in a form that could be used by researchers in any discipline, though his examples tended to favor the agricultural research that dominated the campus.[25]

In 1924, Snedecor and Wallace wrote a pamphlet entitled *Correlation and Machine Calculation,* a paper that would have fit nicely into Karl Pearson's *Tracts for Computers.* It dealt with "practical difficulties" of using least squares techniques for computing correlation problems. The pamphlet showed how Pearson's correlation analysis was actually a form of least squares work and then described how Doolittle's method could be used to do the calculations. The two authors made only a passing reference to punched card technology, but they noted that "where the number of observations runs into the thousands, punched cards should be used with sorting and tabulating machines." The "machine calculation" promised in the title was the calculation done by a human computer with a "commercial form of adding machines."[26]

While he worked on the pamphlet, Snedecor experimented with punched card calculation. He leased a card punch, which he kept in his university office. This was a small device, little bigger than a standard dictionary, and consisted of a wooden base, a frame for the card, and a sliding pad with keys numbered from 0 to 9. He punched each number one digit at a time, advancing the slide as he progressed. Any mistake would ruin the card so that he would have to start afresh. Once he had punched a set of cards, he would send them to Des Moines to have them tabulated.[27] As he gained skill with punched card machinery and began to appreciate what he could do with it, he began to drift away from Wallace. There was no conflict or disagreement between the two, just diverging interests. Snedecor was devoted to statistical research and to the support of scientific study. Wallace was interested in many things and was devoting more time to political causes and new business ventures. His political goals were embodied in a congressional bill that would open new markets for farmers. His new business was a company that produced and marketed hybrid corn seed.[28]

Snedecor's experiments with punched card machinery slowly grew into an organized computing laboratory. He did not have the kind of resources that could be found at the Department of Agriculture in Washington but he was able to hire a human computer in 1925 and acquired a punched card tabulator two years later. When the machines arrived, he discovered that his office did not have the proper electrical wiring for the tabulator, a minor setback that forced him to find a room in another building for his laboratory.[29] Once he had the equipment operating, Snedecor was temporarily overcome with the fascination of new love, the kind of fascination that researchers before and since have bestowed upon new machines. Snedecor demonstrated punched card techniques to any interested researcher and used the tabulators to solve dozens of problems, including many that had no relationship to statistics and a few that were not particularly suited to punched cards. He summarized field trials, factored numbers, built databases, and solved differential equations.[30] His dean grew tired of his lengthy descriptions of punched card computations and suggested that, in the future, Snedecor might simply summarize his accomplishments.[31]

Snedecor named his laboratory the "Mathematical and Statistical Service," a title that defined the role of the office within the college and suggested that Snedecor's machines might be leased by outside clients. It was not the only university computing office in the country. The International Business Machines company was attempting to place its products at colleges and universities. In 1927, the year that Iowa State College acquired its machines, International Business Machines leased tabulators to at least four other schools: Cornell University, Columbia University, the University of Michigan, and the University of Tennessee.[32] Unlike the other sites, the Iowa State facility combined tabulating machines and human computers. The machines summarized data and computed the basic values for least squares problems, just like the machines at the Department of Agriculture in Washington. The computers completed the work, solving the problems by using Snedecor's and Wallace's variation of Doolittle's method.

26. Iowa State Statistical Computing Service

One computer, Mary Clem (1905–1979), rose to play a central role in Snedecor's laboratory. Clem came from western Iowa and had only a high school education. Mathematics, she later claimed, "was my poorest subject," a boring topic that never interested her. By contrast, she thought that computing was something entirely different. "The more I worked with Snedecor," she said, "the more fascinated I became with figures and data. I got to the point where that was my whole life." She became especially good at identifying patterns that could be used for detecting errors, values that she called "zero checks." A zero check was a sum that should equal zero if all the numbers had been correctly calculated. Such sums often escaped the eyes of those with more mathematical training. The first time that she presented one of these checks to Snedecor, he was skeptical. Clem recalled that "he sat there thinking about it a little bit, and he turned around and went through algebraic fiddling around and he said 'that's right they do.' "[33]

Because of her insight into calculation, Clem became the planner for the laboratory, a worker occupying a middle level between the human computers and the research scientists. She was remembered as a strong presence in the lab, a leader who set the schedule, trained new computers, and enforced discipline.[34] She oversaw about six computers and one machine operator. Many of these computers were graduate students or the spouses of graduate students. A few were local high school graduates like herself. The majority were women. Clem felt that men never adapted to the work, judging that "they would rather probably teach or do research work." Male or female, most of her computers were transients, working for only "two or three years" before they left the lab.[35]

Even with the combination of card tabulators and human computers, Clem and Snedecor found that there were problems that were too demanding or too expensive for the lab to undertake. One such problem was presented to the lab by Henry A. Wallace. In 1930, Henry A. Wallace conceived the idea of looking at long-term weather records and comparing them to astronomical ephemerides. He reasoned that the weather was a complex system driven by outside forces. He acknowledged that the radiation of the sun and the gravity of the moon were the largest of these forces, but he speculated that the planets might also exert a physical force upon the weather. Undeterred by what others might think of his plan, Wallace began to develop a framework in which he might search for a relation between the positions of the major planets and the weather on Earth.

Karl Pearson and Frances Cave-Browne-Cave had used correlation analysis to search for connections between weather patterns on the Atlantic Ocean. Wallace borrowed their methodology in order to connect fifty years of climate data with planetary data from nautical almanacs. He began with the ephemerides, punching one card for each of the 18,261 days in the fifty-year period of his study. Each card contained a code for the date, the positions of the moon and the major planets, and a blank spot for weather data, such as the daily rainfall in Des Moines, the high temperature in Cedar Rapids, or perhaps the wind direction in Ames.

Wallace had become a prominent political figure, though he had not quite reached the national stage. When his political opponents learned of his weather research, they mocked it as "weather astrology," a form of research that properly belonged in the misguided world of Jonathan Swift's Laputa.[36] The study was probably naively conceived and certainly belonged to a branch of meteorology that contributed little to the understanding of the weather, but its scientific validity is less important than the scale of the research. To compute a single correlation statistic, Wallace would have to duplicate his set of cards, punch his weather data on each of the 18,261 cards, compute the basic statistics for the correlation, and give those statistics to a human computer to finish the calculation. In this process, the final step was insignificant. The few multiplications and the single division needed to compute the correlation were dwarfed by the other activities. With machines of the 1930 era, a skilled operator would take three hours to create a duplicate set of cards, forty working days to punch the weather data onto the duplicates, and then twenty-one days to compute the basic parts of the correlation. The human computer would need a mere fifteen minutes to complete the task.

Wallace attempted to entice the Iowa State Mathematical and Statistical Service to continue the research. "This study seems to be of such unusual fundamental value that I cannot help but feel if would be a splendid thing if you people at Ames could take over these cards," he wrote to George Snedecor. Busy with his new hybrid seed company, Wallace offered to donate his 18,261 cards to Iowa, suggesting that the database was of "unusual value."[37] Snedecor declined the offer, perhaps sensing that the project was not founded on a firm understanding of meteorology and certainly knowing that such a demanding project would have easily overwhelmed his modest staff and his relatively simple punched card office. His computing laboratory ranked among the most powerful scientific computing facilities of the age, even though it consisted of no more than seven or eight people: Mary Clem, a small staff of computers, and an operator for the tabulator. They could process far more numbers than Pearson's Galton Laboratory with its collection of Brunsviga calculators. Wallace's weather calculations, even if they were based upon valid science, would have consumed all of their time and effort.

The computers at the American Telephone and Telegraph company were veterans of the First World War in the sense that the computing division of the company began to take shape during the summer of 1918. The company had expanded at the start of the war in order to provide the army and navy with radios and telephone equipment. It had a long history of scientific research that could be traced back to the original telephone of Alexander Graham Bell (1847–1921) in the 1870s. Company scientists studied a variety of problems that were related to telephone services. Chemists studied new materials for insulating wires; physicists looked at the propagation of radio waves; and statisticians evaluated different designs for operator stations. The computing division was an offspring of the transmission section, the group that was working to develop reliable and efficient long-distance telephone lines.[38]

The calculations of telephone transmission imposed special demands because they utilized a form of arithmetic that involved complex numbers. Each complex number consists of two values. For historical reasons, one value is call the "real part" and the other is called the "imaginary part." Because of these two parts, complex arithmetic requires more labor than ordinary arithmetic. The sum of two complex numbers requires two ordinary additions. A complex multiplication requires seven ordinary operations: four multiplications and three additions. The most taxing operation, complex division, requires sixteen steps: eight ordinary multiplications, six ordinary additions, and two ordinary divisions.

Though complex numbers increase the amount of calculation, they actually simplify the analysis of electrical circuits, particularly the analysis

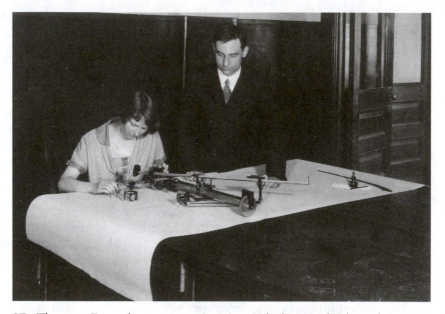

27. Thornton Fry and computer at American Telephone and Telegraph

of vacuum tubes. The vacuum tube amplifier had proven to be the key technology for long-distance transmission. American Telephone and Telegraph had built a prototype amplifier in 1912 and had demonstrated a transcontinental circuit at the 1915 San Francisco World's Fair.[39] In 1916, the company hired a mathematician, Thorton Fry (1892–1991), to assist the electrical engineers with their analyses. In the first year of the war, Fry hired a computer, Clara Froelich (b. 1892). Froelich was a graduate of Barnard College, the women's school affiliated with Columbia University. From what we know about her early years, Froelich was a reserved woman, a student of mathematics who had complained about being isolated among the social circles of Barnard.[40] She shared the computing duties with two other computers, but she proved to be the only permanent member of the computing staff. Fry preferred to hire recent graduates of women's colleges, and few of these workers stayed at American Telephone and Telegraph for more than a year or two.

In 1922, Fry, Froelich, and the other computers were removed from the long-distance transmission division and given their own office, the division of mathematics.[41] Their office occupied one corner of a yellow brick building that sat on the Manhattan riverfront about two miles north of the New York headquarters of American Telephone and Telegraph. As was common in office buildings of the day, no partitions divided each floor to separate the work spaces. The computers could watch the staff of electrical engineers studying the behavior of a vacuum tube or the properties of cables. They could smell creosote drifting up from the laboratories responsible for preserving the wood of telephone poles. From their desks, the computers could look south toward the skyscrapers of Wall Street or watch ships struggling up the Hudson River to the west. In 1925, the building, the computers, and the rest of the researchers were transferred to a new organization, named Bell Telephone Laboratories. Bell Laboratories was the largest of the many industrial research facilities formed in the 1920s. By the end of the decade, "more than 1,600 companies reported that they supported research laboratories," wrote historian Robert Reich, "employing nearly 33,000 people in all."[42]

The mathematics division of Bell Telephone Laboratories "does not regularly supply computing services to other departments," wrote Fry in 1925. He explained that computation was usually "performed by special groups of calculators in the departments where their services are required."[43] Froelich and her peers worked with Fry to develop new methods for calculation, advised other researchers on computational issues, and helped instruct the computers of other departments. Like many scientists of the age, Froelich studied the operation of IBM tabulating equipment, hoping that it might be adapted to the calculations of the company's engineers. The mathematics division did not have its own machines, so she was required to spend evenings in the company's accounting office with the tabulators that were employed for business transactions by day. She "made valiant efforts to use [punched card tabulators] for more purely mathematical problems," reported Fry, "but with little success."[44]

Froelich had greater success with the new desk calculating machines. These machines were the direct descendants of the geared adding machines of the 1890s. In the intervening years, they had acquired electric motors and a fuller set of operations. By 1925, the computing staff could claim to be experts on two machines, one called the Millionaire calculator and the other the Mercedes. The Millionaire had a mechanical multiplication table, while the Mercedes could "perform automatic division, requiring only that the operator set up the divisor and the dividend in proper registers, but not requiring any further supervision." All of these, she found, could be used for the two-part operations of complex arithmetic.[45]

The individual who acquired a reputation for adapting commercial business calculations "with little or no change in construction—to scientific computing" was L. J. Comrie.[46] Though Comrie was intrigued with the idea of building special-purpose machines for scientific problems, he argued that there were many benefits to be gained from commercial machines. First, they were generally flexible. Second, they were usually quite reliable, the product "of groups of experts," rather than a "single and perhaps not too experienced designer." Finally, he argued that they were "economical as compared with the overhead costs of design and construction on a small scale."[47]

After leaving Karl Pearson and the Galton Laboratory, Comrie had moved to Cambridge and had studied astronomy. Slightly older than most students and calloused by his brief experience at the front, he did not easily fit into the society of graduate students. Rather than focusing on the narrow subjects of astronomical theory and the automatic collection of data, he looked to the larger astronomical community and attempted to make a mark. Before completing his doctorate, he had organized a computing section for the British Astronomical Association, a loose network of twenty-four volunteer computers that prepared special tables for the association.[48]

During the early 1920s, Comrie emerged as a restless and ambitious individual. Little in the scientific world met his standards. The computers at the Greenwich Observatory were the first to feel the lash of his criticism. After spending two months in the observatory computing office, Comrie concluded that the computers used inferior methods in their work. Believing that the observatory staff were not listening to his remarks, he made his complaints publicly.[49] In the years that followed, he was often sharp and occasionally angry. While teaching at Swathmore College in the United States, he complained about his teaching duties. When he moved to Northwestern University in Illinois, a school that seemed to be more open to him, Comrie felt that he was unappreciated and underpaid. "I feel that my qualifications and experience entitle me to a better position than the one that I now hold," he wrote to the university president.[50] He found satisfaction only in calculation and looked for a time when he might return to London and take charge of the most sophisticated computing office in England, the computing floor of the *Nautical Almanac*. In 1925, anticipating that the almanac director would soon retire, the British Admiralty offered him a position in the almanac office.[51]

At the almanac office, Comrie experimented with punched card tabulators. His initial experiments, though promising, were not as important to him as a test of a new accounting machine. This device, called the National Accounting Machine, had multiple registers, sets of gears that would store several numbers. These registers were used to keep different sums for accounts. A single action might post a number to the general ledger and accounts receivable. In these multiple registers, Comrie saw a machine that could be operated as a difference engine. It can "be called a modern Babbage machine," he wrote, for "it does all that Babbage intended his difference engine to do and more."[52] For a century, Babbage's difference engine had been a distant but desirable goal, a machine that promised to simplify many almanac computations. Babbage, of course, had failed to complete such a machine. The difference engine built by George and Edvard Scheutz had been sensitive, prone to failure. In a moment, Comrie had discovered a difference engine that was robust, was mass produced, and had "spare parts and expert service . . . readily available." "Above all," wrote Comrie, "others interested may purchase the machines at a moment's notice and at prices that are economical."[53]

The economics of the National Accounting Machine was important to Comrie, as the British Nautical Almanac Office had little money for computing machinery and no funds for developing new computing devices. He would recall the almanac office as a place of "politics and red tape," but this judgment was colored by his later experiences.[54] During the 1920s, he was able to find enough funds to outfit the computing room with ordinary mechanical calculators for production work and was able to purchase a National Accounting Machine with money he acquired from the Mathematical Tables Committee of the British Association for the Advancement of Science. The British Association had been founded in the 1830s by Charles Babbage and his friends as an alternative to the Royal Society. The Mathematical Tables Committee had been formed in 1871, the year of Babbage's death. It consisted of a small group of mathematicians, numbering between six and twenty, who prepared mathematical tables. Since its founding, it had published about fifty tables in the reports of the British Association for the Advancement of Science.[55]

In 1925, the Mathematical Tables Committee was the only professional organization for human computers. It was dominated by the colleagues of Karl Pearson and the students of Pearson's Galton Laboratory.[56] The committee invited L. J. Comrie to join their group in 1928, and Comrie eagerly accepted the appointment. He was "a persistent self-publicist," wrote historian Mary Croarken, and "saw the [Mathematical Tables Committee] as a means to improve his standing in the wider scientific community."[57] Within a year, Comrie had established himself as the leader of the group, even though he officially served the committee as its secretary rather than as its chair. He arranged for the Mathematical Tables Committee to have space in the offices of the Royal Astronomical Society, established a regular schedule of meetings, and outlined an ambitious computing program for the group. Under his urging, the committee agreed to prepare and publish volumes of tables.[58]

The Mathematical Tables Committee had few resources to support its work, but it had two small legacies, which had been left to the committee by former members. The constraints on these legacies specified that the money should be used to compute certain kinds of tables. Comrie was not especially interested in those tables and chose to interpret the legacy requirements liberally. He used the money to purchase a National Accounting Machine and two Brunsviga calculators, reasoning that these machines could be used to prepare the specified tables. Once the legacy tables were complete, he turned to other problems that he thought more worthwhile.[59] He kept the new machines at the offices of the *Nautical Almanac*, even though they were not the property of the British government. In itself, this decision was unremarkable, as the almanac staff included some of the most skilled computers in the country, and Comrie was an acknowledged expert on computing machines. Yet, in combining the resources of the two organizations, he did not always differentiate where the production of the *Nautical Almanac* ended and the work for the Mathematical Tables Committee began. Sometimes the National Accounting Machine prepared tables for the almanac. At other times, the almanac computers helped Comrie prepare tables for the Mathematical Tables Committee. On at least one occasion, the group did some calculations for Karl Pearson, who was affiliated with neither organization.[60]

By 1930, L. J. Comrie had climbed the central pillar of British scientific calculation and taken his place as the country's senior computer, the superintendent of the British *Nautical Almanac*. He directed a staff of about a dozen that used Brunsviga calculators and the National Accounting Machine. His elevation occurred as he was completing a first book of tables for the Mathematical Tables Committee, "a most admirable volume," in the opinion of one reviewer, "which ought to be in every college mathematical library."[61] Like Howard Tolley and Henry A. Wallace, he foresaw a grand future for computing, but the contemporary English landscape was less impressive. Outside of the Nautical Almanac Office could be found the London stock market, whose members had fallen into hard times as the price of corporate shares had fallen At the edge of the city there were factories, gates closed and windows shuttered. Men and women walked the streets looking for a job, a handout, a scrap of food. When the winds blew from the east, one could catch from across the channel the faint smell of mustard gas, the lingering scent of the old battles.

PART III

Professional Computers and an Independent Discipline 1930–1964

> In her mind mathematics were directly opposed to literature.
> She would not have cared to confess how infinitely she
> preferred the exactitude, the star-like impersonality, of figures
> to the confusion, agitation, and vagueness of the finest prose.
> Virginia Woolf, *Night and Day* (1919)

CHAPTER TWELVE

The Best of Bad Times

> She laid down the rules of conduct: self-respect, self-reliance,
> self-control and a cold long head for figures.
>
> John Dos Passos, *The Big Money* (1936)

SPRING 1930. It was the first vernal season of the Great Depression, though the economic collapse was not yet potent enough to touch the annual meeting of the National Research Council. The council was the visible symbol of the First World War's scientific legacy. Formed to coordinate research for the American military effort, it had grown in stature and influence during the 1920s. As part of the National Academy of Sciences, the council occupied a marble-clad building on the National Mall. The entrance to the building stood across the street from the memorial to Abraham Lincoln. Visitors to the facility passed through large, bronze doors into the building's solemn interior. The academy's central auditorium, which stood behind the foyer and the stairs to the executive offices, could be mistaken for the conclave of a mathematically inclined Masonic order. A high balcony of chairs circled the room, and the walls were decorated with allegorical mosaics based upon the symbols of science. The National Research Council, acting as the operating committee of the academy, was charged with the responsibility "of increasing knowledge, of strengthening the national defense, and of contributing in other ways to the public welfare."[1] In practice, this charge meant that the senior leadership of the council would identify outstanding scientific problems and appoint committees to report on those issues. That spring, the problems of the human computers appeared on the council's agenda when one member suggested that something should be done to consolidate the literature of calculation.

The proposal before the National Research Council was simple and straightforward. It would establish a committee to prepare a bibliography of the mathematical tables that had been published in scientific journals. There was, of course, no single literature of computation. Tables and articles on computation could be found in dozens of scholarly publications, including the *Astrophysical Journal*, the *Transactions of the Cambridge Philosophical Society*, *Biometrika*, the *Journal of the Optical Society of America*, and the *Proceedings of the Royal Artillery Institu-*

tion. In addition to a summary of these journals, the proposal asked for a report on "automatic calculating machines, harmonic analyzers," and "special graphical device machines of all kinds."[2] These devices were called, in the language of the day, "aids to computation," hence this compilation would be prepared by a committee with the unwieldy name of the "Subcommittee on the Bibliography of Mathematical Tables and Other Aids to Computation," a title that would quickly be shortened to the initials "MTAC."

"Was it Dr. Veblen who initiated this suggestion?" asked Floyd K. Richtmyer (1881–1939). Richtmyer, a professor from Cornell University and an officer of the council, had been asked to find someone who might organize the MTAC committee. Technically, the answer to his question was no, as the idea had been suggested by a consultant to the Nautical Almanac Office, but the proposed committee clearly bore the influence of Veblen and the Aberdeen veterans. One of Veblen's former assistants within the Army Ballistics Office, Gilbert Bliss from the University of Chicago, served on the National Research Council and championed the proposal. Veblen, though not the instigator of the idea, knew of the plan, approved of it, and suggested several individuals who might serve on the committee.[3]

As Richtmyer sought a leader for this group, he received a letter from the director of the Yale Observatory, informing him that an English mathematician, misidentified as "Karl Pierson," had already prepared a bibliography of tables. The letter reported that one volume of the work had already appeared—the pamphlet on logarithm tables that was published in the *Tracts for Computers*—and that this booklet was one "that astronomers, at least, have frequently used."[4] Richtmyer replied that he was "much indebted" for the information. If Pearson had created a full bibliography, he felt that there would not be a need to create a new one, for "duplication of effort, particularly in a matter like this, is highly undesirable."[5]

After a quick review of Pearson's work, Richtmyer concluded that the *Tracts for Computers* was not a complete bibliography and that there was still plenty of work for the proposed MTAC committee. He then returned to the search for a committee chair. He had two candidates for the position, Thornton Fry of Bell Telephone Laboratories and Vannevar Bush (1890–1974), a professor of engineering at the Massachusetts Institute of Technology.[6] Of the two, Bush was the more intriguing choice. He had recently built a computing machine that could solve differential equations, the equations of planetary orbits, artillery trajectories, and electrical devices. His machine was called a "differential analyzer," though many a commentator would note that it neither differentiated nor analyzed. It used spinning disks and rotating drive shafts to represent mathematical quantities, and it solved differential equations with a technique that was analogous to the method that Andrew Crommelin had used to predict the return of Halley's comet in 1910. However, as the machine worked with motions and not with numbers, it recorded solutions as a graph drawn by a mechanical pen.[7]

Bush's differential analyzer had received much attention from engineers and industrial scientists. The computing division of General Electric had taken an interest in the machine and had used it to do several computations.[8] However, after discussing the merits of Bush and his invention, Richtmyer concluded that the MIT professor was the wrong person to chair the MTAC committee. Bush was not much interested in mathematical tables, and the differential analyzer seemed to be a "special machine and is not likely to be available for general laboratory purposes."[9] With Bush eliminated, Richtmyer turned to Thornton Fry and asked him to lead the group.

"I would like to cooperate if possible," Fry responded, but "I think I had better get a clearer picture of just what this . . . will involve before agreeing to take it on."[10] In part, he was being cautious, as American Telephone and Telegraph had first command on his loyalty, but he was also opening a negotiation, probing the National Research Council to determine what resources might be at his disposal if he agreed to prepare the bibliography of tables. Fundamentally, he did not like the idea of producing a bibliography and argued that it was only "the distasteful but necessary first step in a program of producing the numerical tables." He especially disliked the second part of the committee charge, the review of computing machinery. Such a review, he complained, "would have to contain a certain amount of comparative criticism to which exception would undoubtedly be taken by every manufacturer whose product was adversely mentioned."[11]

There "is no doubt that a very definite need exists for more complete mathematical tables of certain types," Fry concluded. He suggested that the new committee should "map out the need [for new tables] and apportion the work to various people so as to avoid duplication and secure the maximum possible results from the effort expended." He noted that there were many skilled computers who might contribute to such a National Research Council project. He could volunteer the services of Clara Froelich and the other staff at Bell Telephone Laboratories. Karl Pearson, though quite senior, might be willing to contribute. L. J. Comrie, at the British *Nautical Almanac*, would certainly be interested. Fry thought that he might be able to entice some contributions from Aberdeen veterans, such as A. A. Bennett (1888–1971). Bennett had served as the chief mathematical assistant to Forest Ray Moulton and had made substantial contributions to Moulton's revised theory of ballistics. He now held a position at Brown University and was also a special consultant on computation to both the army and General Electric.[12] There were even some new faces that might lend their effort, such as Indiana University professor Harold T. Davis (1892–1974). After stating his vision, Fry indicated that he would be willing to chair the MTAC committee and produce a bibliography if "the committee in question shall carry forward some such program as that which I have outlined."[13]

Richtmyer was sympathetic to Fry's idea, as the National Research Council had undertaken similar cooperative activities in the past. During the 1920s, the council had sponsored a multivolume handbook of data for scientists and engineers. This project, called the *International Critical Tables of Numerical Data, Physics, Chemistry and Technology*, had been praised by scientists in the United States, Europe, and Japan, but it had proven to be an expensive undertaking. The publication had cost $177,000 even though all of the contributors had worked as volunteers. The funds had been donated by "those appreciating its importance and in a position to make the necessary investment," as the National Research Council had no funds of its own. Two hundred and forty-four organizations had donated to the project, although the bulk of the money had come from a single source, the Carnegie Institution of Washington.[14] Richtmyer was confident that the council could find similar funds for Fry's project and asked the mathematician from Bell Telephone Laboratories to describe his idea more fully and prepare a budget for the project.[15]

Fry, engaged in other activities at Bell Telephone Laboratories, delayed his reply for five months. As a consequence, he lost a moment of opportunity. By the time that he presented his plan, the council was feeling the full impact of the economic depression. Richtmyer told him that the National Research Council was "not at the moment in a position to finance a more ambitious program, desirable as such a program obviously is." He asked Fry to consider producing only the bibliography, "in spite of the fact that this proposal falls far short of the plan which you outlined in your letters to me."[16] However, events were moving too quickly. Before Fry could reply, the council withdrew the offer. Feeling awkward about conveying the news to Fry, Richtmyer struggled to find his words. "Partially on account of developments since we began to consider this project," he wrote, "it may turn out that we shall not wish at the present time to form a committee even for the preparation of the Bibliography."[17] Fry did not even bother to reply to this letter but stuffed it in his files.

In the summer of 1931, when Richtmyer withdrew his offer to Fry, neither the National Research Council nor anyone else appreciated the potential opportunities for human computers that would be offered by the Great Depression. The difficult times encouraged new applications of the computing methods that had been developed during the First World War.

The statistics of hog production could analyze the collapse of industrial production or the growing reach of poverty. The mathematics of exterior ballistics could help identify the trajectory of the stock market. Rather than being a hindrance to large computing projects, as Richtmyer feared, the economic collapse encouraged the formation of large computing staffs, since rising unemployment reduced the cost of labor.

The issues that had encouraged the National Research Council to form the MTAC committee did not vanish when they withdrew their offer to Thornton Fry. As computing groups grew and expanded, they looked for the kinds of activities that could be found in the more established disciplines of astronomy or physics or electrical engineering. They desired a unified literature, textbooks, standard ways of training computers, journals to disseminate new ideas, and a professional society that might identify pressing research problems. Such institutions would not appear overnight, as the National Research Council's actions concerning the MTAC committee portended. Human computers would have to make due with interim solutions while they worked to establish computing as a more formal scientific field. Their efforts were complicated by the fact that the same forces that encouraged the expansion of computing laboratories also encouraged the development of computing machinery. At times, it appeared that scientists were caught between two contradictory trends. The first trend was the effort to elevate the status of those who worked with numbers. The second trend pushed human computers to the margins of scientific laboratories and replaced them with precise, unfailing machines.

Harold Thayer Davis, professor of mathematics at Indiana University, shared his last name with the founder of the American *Nautical Almanac*, Charles Henry Davis, but the two had no direct family connection and had little in common. Charles Henry was a naval officer, a member of the Boston elite, and a highly disciplined scientist. Harold Thayer, or H. T., was a reluctant soldier, the son of a western land speculator, and a self-described "ultra-crepidarian," a shoemaker who would not "stick to his last," a workman unable to focus on the tasks he had been trained to do.[18] At different times in his life, he showed the prospects of becoming a classical scholar, a physician, and even a billiards player. The strongest connection between H. T. and Charles Henry was a common interest in scientific computation. Mathematical calculation was the "open sesame," wrote H. T. Davis, "to many undiscovered areas of human knowledge." Though H. T. would never form a computing organization as important as the Nautical Almanac Office of his namesake, he would prove to be adept at performing large calculations under difficult circumstances.

H. T. Davis was born in Beatrice, Nebraska. His grandparents had come west to make their fortune in cattle and gold, but great wealth had eluded them. His mother was the daughter of a farmer. His father was the city treasurer. When Davis was young, his family moved first to Idaho and then to Colorado in search of a more healthful climate for his father, who suffered from asthma. He entered college in 1910, shortly after Halley's comet receded from sight. "As viewed from Cañon City," Davis later recalled, "[the comet] hung just above the western mountains with a brilliant head and a tail that swept across the sky through an arc of 130 degrees." He speculated that the return "might be chosen as the beginning of an epoch in science that has had no parallel in the history of the world." But at the time, he had no interest in studying science. When he entered Colorado College that fall, he intended to study history, literature, and economics rather than mathematics and computation.[19]

Davis became interested in computation after his father fell ill in 1912. Needing to provide for his family, Davis left college and took a job with a civil engineer. The engineer was grading land and needed an assistant to measure newly excavated drainage canals and compute the amount of soil that had been removed. Davis spent several weeks diligently tracing the topography of the land and slowly summing up the volume of dirt. As he grew tired of the drudgery, his "ingenuity was awakened," and he saw a new way to organize the calculations. As he recalled the event, his plan reduced weeks of work to a task requiring "two or three hours."[20]

When his father recovered his health, Davis returned to Colorado College with a new interest in mathematics and computation. He graduated with a degree in mathematics and spent two years teaching the subject at a local high school. When one of his professors took a job at Harvard, Davis followed him in order to study for a master's degree. He arrived in Cambridge in 1917, just as the United States was entering the First World War. Unlike Oswald Veblen or Norbert Wiener or Elizabeth Webb Wilson, Davis originally had no interest in going to war. He believed that America should stay out of Europe's problems and that President Woodrow Wilson was a "dangerous demagogue"; but during the summer of 1918, he had a change of heart and enlisted in the army. It was a conversion of little consequence, for the armistice was declared while he was at a training camp in South Boston.[21]

Returning to Harvard, Davis was drawn to empirical subjects, rather than the more theoretical topics that had been introduced by German mathematicians some twenty-five years before. While studying theorems and proofs, he complained that "one grew sick, indeed, at the torrent of abstract symbolism which was poured forth at every [mathematical] seminar." His studies combined the two central methods of scientific calculation, statistical methods and calculus-based methods. He took classes with a mathematical statistician and worked in a Harvard statistical laboratory. "It was in this laboratory," he remembered, "that [I] first saw [a] multiplying machine, a nice, black, shiny Monroe calculator, which operated with a crank." When the time came for his master's exam, his professors asked him to undertake a broad survey of the computational methods and to show how to compute a class of functions called elliptic integrals. Writing of the experience in the third person, as he often did when describing his life, he said that "the subject suited his taste and he undertook it avidly."[22]

From Harvard, Davis went first to the University of Wisconsin to begin doctoral studies and then to Indiana University to teach. He had not finished his doctorate when he arrived at Indiana, but he had taken the job in order to support his new wife and family. The country was in the midst of an economic recession, the difficult time that followed the end of the First World War. Davis was grateful for a university job, but he was surprised at the relative poverty that he found on the campus. Once the university had been the promising new school of the Midwest, but its fortunes had fallen. Davis reported that the department of mathematics "was housed in a single dingy room in an old building used mainly by chemistry. Desks were crowded together and the author found space for only a small table in the place assigned to him."[23] The university president admitted that young faculty members worked so hard that they had "little leisure, little energy left. [They] can not brood by the hour over [their] own studies as a man must to grow rich in them."[24]

In spite of the conditions at Indiana University, Davis remembered the school as a place of great freedom, an opportunity to "go his own way and explore those paths into which his own interest and his own imagination" directed him. In 1927, with his doctoral work behind him, Davis decided that he wanted to build a statistical computing lab like one that he had once used at Harvard. The national economy had recovered, but Indiana University had no money to support his research. Davis was able to acquire some funds from a local charity, a foundation that had been created by a grateful alumnus of the university. The grant was small, a "few hundred dollars at most," but it allowed him to acquire "a battery of electrically driven Monroe calculators." The dean of the business school found a room for the new laboratory in the attic of the library, a space previously used as an artist's studio. The university offered $86 to give the walls a fresh coat of paint, install electric lights, and provide some tables for the machines.[25]

Thornton Fry once quipped that H. T. Davis computed "various things as they occurred to him."[26] The first test of the new computing lab came from the physics department. One of the physicists was experimenting with beams of light. Davis was attracted to one experiment that produced a "beautiful circular pattern with seventy rings, which broadened as they neared the circumference." The physicist wanted to compute the amount of energy in each ring as a way of testing the wave theory of light. The problem had nothing to do with statistics, but Davis saw that the calculation would provide "an immediate use for his machines" and volunteered to do the work.[27]

The calculation required Davis to use some values of the Bessel function, the function that had been tabulated by the Edinburgh Mathematics Laboratory and the British Mathematical Tables Committee. Finding none of these tables in the Indiana University library, Davis decided that he would create a new table for his own use. He became absorbed in this extra task, preparing pages of Bessel functions that would not be needed for his calculation. Only after the table was complete did he return to the original physics problem. When all the work was done, Davis and his colleague discovered that the final values for the energy in the light did not match the empirical measurements from the experiment. The two reviewed the figures, looking for an error in the calculations, but found nothing of significance. They eventually realized that Davis had used an incorrect value for the frequency of the light. At this point, they had two ways to adjust their results. They could either recalculate the energy levels using the correct frequency or redo the experiment using light that matched the frequency from the calculation. Davis conceded that repeating the experiment would be "arduous," but he claimed that it would be easier to adjust the experimental mechanism than to perform the calculations a second time. His colleague eventually agreed with this reasoning, returned to the laboratory, and performed the experiment with the new frequency of light. This time, reported Davis, "the experimental evidence and the mathematical values coincided exactly."[28]

Davis emerged from the calculation with a new table of the Bessel function and the notion that he should create a compendium of mathematical tables.[29] He recognized that there was an epic irrationality in such an idea. "Although the machines were present in the laboratory there were no funds with which to operate them," he wrote, "no trained personnel to hire even if funds had been available, and not the slightest chance to print and publish to the world the fruits of the heroic computation could it actually be achieved."[30] Under such circumstances, most mathematicians would have not even attempted the project, but Davis believed in the idea and had the ability to make others believe in it as well. He recruited volunteer computers from among the university's mathematics students and convinced them that a book of mathematical tables was a grand project, a goal worthy of their noblest efforts. He reported that the computers arrived in the laboratory "fired with enthusiasm" and that their calculations "filled the laboratory with their music."[31]

The computers finished their work in 1932, three years after the stock market crash and one year after the National Research Council had decided that it was not a good time to publish a bibliography of tables. Davis estimated that it would cost $2,100 to typeset his manuscript of tables, print the pages, and bind the book, a sum that far exceeded the annual salary of a professor, yet he seemed undeterred by the economics of his project. He was a partner in a small scholarly publishing firm, a company that he had named Principia Press, after Isaac Newton's great work. Davis confessed that Principia Press was a "reckless adventure," as there seemed to be little demand for their eclectic list of scientific and philosophical books, but the firm had done surprisingly well in spite of the poor economic conditions. "Any one who entered a printing plant with a thousand dollars in real money," he observed, "was a person to be met with open arms."[32] The press was able to raise enough money to publish his manuscript, which he entitled *Tables of Higher Mathematical Functions*.

After releasing this first volume, Davis started work on a second collection of tables. For this calculation, he paid his computers with money that he received from one of the New Deal agencies, the National Youth Administration, or NYA. The National Youth Administration was created as a "relief program for the middle class." It provided training and educational activities for young men and women who were working to support their families. Most of the National Youth Administration activities were designed for high school students, but one of its program provided funds to employ college students. These NYA grants paid students to serve as part-time administrative aides, to work as teaching assistants, and to perform research for faculty. Like every other program of the New Deal, the National Youth Administration was a controversial activity. Many educators refused to take NYA funds, arguing that it was "the first step in the establishment of a federal education system competitive with the schools."[33] Even at schools that could easily employ research and teaching assistants, deans often had to push their faculty to find some use for an NYA grant.[34]

By nature, Davis should have been among the critics of the National Youth Administration. He was a political conservative and once described President Franklin Roosevelt as a "violent man, who showed the nature of his spots immediately after his election." Yet, when Indiana University announced that it would start a National Youth Administration program, he put his political objections aside and moved "with a boldness that verged on rashness" to secure NYA funds for his computers. He suspected that "the University administration, caught without definite plans for the employment of these young people, would set them to menial tasks about the campus." With a clear program in hand, he was able to acquire a substantial amount of the National Youth Administration funds allocated to Indiana University, though he confessed he gained "few friends by his boldness."[35] In all, the National Youth Administration funds helped him produce two more volumes of tables.

The quality of Davis's calculations was tested when *Tables of Higher Mathematical Functions* reached the British Nautical Almanac Office and the desk of L. J. Comrie.[36] "With an unknown author," wrote Comrie, "it is desirable to check considerable parts of the tables, to see if he has the reliability of Andoyer and Peters, the proneness to error of Steinhauser, Gifford and Hayashi, or the plagiaristic tendencies of Duffield, Benson and Ives."[37] These other human computers had been the targets of pointed reviews by Comrie; and Davis was to join their number. Comrie worked through the book, table by table and value by value. By recomputing 2,000 entries, he reconstructed Davis's original computing plan, uncovered a substantial number of errors in the book, and identified the source of each mistake. Comrie noted that the pattern of errors indicated that the computers had verified their results by repeating the calculations a second time, "the poorest possible check" in his eyes. They had made mistakes in rounding the numbers and had introduced errors when they transcribed the values. Even the design of the book did not escape his attention. "The general lay-out of the tables," Comrie wrote, "shows a lack of acquaintance with many elementary principles of tabulation, lack of consistency and lack of consideration for the user." The review was not entirely negative, though its conclusion, that "table-lovers are assured that they should possess this work," seemed faint praise.[38] Davis apparently paid due attention to Comrie's criticism, for the second set of tables, those produced with NYA funds, had considerably fewer errors than the first.[39]

During the first years of the Great Depression, H. T. Davis spent summers at the family home in Colorado Springs. It was a time to get away from the demands of university life, a chance to see old friends, and an opportunity to meet those who had come to southern Colorado in search of a better life. Town society included wealthy miners and poverty-stricken farmers. There were ministers hoping to find a land more spiritual and patients who hoped that the dry air might free them from tuberculosis. Among this group was a businessman named Alfred H. Cowles (1891–1984), who had survived the stock market crash with most of his fortune intact and had come to think that he might use his resources to improve the national economy.

Cowles was not formally trained in economics or in the methods of research, but he came from a successful newspaper family. The Cowles family owned the Cleveland *Leader*, had a substantial share of the Chicago *Tribune*, and also published farm papers in Washington, Oregon, and Idaho which were similar to *Wallace's Practical Farmer*. Alfred Cowles had begun his career at the family paper in Spokane before deciding that he wanted to form his own business. He established an investment firm in Chicago and specialized in acquiring and restructuring small railroads. For a time, he ran a small conglomeration of southern lines that was known as the Alfred Cowles Railroad,[40] but he never entirely broke his ties with the world of journalism. His firm published a stock market newsletter that analyzed the state of the market and recommended stocks to buy or sell.

In the late 1920s, Cowles was diagnosed with tuberculosis, and as others had done before him, he moved west in search of better health.[41] He withdrew from business shortly before the stock market crash of 1929, an event that caused him to think about the health of the economy as well as his own physical well-being. The initial prognoses for the market were optimistic and forecast a quick return to prosperity, but stock prices continued to fall. With time to ponder the situation, Cowles "began to feel that most of the forecasters were just guessing, himself included."[42] Living a life of enforced leisure, he began to sketch ideas for studying the stock market. His financial work had taught him something about correlation and statistical least squares. His first analysis was a regression model, one that simultaneously compared the predictions of twenty-four different market newsletters with the actual stock market prices. The equations required him to use least squares and demanded a substantial amount of calculation, calculation that Cowles did not know how to do.

With the resources at his command, Alfred Cowles could have easily found university researchers willing to study stock market predictions. He had connections at Yale University, where he had gone to college, and at the University of Chicago. Yet it seems that Cowles wanted to do the work himself, that he wanted to be the gentleman researcher, the amateur scholar working in semiretirement. Believing that he could do the work if he had assistance with the computations, he went in search of calculating help. He was led to H. T. Davis through a mutual friend, the director at a local tuberculosis foundation. In many ways, Davis was a good match for Cowles, for he never questioned the businessman's approach to research. Had he been more active in the economics community, Davis might have tried to push Cowles toward a certain type of research, or he might have tried to make Cowles a silent partner in his own research. Instead, he tried to be the best help he could. "As far as I know, such a regression equation has never been made," Davis replied, apparently unaware that Myrrick Doolittle had solved such equations by hand at the Coast and Geodetic Survey, that Howard Tolley had done similar work at the Department of Agriculture, and that George Snedecor was doing the same at Iowa State College.[43] "It happens, however," Davis continued, "that a new machine is now available, a Hollerith Calculator, by means of which such a problem can be solved."[44]

In fact, Davis had little, if any, experience with "Hollerith Calculators," as tabulating equipment from International Business Machines was sometimes called. His opinion was deduced from what others had told him or from what he read about the machines in the scholarly literature. While card tabulators could ease the labor of computing a large regression problem, they were unable to do some of the most difficult and time-consuming work, as the workers at the Iowa State Statistical Laboratory had learned. The Iowa State computers were able to use the machines only to calculate correlations and do the preliminary work for a regression problem. The final solution had to be calculated with an adding machine. To solve a regression problem with twenty-four elements could require eight or ten months of labor.

Though he knew little about tabulating equipment, Davis agreed to help Cowles build a computing laboratory in Colorado Springs. Later that summer, the two of them took the train north to Denver and visited a company that used punched card tabulators in its business. Cowles was satisfied with what he saw and leased a full set of equipment for his office.[45] Only after the tabulators arrived in Colorado Springs, perhaps as late as August, did Davis realized that the tabulating equipment could not compute a correlation or solve a regression equation. To compensate for this deficiency, he pressed Cowles to hire some of the computers that had prepared the *Tables of Mathematical Functions* in Indiana.[46]

Cowles eventually adjusted his research plans, replacing the massive twenty-four-term regression with a series of twenty-four smaller calculations. Much of the work for these smaller calculations could be handled by the punched card equipment, leaving only a little arithmetic for the computers to do by hand. As he gathered data, he also developed a simpler way of analyzing the stock market forecasts, a method that compared the forecasts to random guesses. The calculations for this method were easily handled by the punched card machines. It was not the work that Cowles had planned to do, but it allowed him to answer the question "Can stock market forecasters forecast?" with the quick summary "Not very well." He elaborated that "the best individual records failed to demonstrate that they exhibited skill and indicated that they more probably were the results of chance."[47]

Cowles must have known that he would never be a major economic researcher, but he wanted to be a part of the economics community and guide its research. He incorporated his office into a private research foundation, the Cowles Commission for Economic Research, and he actively reached out to scientists who were interested in economic research and organized computation. Using H. T. Davis as an intermediary, he met

28. Alfred Cowles (center) with H. T. Davis (right) at Cowles's home in Colorado Springs. Photo taken by Elizabeth Webb Wilson

with James Glover, who taught actuarial mathematics and computation at the University of Michigan, and Thornton Fry of Bell Telephone Laboratories.[48] He also approached the Econometric Society, a new professional society devoted to the mathematical study of the economy, and offered to collaborate with them, to provide them with computing services, and to fund their publications. Initially, his proposal produced hesitation among the economists. One member reported that a few "became alarmed lest the Society's good name be harmed by its implication in a venture with a man who was willing to spend a considerable sum of money in order to accomplish they knew not what purposes of his own." The group debated the proposal among themselves before sending one of their number to meet with Cowles and to inquire about his intentions. The emissary spent a week with the businessman and concluded "that Cowles was sincerely interested in econometric research." He urged the society to take Cowles's money, work with his research organization, and utilize his computing staff without fearing that he would attempt to influence them or their work.[49]

At the Econometric Society, Cowles crossed paths with Elizabeth Webb Wilson, the ballistics computer of the First World War.[50] Wilson was living in Cambridge, Massachusetts, and pursuing a doctorate in economics at Radcliffe College. She was the only woman among the First World War computers to find some kind of scientific role, but her success had been shaped more by her wealth than by her wartime accomplishments. She had taught high school after the war until an inheritance from her parents left her a substantial income. Freed of the need to work, she had pursued graduate study, first at the University of Michigan and later at Radcliffe. Like Cowles, she tried to move beyond the role that wealth had brought her, though she was limited by her gender in a way that Cowles was not. Instead of financing industry, she had done calculations for the Michigan Teachers' Annuity. Instead of commanding the Econometric Society, she had represented Michigan professor James Glover at actuarial congresses. Though she might yearn to be the leader, she would accept the role of observer and commentator.[51]

If "business principles were quite free to work out their logical consequences, the outcome should be to put the pursuit of knowledge definitely in abeyance," quipped the economist Thorsten Veblen.[52] Veblen's nephew, the mathematician Oswald Veblen, rarely agreed with his uncle's pronouncements, but he might have conceded the idea that commercial businesses rarely supported unfettered scientific research. The younger Veblen had spent part of the 1920s in a frustrating attempt to convince industries that supporting mathematical research was in their interest, a campaign that he called "our Debt to Mathematics."[53] It proved to be a much harder task than organizing the computing staff at Aberdeen. Few industries, save those involved in power generation or aircraft production, saw any value in pure mathematics. Mathematicians found it easier to raise money for computation, as computation could support a corporate goal. Both Westinghouse and General Electric expanded their computing staffs during the 1930s, as they were involved in the construction of the large New Deal electrical projects: the Rural Electrical Association, the Tennessee Valley Authority, and the dams on the Columbia River. The two companies investigated Vannevar Bush's new computing machine, the differential analyzer, in order to handle their engineering calculations. One of the few examples of a company funding an independent computing lab, a lab with little direct connection to company production, was the scientific computing facility at Columbia University, which was financed by International Business Machines.

The Columbia computing facility, originally known as the Columbia University Statistical Bureau, was the joint creation of a scholar and a business leader, Benjamin Wood (1892–1984), a professor at Columbia's education school, and Thomas J. Watson, president of International Business Machines. Wood was a Texan, described by a biographer as a "towering and disciplined intellect against which sparks of imagination impinged with the force of ignition."[54] Watson was a salesman by training, enthusiastic yet focused, and strategic.[55] The two met when Wood was

29. Columbia University Statistical Bureau. Ben Wood (in dark suit) stands in the center

studying the Alpha and Beta intelligence tests, the exams that the army had used during the First World War to screen recruits. Wood had collected hundreds of tests as his raw data. Each of these tests was graded by hand, and the results were analyzed with correlation statistics. By his estimate, it cost him five dollars in labor to process each batch of tests.[56] Thinking that there must be some way to grade the tests with a machine, Wood wrote to the presidents of ten companies, described his problem, and asked their assistance. Of the ten, only Thomas Watson responded. He had no solution to the problem, but he was intrigued with Wood's letter.[57]

The story of the initial meeting between the professor and the businessman comes from Wood, who is not the most reliable witness. He liked to boast and would later attribute quotes to Watson that could not possibly have come from the leader of IBM.[58] Wood stated that the two met in a downtown business club and that he began the meeting by flattering the IBM product line, telling Watson that the tabulators could "do marvelous things in science and education, civil service, government, logistics, military, aviation, astronomy, every science in the world."[59] No matter what Wood may have said, something in his presentation appealed to Watson. Watson's son later wrote that the Columbia professor described a vision that was "music to the ears of a tabulating machine maker."[60] During the meeting, Watson offered to provide Wood with punched card equipment. As Wood would later recall, "two or three weeks later, or maybe one week later, two or three huge trucks arrived at my office" and delivered what he remembered as "every IBM machine they had."[61] Of course, the installation was not quite that easy. Wood had to get permission from his dean, find space on campus, install the equipment, and find a trained operator.

Wood outlined a grand vision for the tabulating machinery. "I was planning to make tests over the whole gamut of the curriculum," he recalled. Wood provided Watson with a basic understanding of statistical calculation as employed by social scientists and suggested new equipment that might be useful in statistical research. He also worked with IBM engineers to create a progressive digiting machine that could automatically perform multiplication and helped IBM identify a mechanism that would score standardized tests. When called upon to give a speech or demonstrate his equipment or even write a letter of recommendation for Watson's teenage son, Wood was glad to comply.[62] Perhaps his most important contribution was to admit into his laboratory a young astronomy professor named Wallace J. Eckert (1902–1971).[63]

Unlike social statistics, astronomical calculation was not a growing field, and unlike the boisterous Wood, Wallace Eckert was a "small and retiring man," characterized by one IBM historian as "so soft-spoken that he was scarcely audible."[64] A Columbia graduate student went so far as to call him one of the "passé guys," a scientist who looked to the past rather than anticipating the future.[65] Passé or not, Eckert came to the laboratory wanting to see how the punched card machines could handle a classic and difficult calculation, the three-body problem. The specific version of the problem was the system involving the Earth, the Sun, and the Moon. Eckert had an unusually detailed analysis of the problem, which he had obtained from Yale professor Ernest W. Brown (1866–1938). Brown had extended his analysis beyond the three bodies of the Earth, the Moon, and the Sun and included the tug from the giant planets of Jupiter, Saturn, Uranus, and Neptune as well as the slosh of the Earth's oceans as they dragged behind the Moon. He had summarized this solution in a 660-page volume with 180 tables.[66]

With Brown's tables, a computer could determine the position of the Moon for any day and any time, but the labor was almost overwhelming. Writing from the British Nautical Almanac Office, L. J. Comrie complained that the task of preparing a lunar ephemeris from Brown's tables required "the continuous work of two skilled computers."[67] Comrie transferred this calculation from human computers to the punched card equipment. This new approach to preparing an ephemeris required Comrie to think in terms of operations that were not obvious to the astronomical computer. He selected the tables he needed and punched them onto a separate deck of cards. Each table represented a different force on the moon. Next, he duplicated the cards and shuffled them together, using a card sorter in a downtown London office.[68] When he was done, the first cards had the values for January 1, the next cards had the values for January 2, and so forth. In the final step of the calculation, he put the cards for each day through a tabulator in order to sum the values.[69] The process was relatively quick and possessed the added benefit that "a small change may be made in the elements and the new values of the coordinates obtained with almost no additional work."[70]

Comrie was not in a position to further develop the methods of punched card computation, as the British *Nautical Almanac* did not have its own tabulator. However, he sent a copy of his computing plan to Eckert, who had access to Wood's tabulating facility at Columbia. Eckert studied Comrie's plan and carefully duplicated the English computations.[71] From this work, he slowly began to expand his skill with the tabulating equipment, learning how to handle complicated analyses and difficult computations. By the spring of 1934, he had become the expert in scientific computation with punched card equipment and had supplanted Benjamin Wood as the faculty contact with International Business Machines. That spring, IBM recognized Eckert's growing prominence by helping him create a new facility, the Columbia University Astronomical Computing Bureau. This organization was not really a separate laboratory, for it used the same machines that were being used to tabulate educational statistics. It did have one new piece of equipment, an IBM 601 multiplying punch, which eliminated the need to do progressive digiting.

For all of the attention that International Business Machines gave to Columbia University, Benjamin Wood, and Wallace Eckert, the company considered scientific computing to be only a minor application of their equipment. In 1935, the company prepared a book to promote the use of tabulators in higher education, entitled *Practical Applications of the Punched Card Methods in Colleges and Universities*. The volume was edited by a company employee and was published by Columbia University Press. "So numerous are the uses of the punched card method in colleges and universities," read the preface, "and so great the interest shown by these institutions that the creation of this volume was a logical development."[72] Fully four-fifths of the contributions to the book were business applications: class records, patient histories, student accounts, course registration, resource scheduling. Of the remaining fifth, most dealt with

social statistics. Three entire chapters had been submitted by the statisticians at Iowa State College. Only Eckert's chapter, buried at the back of
the book, dealt with astronomy.

Eckert began to record the methods of scientific computation in a notebook which was known locally as the "Orange Book." In the pages of
this book, he described how to reduce data, create a star catalog, compare observed positions with theoretical calculations, and mechanize the
solution of differential equations.[73] His descriptions showed how to prepare the cards and how to work with the sorters, tabulators, and
punches. Through 1935 and 1936, scientists began to appear at the door
of the laboratory. Some had written to ask for permission to visit; others,
who may have been passing through New York or attending a conference
down by one of the railroad stations, simply arrived unannounced. They
talked with Eckert, handled the cards, watched the machines in operation, worked with one of the machine operators, and usually inquired if
they might make a copy of some page of the Orange Book.[74]

In 1936, one of Eckert's visitors was an astronomer from the Soviet
Union, the aptly named Boris Numerov. Numerov walked freely among
the equipment, listened to Eckert talk about the benefits of punched card
calculation, and, most likely, took copies of pages from the Orange Book.
When Numerov left the lab, he apparently told Eckert that he would
keep in touch or that he might write later or that perhaps he would need
Eckert's assistance in building a punched card facility in the Soviet Union.
Eckert accepted the farewell, watched Numerov depart for Moscow, but
never heard from him again. It was the era of the Stalinist purges, and
Eckert came to believe that the Soviet astronomer had been punished or
killed for inquiring about punched card tabulation. It was a misguided
concern, as Numerov almost certainly had the permission of the Communist Party for his visit, but it was not without basis in fact. The threat
to the astronomer came not from a visit to an American computing laboratory but from a German name for an asteroid. Numerov was arrested
after the Soviet secret police learned that German astronomers had
named a small planet Numerov. Concluding that anyone who received
such an honor from Germany was likely a spy, the secret police had him
executed.[75]

Eckert may not have appreciated the politics that pulled Numerov to
his doom, but he was able to recognize the forces pulling on his computing facility. By 1936, the Columbia Computing Bureau was the prominent facility for scientific computation with punched card equipment, a
laboratory far more visible than the Iowa State Statistical Laboratory or
the computing office at the U.S. Department of Agriculture. As the leader
of the astronomical computing bureau, Eckert was increasingly identified
with International Business Machines, even though he remained a mem-

ber of the Columbia faculty. In his writings, however, he made it clear that he was not a blind advocate for IBM, an unquestioning promoter of tabulation equipment. "The main question in any case is not 'can the problem be solved by these machines,'" he wrote, "but rather 'have I enough operations of this type or that, to justify such powerful equipment.'"[76] Still, his ties with IBM were close, and when he decided to publish his Orange Book of computational methods, he used the publication services of International Business Machines rather than Columbia University Press or some other university publisher.[77]

In 1933, when organized computing had taken root at Indiana University, at the Cowles Commission, and at Columbia University, the National Research Council returned to the idea of preparing a general bibliography of mathematical tables. Some on the council, including Thornton Fry, believed that Davis might be the appropriate person to head the unfinished Subcommittee on the Bibliography of Mathematical Tables and Other Aids for Computation. Others had reservations. Davis would not "be the right man to head up a committee along this line," wrote Henry Rietz (1875–1943), a professor at the University of Iowa. He observed that Davis expressed more enthusiasm than discipline and that "Professor Davis very strongly believes that the real problem of aids to computation rests in table-making itself rather than in a comprehensive bibliography."[78] The rest of the council reluctantly concurred and continued its search. Finally, they returned to the Aberdeen veterans and selected A. A. Bennett of Brown University, a mathematician who had worked with both Oswald Veblen and Forest Ray Moulton. "Bennett seems especially well-fitted for the work," argued Rietz, "not only because he would probably do a scholarly piece of work, but because Brown has . . . the best collection of mathematical tables in connection with any university."[79]

Bennett accepted the chair of the MTAC committee in 1935, but the appointment came at a poor time for him. Along with Oswald Veblen and Gilbert Bliss, he was a consultant to the Aberdeen Proving Ground. In 1935, the proving ground had reorganized the division engaged in ballistics research and had increased the number of test firings on the artillery ranges for the first time in a decade.[80] In the years since the First World War, ballistics research had been divided into three distinct fields. The traditional computation of trajectories was now identified as external ballistics, which was contrasted with internal ballistics, the study of the stresses and pressures within a gun. The final division dealt with the physics of exploding shells and was called, appropriately, terminal ballistics. The new Ballistics Research Office had sections devoted to each aspect of the research as well as a new central computing office.[81]

Bennett had some initial success organizing the MTAC committee, but he was unable to give the work the kind of effort and attention that it deserved. He wrote to the Galton Laboratory at the University of London to enquire about the plans of Karl Pearson. Pearson's son replied that his father had died the year before but that he had left a great deal of material which might be used by the new committee. Bennett also asked a half dozen individuals, including H. T. Davis and L. J. Comrie, to join him on the committee and help with the bibliography. Writing from Colorado, Davis exclaimed that "I am very much pleased to accept membership on this Committee because I believe sincerely in the importance of the project."[82] Comrie, however, was more circumspect. "Am I right in interpreting 'Aids to Computation' as meaning calculating machines?" he asked.[83] Only after Bennett assured him that such machines would have a place in the bibliography did he agree to serve.[84]

Comrie's ability to contribute to MTAC would be limited. The problem was not so much the economy as it was a fall from grace, a small lapse of judgment. In the winter of 1935–36, the British Admiralty discovered that Comrie had not been accurately reporting the activity of the Nautical Almanac Office. Comrie was convinced that his staff was too small and that he was unable to retain the best computers because of Admiralty personnel regulations. Unable to make his point through memos and arguments, he was trying to impress upon the Admiralty the shortcomings of their policies by delaying the release of important work. He told his supervisors that the almanac computers were overworked and unable to do certain computations when, in fact, those computations were already finished and residing in Comrie's files.[85] The flaw in this strategy was exposed when an investigative board arrived unannounced at the almanac office. The board discovered the missing computations, charged Comrie with obstructing Admiralty work, and dismissed him.[86] "Comrie's often-expressed complaints that the civil service regulations were petty and restrictive are given some justification by [the record]," observed historian Mary Croarken, "but it is also clear that Comrie was inept at 'playing the game'" and working effectively within a large government organization.[87]

After his departure from the British *Nautical Almanac*, Comrie rented a building in London and formed a private computing laboratory, the Scientific Computing Service Ltd. The company grew quickly, sustained in no little part by computing contracts from the British government, including his former employer, the Admiralty. In less than a year, he employed a staff of sixteen computers, "most of whom have academic training," he boasted. The company handled the same kinds of calculations that were done at the Nautical Almanac Office: navigation tables, astronomical calculations, statistical summaries. Looking for commercial business, it also advertised a specialty in the analysis of questionnaires and "an advisory and investigational service relating to the purchase and use of calculating machines."[88]

Though Comrie's new company prospered, his fall from grace had consequences. His scientific life was burdened with the need to satisfy bank officers, customers, and investors. He could no longer freely volunteer his time to scientific organizations, either the Subcommittee on the Bibliography for Mathematical Tables and Other Aids for Computation or the Mathematical Tables Committee of the British Association for the Advancement of Science. He could remain a member of both groups and could contribute to the work of each, but he could not afford to take a leadership role. In his new position, he had to earn his own way and support the economic prosperity of his company.

CHAPTER THIRTEEN

Scientific Relief

> Get that human adding machine out of my way. . . .
> Clare Boothe Luce, *The Women* (1937)

MALCOLM MORROW (1906–1982) was the faceless bureaucrat of computation, the government worker who created the largest human computing group of the 1930s but left little record of himself. He lived in a working-class district of Washington, D.C., only a few blocks from the original Naval Observatory and the building that had once housed Simon Newcomb's Nautical Almanac Office.[1] He held jobs in several of the New Deal agencies and eventually settled into the executive office of the Work Projects Administration (WPA)[2] as an assistant statistician. His title gives us little information about his mathematical ability, as the WPA employed many assistant statisticians. Some were nothing more than clerks. Some were project managers. Some prepared questionnaires. Only a few were employed in the analysis of data.

Morrow's WPA was the largest and most developed of the New Deal's relief agencies. Formed in 1935, it was Franklin Roosevelt's third attempt to reduce unemployment through the construction of public works. The WPA operated offices in each of the forty-eight states, as well as an extra office in New York City, that provided funds for projects judged to be in the public good. The WPA paid for parks and bridges in New York City, sidewalks in Michigan, dams in Texas, city offices in Los Angeles, 226 hospitals, 1,000 libraries, 1,200 airport buildings, 2,700 firehouses, 9,300 auditoriums. WPA workers renovated the army's research facility at the Aberdeen Proving Ground, repaired the Naval Observatory in Washington, and constructed a new building for the Iowa State Statistical Laboratory.[3]

In general, the WPA did not manage the projects that operated under its name. It provided only wages for workers. Local organizations would identify appropriate projects, develop plans, and pay for any necessary materials. Only rarely would the WPA staff in Washington plan and manage a project without the assistance of a local sponsor. Those projects that were directly overseen by the WPA tended to be national in scope, such as the Federal Theater Project and the Federal Writers' Project. The theater project supported performing arts across the country and proved to be a controversial activity. While only a few of the theater productions had any political content, notably Orson Welles's *The Cradle Will Rock* and Sinclair Lewis's *It Can't Happen Here,* the project was the target of conservative critics who claimed that the government-supported artists were promoting liberal ideas and undermining American society. Eventually, the critics found enough support within Congress to terminate all WPA theatrical productions.

The writers' project generated less controversy, but it experienced serious operational problems. The project was intended to support regional authors by having them create guidebooks for each state of the union. The Washington WPA office provided a style manual that described how the "geographic, historic, cultural, social, recreational, industrial and commercial information should be assembled." In spite of the best intentions, the project was unable to recruit sufficient regional talent to prepare the guides. Not wanting to see a visible project fail, the WPA administrators had many of the guidebooks rewritten "by experienced writers drawn from New York City and other centers, who were paid for their work on a non-relief basis."[4]

Among the tens of thousands of WPA projects, there were about 2,400 that supported scientific research. Future Nobel laureates Glenn Seaborg and Luis Alvarez at the University of California are on the list of those who received WPA assistance.[5] WPA funds paid the salaries of laboratory workers in much the same way that the National Youth Administration had provided funds for student assistants. Most WPA-sponsored science involved large statistical studies of social or economic problems.[6] Malcolm Morrow's division of the WPA, the Central Statistical Office, oversaw most of these projects. This office suggested possible studies, reviewed proposals from universities and state agencies, and granted funds that would pay for workers to conduct surveys, collect information, tabulate data, and analyze the results. The Iowa State Statistical Laboratory was a major beneficiary of the Central Statistical Office. It received WPA money to summarize harvests, measure farm income, and analyze the extent of rural poverty.[7]

Sometime early in the operation of the WPA, perhaps in the summer of 1935 or the winter of 1936, the staff of the Central Statistical Office became engaged in a discussion of mathematical tables. As they were starting to oversee a large number of statistical studies, they thought that it might be useful to provide a standard set of mathematical tables to each of the projects and to each of the regional WPA offices. If all of the statistical projects had the same tables of logarithms and probability functions, then the Central Statistical Office could compare the results of the different studies with a little more confidence. This idea produced no immediate action, as the office had more pressing things to do, but it was revived in the fall of 1937 when the nation entered a new period of economic hardship. After three years of tentative growth, the economy had begun a contraction that was quickly labeled the "Roosevelt Recession." Faced with growing unemployment and greater labor unrest, the WPA began looking for new projects in order to expand the number of relief jobs. With winter approaching, they needed to identify activities that could operate during the cold months.[8] The construction jobs, which constituted 75 percent of the WPA projects, were curtailed between late fall and early spring. As WPA officials evaluated different ideas, they recognized that a computational project could employ a large office staff and assigned Malcolm Morrow the task of organizing such a group.

The mechanics of starting a new project were fairly simple. Morrow needed to find office space, acquire the necessary furniture, and notify the local WPA office that he could accept a staff of qualified workers. The local office would send him as many people as he needed from those who had applied for relief jobs. The difficult part of Morrow's assignment was the task of finding a sponsor, a scientific organization that would accept the project, oversee the computations, and provide the appropriate mathematical expertise. He started looking for a sponsor at the institution that considered itself to be at the center of American scientific research, the National Academy of Sciences. The academy building stood just a few blocks from the mammoth headquarters of the WPA, so one day in late October, Malcolm Morrow left his desk early and called upon Albert Barrows, the permanent secretary of the National Academy, and asked for assistance.

Morrow began his presentation slowly, stating ideas that were familiar to many of the scientists who worked with the academy. He reported that the statisticians on the WPA staff felt that "there was a very large amount of work, which can be done to improve the mathematical basis on which a great deal of other scientific work depends." He noted that there were substantial errors "in a number of the standard mathematical tables" and that the "tables already in use ought to be greatly extended," as more scientists were employing them in their research. Reaching his central point, he told Barrows that the WPA was planning to build a large computing center in New York City. The center would prepare mathematical tables for use "not only by mathematicians and astronomers, but also by surveyors, engineers, chemists, physicists, biometricians, statisticians, etc." If they were successful, it would be the biggest computing organization in history, a group that might "employ a thousand people, as well as experts to advise." All of the computers, of course, would be drawn from New York's unemployment rolls, but he assured Barrows that there would be calculating machines and a "training school to initiate employees." The new center would be an expensive operation, Morrow conceded, but

New York City had $2.3 million available for relief. He implied that a large part of this money "could be devoted to this project."[9]

After describing a computing facility nearly as fantastic as Lewis Fry Richardson's weather computing room, Morrow retreated to a more practical realm and admitted that the project would begin on a small scale, perhaps "50 employees under the supervision of technical directors and office managers." Coming to the key question of the meeting, he asked Barrows if the National Academy of Sciences would be willing to appoint an advisory committee to the computing center, a group that would identify projects for the computers and help prepare computing plans.[10]

In general, the members of the National Academy of Sciences belonged to a class that distrusted Franklin Roosevelt and disliked the idea of government-sponsored work relief. They joked that the initials "WPA" stood for "We Poke Along" and claimed that the public works projects were nothing more than an attempt to buy votes for the Democratic Party. This view of work relief was so pervasive among the secure classes of American society that twenty years after the end of the Great Depression, the novelist Harper Lee could damn a character by claiming, "He was the only man I ever heard of who was fired from the WPA for laziness."[11] If Barrows held such impressions about work relief or the WPA, he was able to put them aside as he summarized the meeting for academy president Frank Lillie (1870–1947), a University of Chicago zoologist. "Mr. Morrow (a man I should judge of 40 or 45 years of age) impressed me as competent to discuss this project," he wrote.[12]

Lillie responded to Morrow's visit by sending a note to the chief statistician of the WPA. "I wish to assure you that we shall be most happy to render any service within our power," he wrote; "the National Academy of Sciences in its part would be glad to appoint a committee of mathematicians to advise on specific undertakings."[13] Lillie named no names, but he and Barrows had already discussed possible candidates for such a committee, including Oswald Veblen, Vannevar Bush, Harold T. Davis, and James Glover.[14]

When Malcolm Morrow ended his meeting at the National Academy of Sciences, he had told Barrows that there was "an element of urgency in organizing the matter because of the impending election of mayor in New York City," and he had asked that "no intimation of consideration of the project be permitted to be made publicly or find its way into the channels of the press."[15] The WPA computing office, tentatively called the Project for the Re-computation of Mathematical Tables, was not an ordinary scientific project that was supported by a university or funded by industry. It was going to be a federal project in a city where Democrats and Republicans fought over federal funds and where the outcomes of such battles could have national repercussions. Morrow was probably concerned about protecting his project, getting the computing office established before some city agency attempted to capture its budget for another purpose. The sitting mayor of New York City, Fiorello La Guardia, was an ally of Franklin Roosevelt and willing to have projects managed from Washington.[16] His opponent, a member of the city's Tammany Hall political organization, was not as sympathetic to the Roosevelt administration and might be able to redirect the project funds to some organization with no qualifications other than ties to Tammany Hall.[17]

After his meeting at the National Academy of Sciences, Morrow traveled to New York City, hoping to persuade one of the city's universities to sponsor the project. Columbia University was the most likely choice, as they had a strong mathematics department and housed the offices of the American Mathematical Society. However, its faculty already operated the Columbia Computing Bureau with its tabulating machines. They showed no interest in sponsoring a large group of human computers. Morrow then approached New York University and proposed that the school provide space for the computers and mathematicians to manage the calculations. Though university officials were interested in the project, they felt that they could not provide the necessary funds and declined the offer.[18]

As the date of the mayoral election approached, Morrow had no sponsor for the project, no site to house it, and no plans for its operations. At the National Academy of Sciences, which had received no report from Morrow after his initial visit, secretary Barrows voiced a growing skepticism about the idea. "I still have the feeling," he complained, "that this is one of those matters which they would not be able to complete if started."[19] Morrow found his sponsor three days before the November election. After finding no help in New York, he returned to Washington and convinced the director of the National Bureau of Standards, the physicist Lyman Briggs (1874–1963), to manage the new computing organization.[20]

Briggs had both the technical background and the political experience for such an assignment. As a physicist, he knew how hard it could be to prepare a useful, error-free table. From his years with the Bureau of Standards, he understood that government support for science was unstable in the best of times and hostile in the worst. Since its founding in 1901, the bureau had been repeatedly attacked by congressional critics, who had argued that the government should not be involved in research and should not set standards for private industry. In 1933, Congress cut the budget of the National Bureau of Standards in half. "It was a bitter experience for us," Briggs recorded. "More than one third of our staff was dropped on a month's notice."[21] Attempting to sustain the bureau, he discovered that he could obtain relief grants to replace some of his lost budget. Money from the Civil Works Administration and the Public Works Administration, predecessors of the WPA, was used to clean and repair bureau buildings.[22] "Several of the abler mechanics and technicians of the Bureau," wrote one historian, "let go earlier, found their way into [maintenance work] and tarried there until they could be restored to the Bureau payroll."[23]

In agreeing to sponsor the WPA computing project, Briggs extracted concessions that New York University was unable to get. He would manage the project, recruit senior personnel, and provide scientific expertise, but he would provide no money for the group. The WPA would fund the entire budget for the project. Furthermore, Malcolm Morrow, who had identified himself as the project leader, was relegated to administrative issues, such as finding a suitable office, handling WPA paperwork, and communicating with the New York City relief agencies. This agreement was consummated so quickly that Morrow apparently failed to tell Lyman Briggs of his contacts with the National Academy of Sciences, for the politically astute Briggs made no effort to contact Albert Barrows or Frank Lillie. With no information from either Briggs or the WPA, the senior members of the academy naturally concluded that the relief agency had been unable to organize the computing project. "Since the day Mr. Morrow was in here saying we were to get a letter the next day, I have wondered what has happened to him," commented one member of the Academy staff. "He seems to have disappeared off the horizon."[24]

It would have been best for the Mathematical Tables Project if Lyman Briggs had been able to recruit an established computer to be its leader. If the project had been led by Oswald Veblen or even H. T. Davis, it would have started operations with strong connections to the scientific world. However, the idea of leading a work relief project appealed to few scientists, and Lyman Briggs had to turn to the ranks of the underemployed, the scientists who were not able to find a position at a top college or university. The path that led him to choose Arnold Lowan (1898–1962) as project director is no longer well marked. It may have involved Oswald Veblen, probably passed through Columbia University, and almost certainly involved thermodynamic calculations, the computations of heating and cooling.

In the fall of 1937, Arnold Lowan was holding two part-time teaching positions. By day, he taught physics and mathematics at Yeshiva College in Manhattan. At night, he taught the same subjects at Brooklyn College.[25] He had immigrated to the United States in 1924, fleeing anti-Semitic pogroms in his home country of Romania and rushing to reach New York before a new law restricted immigration from Eastern Eu-

30. Arnold Lowan, director of the Mathematical
Tables Project

rope.[26] He was a desirable immigrant, a chemical engineer, but he had
waited four years before he found professional employment, a period he
would never describe on any of his resumes or reminiscences of the time.
He likely spent the time among the immigrant Jews of Brooklyn or Manhattan's Lower East Side, learning English and becoming acclimated to
the United States. If he was employed during this time, he was likely
doing manual labor or other work that he considered subprofessional.
The first job he was willing to identify was one he took in 1928 with a
utility contractor, working as what he called a combustion engineer. He
described his job in grand terms, claiming that he was researching the
"thermodynamics of gaseous hydrocarbons," while he was actually working with coal gas furnaces and boilers.[27]

The salary he gained from tending furnaces by day allowed him to
study physics by night. He took a master's degree from New York University and then transferred to Columbia University in order to earn a

doctorate. "He certainly has great industry and perseverance," remarked a Columbia physicist, and "in the main [he] follows out his own ideas." At a time when the brightest students were looking at problems that would lead to the subjects of relativity and quantum physics, Lowan turned to conventional thermodynamics, the study of heat. Though he rarely asked the Columbia faculty for guidance, he did become acquainted with one of the veterans of the Aberdeen Proving Grounds, a physicist who had served as Oswald Veblen's assistant during the First World War. This connection gave Lowan a brief entry into the senior ranks of American scientists, a postgraduate fellowship at the newly formed Institute for Advanced Study in Princeton, where Veblen was on the faculty.[28]

At the institute, Lowan seems to have kept to himself and rarely interacted with the other researchers. He spent his year developing his ideas on thermodynamics and completing a large calculation that described the cooling of the earth. In all, he did the research for ten scientific papers, each of which involved extensive numerical work. He may have been overawed by scientists like Veblen and Albert Einstein, or he may have had difficulty communicating with others, or he may simply have lacked the background to deal with research problems in the new fields of physics.[29] Judging from his later letters, he clearly spent some time with Einstein, Veblen, and the Hungarian mathematician John von Neumann (1903–1957), but not enough to be their friend or even their familiar. In 1934, when Lowan's fellowship ended, they were unable or unwilling to help him find a job at a research university. With the lingering depression, budgets were tight; Lowan could not find any full-time job in physics. Instead, he took his two part-time teaching positions.

There are three plausible routes that might have led Lyman Briggs to Arnold Lowan. The shortest begins with Oswald Veblen, the great computer of the First World War, who would have recalled Lowan's computational work at the Institute for Advanced Study. The second path is more complex, but it was suggested by one of the WPA workers. This path begins with one of Lyman Briggs's assistants, a geologist who had prepared mathematical tables for the Smithsonian. For this path to work, the geologist would have to have read Lowan's articles on the cooling of the earth and appreciated the extensive computations. The final path leads through Columbia University. Briggs may have simply asked the Columbia faculty or the American Mathematical Society if they knew of anyone qualified to run a computing organization.[30] No matter how Briggs was introduced to Lowan, he could not have found a leader better suited for the Mathematical Tables Project. Lowan was ambitious to make a reputation as a scientist and was not repulsed by a work relief job.

Lowan joined the Mathematical Tables Project in late November

1937 and was told to begin operations in the first months of the new year. The project office would be one floor of an industrial building on the west side of Manhattan, not far from Times Square. Arnold Lowan found a used desk to serve as his work area, placed it in a corner of the empty space, and started to make his plans. He would be the executive of the group, the administrator who would deal with budgets, personnel, correspondence with the public, and any communications from the Bureau of Standards. His first problem was to find more mathematical talent, a technical director to analyze the computations and prepare the detailed computing plans. Almost immediately, he hired two graduate students from Brooklyn College, but they acted as personal assistants, not mathematical leaders. His technical director needed to hold a doctorate and should know something about the operations of a large office. He probably had no idea where to search for such an individual and must have been surprised to find one in his night class at Brooklyn College.[31]

That fall, Lowan was teaching a course on the subject of relativity, an elementary presentation designed for individuals who were trying to expand their horizons. Most of the students came from day jobs in Manhattan. Sitting near the back was a short woman who wore the kind of business dress that could be washed in a bathroom sink and dried overnight. On most nights, she seem a little tired and not especially engaged in the subject. When it was cold or raining, she did not attend the lecture at all. Lowan paid little attention to her until he began to grade the homework papers. Her mathematical reasoning was the work of a professional. It identified all the assumptions, moved through each step of the analysis, and presented the solutions in a clear manner. As the student used mathematical concepts that had not been presented in class, Lowan suspected that she was receiving outside assistance. He traveled to and from campus on the same bus line as this student, so one night after class, he sat next to her and tried to start a conversation. At first, the student was reluctant to speak, as if she was unsure of Lowan's motives. After some coaxing, she began to relax and tell her story.

The student's name was Gertrude Blanch (1896–1996), and she confessed to Lowan that she held a doctorate in mathematics. She had been born Gittel Kaimowitz in Kolno, Poland, a Jewish settlement near the Russian border. Like most such communities, it had suffered the czarist pogroms, and her family had fled to the United States. Her father had come first, followed by her mother and finally, in 1907, by Blanch and her sister. The family had settled in Brooklyn, which was considered a "pastoral neighborhood" compared to the tenements of the Lower East Side.[32] Blanch, who had the opportunity, rare for a Jewish girl, of attending school in Poland, settled easily into the public school system. She

31. Gertrude Blanch, lead mathematician of Mathematical Tables Project

completed the primary curriculum in three years and gained admittance to Brooklyn's Eastern District High School.[33]

Blanch rushed through her high school studies. "It was up to me to get a job as soon as possible," she recalled. Her father's health was declining, and when he died, just at the time of her graduation, all hope of attending college died with him. To support her mother, she took a job with a Manhattan hat dealer named Jacob Marks. "He would export things," she said, "and I would organize the transactions for them. It was paperwork. They paid me very well."[34] She mastered the computations for foreign currency payments, gained added responsibility, and acted as Marks's office manager. While she was working, she assembled a small library of mathematical books, reminders of the subject she most loved in school. Many of them were commercial tracts, the kind that were distributed by adding machine manufacturers or sold by business teachers trying to improve their income. They explained various accounting computations, such as the cost of money or the depreciation of stock. They showed how to simplify calculations, check values, and reuse results.[35]

For fifteen years, Gertrude Blanch worked for Mr. Marks. She watched the men of her generation march off to war and heard of the new opportunities for college-educated women, but she was unable to leave her office and her responsibilities to her family. Only after her mother died in 1927 was she able to begin a new life. She cleaned her parents' apartment for the last time, accepted an invitation to live with her sister's family, and enrolled in her first courses at New York University. When she announced to Mr. Marks that she was quitting her job in order to attend college, he countered that he would pay her tuition if she would attend night school and spend the days at his company. She accepted the offer and graduated four years later with a degree in both mathematics and physics. Her diploma was awarded with highest honors, "summa cum laude," and she was inducted into the national Phi Beta Kappa honors society.[36]

Blanch wanted to attend graduate school, but she knew that it would be a gamble, a risk that would hazard all her resources. In the United States, there were little more than a hundred female mathematicians, most of whom were relegated to limited roles in the profession.[37] The odds were lengthened by her Jewish heritage and her age. At thirty-six years, she was a full decade older than most graduate students. "Science is a young man's game," wrote the mathematician G. H. Hardy. Citing Isaac Newton as an example, Hardy had claimed that the great mathematician "recognized no doubt by the time that he was forty that his great creative days were over."[38] As Blanch prepared for further schooling, she recognized that her scientific career would have barely started when she reached her fortieth birthday.

Before she departed for graduate school, she made her immigrant background less obvious by Americanizing her name. She had long been known as Gertrude instead of Gittel. Informally, she had occasionally used Cassidy instead of Kaimowitz for a last name.[39] When she came to choose a permanent last name, she selected the birth name of her mother, Dora Blanch.[40] She was not a feminist, as modern scholars would define the term, and later in life she would dismiss the idea that she had been limited by her gender. As she pondered her choice of graduate schools, she was aware that many of the best graduate programs were closed to her. Princeton University, where Oswald Veblen taught, did not admit women. Harvard educated women only through the back door of Radcliffe College. Even the most liberal of graduate programs, such as the graduate school of the University of Chicago, had their limits. Chicago gave more PhDs to women than any other school. Still, Blanch observed, "All things being equal, they would choose a male student."[41] The university offered her admission to the mathematics program, but when she asked for a fellowship, they told her that "scholarships to women were given out only in very rare circumstances."[42] Instead, she chose to attend Cornell University, which also offered no scholarship but welcomed women and also had a substantially lower tuition.

Cornell proved to be a good choice for her. "They appreciated me," she recalled, "to the extent that I contributed as much as any other student. I can't claim discrimination while I was studying."[43] Her doctoral advisor was Virgil Snyder, a past president of the American Mathematical Society. Snyder, like James Glover at Michigan, advocated mathematical education for women and guided the graduate study of several female students. Blanch would later write that she was "deeply grateful to him for unfailing encouragement."[44] She also found support in a small club for women graduate students. Every few weeks, these women met to socialize, talk about their lives, and share the lessons of graduate school.[45] She enjoyed graduate school, with its seminars and discussions and parties, even though the failing economy made study increasingly difficult for her. As her resources began to wane, the university found a fellowship that allowed her to continue her studies uninterrupted. During the most difficult times, she had to purchase food on credit from the university's agricultural school.[46]

When Blanch completed her degree in 1936, she found a temporary position at Hunter College for Women, a job which paid the "munificent sum of thirty dollars a week."[47] It proved to be only a brief respite from the problems of the Depression and the limitations imposed upon women scholars. She spent much of the year searching for a permanent university job and completing dozens of applications. She stopped only when Cornell University refused to issue any more transcripts, claiming that her "requests have been excessive."[48] No school offered her a position, so she turned to the employment ads in the *New York Times*. She applied to be the office manager for a company that made cameras for color photography. In her letter to the firm, Blanch explained that she was a Cornell graduate but never stated that she had a doctorate in mathematics. She was invited for a job interview in a pleasant mid-Manhattan office with carpets, "wall paneling and wainscoting."[49] The senior manager, who was conducting the interview, remarked that the company had received fifty letters of application for the position but that Blanch's was one of two "written in good English." The manager also said that she was grateful to see that Blanch "was carrying *The Nation* under her arm," a sign that Blanch was not Catholic and hence not Irish.[50]

When the company offered her the job, Blanch accepted it and settled into the familiar routine of correspondence, scheduling, and bookkeeping. As the work demanded nothing of her mathematical skills, she decided to take a course at Brooklyn College. Scanning the list of what was offered, she decided that Lowan's class would be the most interesting, even though she recalled, "It was very elementary," an assessment that must have made Lowan wince.[51]

Blanch finished her story just as the bus trip ended, and the two of them started on their separate ways. Nothing more was said that evening except a pleasant farewell and a hope that Blanch might attend the next lecture. One week later, when she arrived at class, Lowan asked if he could again accompany her home. On this trip, he told her about the WPA and its plans for a computing laboratory. He explained that the computing project was being sponsored by the National Bureau of Standards and that he was the executive director. As the journey came to an end, "he asked me if I would join the project,"[52] Blanch recalled. Two bus trips through the night did not provide her with enough information to make a decision, so she asked Lowan if she could delay her answer until she had had an opportunity to visit the office of the Mathematical Tables Project.

The following week, with the weather growing cold and the class term coming to an end, Blanch and Lowan took the train to Manhattan and walked past empty warehouses and closed machine shops to the building that housed the project office. Lowan showed her to the elevator, which had been designed for moving equipment. It had an open cage and rose slowly past exposed concrete beams and dangling light fixtures. When they reached the top and Lowan opened the gate, Blanch could see that dust covered the concrete floor and that the only furniture, beyond Lowan's corner desk, was a mismatched collection of battered and weary tables. The windows near the staff desks were streaked with dust and dirt. Ceiling lamps gave a harsh and unpleasant glare. Permeating the air was the lingering odor of machine oil, yet something in this scene told Blanch that the WPA might represent her best chance to become a mathematician, the final gamble that might give her a place in the world of science. Before she left the building, she had "decided to resign from my job in my beautiful office and to join Lowan and go up the freight elevator."[53]

It would take more than simple determination to get the project going. Lowan needed to find more furniture for the office and purchase supplies. Blanch needed to learn the literature of computation and begin preparing computing plans. Most important, both of them needed to complete the work that Malcolm Morrow had begun three months before. From the scientific community, they needed to solicit a list of tables that could be used by practicing scientists and yet be prepared by a large staff of untrained computers. Morrow had asked the National Academy of Sciences to appoint a special committee to provide this advice, but Lyman Briggs, the new sponsor of the project, knew that such a committee already existed, the Subcommittee on the Bibliography of Mathematical Tables and Other Aids to Computation, MTAC. Briggs contacted A. A. Bennett, the committee chair, and scheduled a meeting for the group in the offices of the National Bureau of Standards. The only member who declined to come was L. J. Comrie, who informed Briggs that he was interested in the project but could not afford to travel from England.

Briggs convened the meeting on January 28, 1938, only three days before the start of operations. H. T. Davis was present, as were A. A. Bennett and a second Brown University mathematics professor, Raymond Claire Archibald (1875–1955). Briggs began by introducing the committee to Malcolm Morrow, even though it was the last time that the WPA statistician would have anything to do with the project. Arnold Lowan, who had traveled from New York for the meeting, sat quietly in the room and listened to Briggs present the goals for the group. When his turn came to speak, Lowan described the plans for the first calculation, a table of the first ten powers of the integers from 1 to 1,000. This table was a relatively simple project that involved none of the difficulties that would be found with more complicated functions. It extended a table that had been created in 1814 by Peter Barlow (1776–1862), the nineteenth-century computer who claimed that calculation was nothing but "persevering industry and attention."[54] Each entry in the table required only a single multiplication. The square of a number was computed by multiplying one number by itself. The cube was computed by multiplying the square by the original number, and so on.[55]

The committee accepted Lowan's proposal and moved to consider other tables for computation. They recommended that the second project be a detailed table of the exponential function, e^x. The committee gave Lowan a rough idea of how the table should be structured and suggested ways of computing it. By the end of the day, they had identified twenty functions for Lowan, a list that could keep the project busy for two or three years. The meeting closed with a final recommendation that Lowan coordinate his efforts with the work being done by the Mathematical Tables Committee of the British Association for the Advancement of Science.[56]

Initially, the discussions with the Subcommittee on the Bibliography of Mathematical Tables and Other Aids to Computation seemed to give the new WPA project a good connection with the scientific community through the National Academy of Sciences and the National Research Council. When the academy leadership finally learned of the meeting, they gave their blessing to the project and expressed their approval of the advice given by the MTAC committee.[57] However, such generosity was not as strong as Arnold Lowan might have liked, and it certainly did not extend to all levels of the organization. Later that winter, when the National Research Council reviewed the progress of the MTAC committee, some members were distressed to learn of the January meeting. When one council member gave a quick description of the meeting in Lyman

Briggs's office, another snapped that the WPA had "no connection with work assigned to committee" and that "the Committee work should be limited to a bibliography."[58]

On February 1, the WPA began to send workers to the Mathematical Tables Project office. Those who took WPA jobs were desperate for work. They lived at the edge of poverty and usually had held no stable job for a long time. The WPA's figures suggested that 90 percent of them lacked the skills that would gain them employment with a private firm.[59] Gertrude Blanch, who had a tendency to see the best in anyone, recalled that "among them we found some very good material. Most of them were willing to learn, but we knew that we couldn't expect too much."[60] Though such native grace made her job a little easier, it did not allow her to be complacent or to feel sorry for her workers. She had less than five months to turn these workers into human computers. The WPA had authorized funds for the Mathematical Tables Project only through June 30. By then, if she could not demonstrate that her computers were producing useful public works, funds would be terminated and the project ended.

The most detailed picture of the computing floor in its first year comes from Blanch's closest friend at the project, Ida Rhodes (1901–1986). Rhodes did not join the project until 1940, but she would be Blanch's confidant for nearly forty years. She was outgoing while Blanch was retiring, flamboyant when Blanch was reserved, and critical when Blanch might have been gentle. Blanch generally approved Rhodes's accounts of the Mathematical Tables Project. "She had a way of getting across a point," Blanch recalled, "in a way that no one else could, all with a sense of humor."[61] Some of Rhodes's stories contradict the administrative record of the WPA, but those that agree with other accounts of the project offer a bleak portrait of that first winter. "Many of our workers were physically ill," Rhodes reported; "arrested TB cases, epileptics, malnourished persons abounded." She seemed to delight in remembering that some of the computers had engaged in "several types of perversion and vice." Those who were hardest to engage, those who were "most pitiful," in her words, were the workers who had "lost their self respect in that horrible year."[62]

By early spring, Blanch was working with a computing staff of one hundred and twenty-five, a number far short of the thousand Morrow had promised, but it was as large a computing force as had ever been assembled. The operations of the Mathematical Tables Project were overseen by a planning committee, a group of six mathematicians who prepared the computing plans. In theory, the planning committee was chaired by Lowan, but Blanch generally ran the group. Each member of the committee would take responsibility for one computation, research-

32. Computing floor of the Mathematical Tables Project

ing the background for the table, recommending a certain mathematical approach, preparing worksheets for the computers, and checking the final results. One committee member oversaw the operation of the "computing floor," as the bulk of the computers came to be called. In addition, the planning committee worked with two smaller computing groups, the special group and the checking group. The special group tested methods, calculated initial values, and did other work for the planning committee. The checking group worked with the finished worksheets and with the final proofs for the tables. In 1938, the computers of these last two groups were the only ones who had access to the project's three adding machines.[63]

Blanch divided the computing floor into four groups, one for each of the arithmetical operations. The largest group, identified as group 1, did addition only. A slightly smaller group, group 2, did subtraction. Group 3, which had about twenty computers, multiplied numbers by a single digit. The elite of the computing floor was the tiny group 4. Its members did long division. She isolated each group within the Mathematical Tables Project office. She placed group members at long tables facing a wall. On the wall she put a poster to remind the computers of the basic rules of arithmetic. Few of them had completed high school, and fewer still could be trusted to work without direction. Since most did not know how to manipulate negative numbers, she devised a scheme that used black pencils to record positive quantities and red pencils to record the negatives. The wall posters described how to handle numbers of different colors. The poster for the addition group read:

Black plus black is black.
Red plus red is red.
Black plus red or red plus black, hand the sheets to group 2.[64]

The worksheets had been duplicated on a mimeograph machine. "They generally had 100 lines and were on graph paper," explained a veteran of the project.[65] Whenever possible, Blanch tried to replace complicated operations with repeated additions. It was an approach that mimicked her driving habits. Family members remarked that she would go to great lengths to substitute three right turns for a single left-hand turn across traffic.[66] Because the computing floor had one section for each operation, she could leave some multiplications and divisions on the worksheets. The sheets circulated through all four groups on the floor, often wandering through each section several times before the work was complete. For example, a sheet might start in the addition group and then move to the multiplication group after all the initial additions were done. "The human computers who liked boring work, did calculations vertically," observed a planning committee member. "They did 100 operations before they moved to the next column."[67]

When the worksheets returned to the planning committee, they were checked for errors. Blanch and Lowan went to extraordinary lengths to find every possible mistake, including errors created in transcribing the values to the final table. Behind their efforts was the knowledge, rarely mentioned, that one poorly computed table could permanently damage their reputation and limit their acceptance within the scientific community. "We had very little equipment," reported Gertrude Blanch, "and a lot of supervisors who thought that this was another boondoggle project."[68] The first defense against errors was found within the worksheets. Blanch designed them in such a way that elementary mistakes in calculation could be quickly identified. The computers filled in a grid of numbers. The last column of this grid had to match a set of predetermined values, numbers that had been computed by the planning committee or one of the special computing groups. If they did not match, the sheet contained a mistake in calculation.

Some computers attempted to avoid the discipline imposed by the worksheets. Recognizing the structure of the sheet, a worker could put the predetermined values in the last column and then fill the remaining blanks with any values that seemed appropriate. This did not happen often, given the quality of the final tables, but each member of the planning committee could recall at least one incident of cheating. Blanch remembered two women who submitted falsified worksheets. "You can't cheat in those things," she said flatly. "It comes out almost immediately. Well, these girls were fired immediately."[69] The other examples of cheating or carelessness also ended in dismissals. One resulted in the departure of a planning committee member.[70]

Blanch checked each calculation with six or eight different procedures. Some of the procedures were based upon the mathematical properties of the function they were tabulating. All tables were checked with the technique known as differencing. Differencing had been used by the computers of de Prony and Maskelyne, but it reached a new level of sophistication in the hands of Gertrude Blanch. It reversed the method on which Charles Babbage based his difference engine. Babbage had shown that many functions could be reduced, through the method of finite differences, to a series of repeated additions of constant terms. Blanch used this technique to check tables by repeatedly taking the differences of adjoining terms. In a table of cubes, the first five values are the numbers 1, 8, 27, 64, and 125. The first differences are $8 - 1$, $27 - 8$, $64 - 27$, and $125 - 64$, or 7, 19, 37, and 61. If we repeat the process and take adjoining differences of these values, we get $19 - 7$, $37 - 19$, and $61 - 37$, or 12, 18, and 24. In one final set of differences we get $18 - 12$ and $24 - 18$, or the numbers 6 and 6. If one of these third differences were not 6, then one of the original values in the table would be in error. For most functions tabulated by the Mathematical Tables Project, the sixth or seventh repeated difference would produce a list of nearly constant numbers. Any significant variation would indicate an error in the calculations.

By WPA regulation, the human computers worked thirty-two hours a week. They were supposed to spend the remaining time looking for permanent employment. This limitation on the program gave Blanch the opportunity to study the worksheets when the computing floor was silent. She spent many an evening hour checking calculations and preparing new worksheets for the next day. She seemed to enjoy the quiet time after the computers left for the day, when there were no workers to encourage, no questions to answer, no formulas to explain. The office was quiet, and she could concentrate on mathematics.

The Mathematical Tables Project finished the table of powers in early April 1938, printed the manuscript with a mimeograph stencil on rough sheets of paper, and bound the pages with a cardboard cover. A few of the numbers were a little difficult to read. Prominently displayed on the face of the book was a WPA seal, a small rectangle that identified it as the product of a relief effort. Lowan and Blanch claimed that there were no errors in the table, a claim that was tested when a copy reached the desk of L. J. Comrie in England. Comrie reviewed the entire work, checked every value with a difference test, and concurred that the table was entirely free of errors.[71]

Finishing the first table was a relief to the entire group and created what Blanch called a sense of "heartwarming camaraderie" among the staff.[72] It steeled their courage for their second project, creating tables of the exponential function, e^x. The exponential function is harder to compute than a simple power of an integer. It is the inverse of the logarithm and can be expressed as the sum of an infinite number of terms, $e^x = x/1 + x^2/2 + x^3/6 + x^4/24$ and so on, continuing in the same pattern. In practice, a computer does not need to deal with an infinite number of operations, as the terms eventually become so small that they can be safely ignored.

The computers began working on the exponential calculations in the early spring, while they were still working on the powers of integers. It seemed important that this project be under way when the WPA reviewed the progress of the group in June. For all the anxiety among the planning staff, the review proved to be quick and perfunctory. The WPA officials reported that they were satisfied with the accomplishments of Blanch, Lowan, and the human computers and approved another six months of budget for the group.[73] With the future of the project on a stronger financial foundation, Lowan was able to arrange for Columbia University Press to publish the tables from the project.

While Blanch was preparing computing plans and directing the daily operations of the human computers, Lowan was attempting to bring the Mathematical Tables Project to the attention of American scientists and engineers. His first problem was to overcome the stigma of the WPA. His second was to show that his group could equal the accomplishments of the two other scientific computing organizations in New York City. Thornton Fry, Clara Froelich, and the other Bell Telephone Laboratory computers worked two miles south of the Mathematical Tables Project office. Wallace Eckert and the Columbia University Astronomical Computing Bureau sat a slightly greater distance to the north. Both had reputations that Lowan would have liked to equal.

Beginning in the fall of 1938, Lowan began to promote his project as a general-purpose computing office for the nation's engineers and scientists. He prepared mimeographed circulars, using the same machines that reproduced his computing sheets, and mailed them to universities, government offices, manufacturers, and every scientist he could name.[74] In general, these efforts at self-promotion were frustrating affairs, as he might have learned from anyone who dealt with direct mail advertisement. He often posted one hundred or two hundred letters in a month and received not a single reply. When he did receive a reply, he was usually disappointed. Lowan received requests to prepare salary tables, to tabulate data, and to do simple calculations for other WPA projects. In general, he tried to decline such work by citing the WPA requirement that all products needed to be "widely usable."[75] However, he quickly learned that he was part of a government organization that could be subject to political pressure. More than once, a rejected request returned to his office in the form of a letter from Lyman Briggs. These letters reminded Lowan that the Mathematical Tables Project received government funds and that it would be well advised to honor the request, no matter how uninteresting it might be.[76]

The few letters from well-reputed scientists could be tantalizingly sad, like the gentleman caller who offers a moment of hope and happiness and promise before disappearing. In response to a circular, John von Neumann wrote, "Many thanks for the announcement of your project. I am much interested in your program and should like to get your material." He ended the letter, "I may have some remarks and suggestions in connection with these things and will write to you concerning them before long."[77] As Lowan had known von Neumann at the Institute for Advanced Study, he had some reason to expect that the mathematician would act on his promise, but the weeks passed, and the Mathematical Tables Project received no further correspondence.

The one scientist who offered more than hope was Philip Morse (1903–1985), a professor of physics at the Massachusetts Institute of Technology. "I have been in charge of a number of similar projects," he wrote, and felt "that I should get in touch with you." With no more introduction or pleasantries, he then described a lengthy calculation, acknowledging at the end that "this calculation has no particular difficulties but requires considerable man-hours and if you are in a position to be of help, this help would be welcomed." Such interest was well received by Lowan, Blanch, and the rest of the senior staff, but it was accompanied by an embarrassing query. Morse ended his letter by stating, "I will be very pleased to know just what sort of arrangements you make on calculating facilities and in general how you operate."[78]

Lowan responded quickly, assuring Morse that his group would be happy to compute for anyone at the Massachusetts Institute of Technology. He gave a brief overview of the Mathematical Tables Project, telling Morse that "we are operating with a staff of 110 workers under the supervision of a planning section of which I am in charge." He mentioned nothing about worksheets or relief workers. When he reached the topic of machinery, he equivocated. "Most of the work is, of course, done with the aid of calculating machines."[79] Strictly speaking, the statement was true, as the project's three adding machines contributed indirectly to everything that the staff did. However, the bulk of the calculations were done with only paper and pencil.

We do not know when Morse learned the true nature of the Mathematical Tables Project computers. He was unusually well connected in the scientific community and may have known from the start that Lowan was stretching the truth. He never criticized Lowan, never suggested that there was anything wrong with the project. In fact, he seemed to grasp innately the nature of the group. In his second letter to Lowan, he said that he had "one or two other sets of calculations which I want to suggest to you in the next month or so" and that he would "consult with a number of the other members of the Physics and Mathematics Departments here and at Harvard in order to submit to you a program that is not a set of personal interest."[80]

The last phrase in Morse's letter was important. Lowan had to establish that every calculation was of general interest and would benefit more than a single scientist or engineer. None of the tables could be copyrighted, as they had been prepared with public money. It was one of the WPA policies that governed his actions as long as he accepted their money. The WPA also required that he use labor-intensive methods, in order to employ the greatest number of people, and restricted the number of female employees to about 20 percent of his staff, an attempt to prevent one household from receiving two sets of benefits.[81] Every request for scientific calculation had to follow the WPA rules, even if that meant using methods that seemed archaic or unsuited to 1930s research.

Morse never seemed to be bothered by the WPA restrictions. He proved to be a tireless ally and a key link to the scientific community. He helped bring the group its first important scientific computation, a problem posed by the Cornell physicist Hans Bethe (1906–). Bethe was studying the processes that produce solar energy. In order to understand these reactions, he needed a table that would show the internal temperature of the sun, beginning at the outer edges and moving inward to the core. Since these measurements could not be taken directly, they had to be computed from a mathematical model.[82] Bethe claimed that the calculations were not "prohibitively laborious" but were so difficult that no one had attempted the work without first substantially simplifying them.[83]

Lowan eagerly agreed to undertake the project, for the work would bring the Mathematical Tables Project to the attention of astrophysicists and might also show that the group was as capable as the Columbia Astronomical Computing Bureau. If he expected that the calculations would be done quickly, he was mistaken. After the planning committee reviewed the plan, they had a half dozen questions for Bethe. They asked Bethe to verify equations and check key values. The calculations involved a three-dimensional system of differential equations, the form of equations found in Richardson's weather model. Though the equations were simpler than those posed by Richardson, they still demanded a great deal of work. If the starting values were wrong, the effort would be wasted.

Not even Hans Bethe could adjust the sun and bring it into agreement with a mistaken calculation.

In the end, Gertrude Blanch decided to do all the computations herself. She concluded that she could calculate the tables in the time that it would take her to prepare worksheets for the computing floor.[84] The calculations took about three weeks of effort. Bethe was impressed with the results, calling them the "first modern determination of the temperature of a star."[85] The table was published in the *Astrophysical Journal*, which must have pleased Lowan. It was the first major publication to come from the project. Perhaps more important, it brought attention from a major physics department. Bethe reported that his colleagues had taken a look at the group's first tables and that "the department has decided to order copies of them."[86]

By this time, the project's second volume, *Tables of the Exponential Function*, had appeared in print. Compared to Karl Pearson's *Tracts for Computers* or H. T. Davis's *Tables of Higher Mathematical Functions*, this new volume received serious attention from the scientific press. Though book reviewers inevitably mentioned that the volume had been produced as a relief project, they generally treated it with respect. The review in the widely circulated *American Mathematical Monthly* was positive, though perhaps not as enthusiastic as Lowan might have liked. The reviewer, John Curtiss (1909–1977) of Northwestern University, noted that the book claimed to be "entirely free from error." If he had been L. J. Comrie, he would have tested this assertion. As he was not, he was willing to concede that "the precautions taken seem to give considerable weight to this claim."[87]

John Curtiss's review was an important milestone for Blanch and Lowan, evidence that they had done their job well, had met the requirements of the WPA, and had created an organization that could produce something of value to at least a few scientists. Nevertheless, the project was still a work relief effort. It still occupied a dirty industrial loft in an unattractive part of Manhattan. Each week, the members of the computing floor collected their wages hoping that by the next payday they would have a full-time job with some employer that was not the WPA.

CHAPTER FOURTEEN

Tools of the Trade: Machinery 1937

> Some of them hated the mathematics that drove them, and
> some were afraid, and some worshiped the mathematics
> because it provided a refuge from thought and feeling.
> John Steinbeck, *The Grapes of Wrath* (1939)

WHEN MATILDA PERSILY came to work at the Mathematical Tables Project, she followed a little ritual to prepare her tools for the day. She was part of the special computing group, one of the few who regularly used an adding machine. Her machine was a old Sunstrand, an inexpensive device that had ten keys on the top and a crank on the right side. She would first add a few meaningless numbers while sharing a moment of conversation with other members of the staff. As she talked, she would listen for the sound of grit in the gears and try to feel any catch or slip in the mechanism. She could apply some lubricating oil from a can that sat on a small shelf, but too much oil would attract the very dust she was trying to avoid. She had found that the machine was best cleaned with an orangewood stick; such sticks were sold at drugstores to groom fingernails and cuticles. Once convinced that the machine was ready, she would sit at her desk, take out her worksheets, and begin to compute.

Most of the computing machinery acquired by the Mathematical Tables Project during its first years of operation was scavenged from terminated WPA offices and other government agencies.[1] Ida Rhodes described this equipment as "the most broken down, the most ancient of contraptions."[2] The majority were inexpensive, hand-cranked Sunstrand calculators. They were not designed for the repetitive work of the computing floor, nor was their crank mechanism easy for project computers to pull. Rhodes would complain, "Oh, how my arm ached by the end of the day."[3] Few of these machines were in operating condition when they arrived at the project's office. A couple of the senior computers had learned how to transplant gears and levers from one machine to another and were able to produce two or three working machines for every four that came to their door. Through 1938 and 1939, the group had no other way to acquire calculators. A new mechanical calculator cost $400, almost as much as the $560 annual salary paid to a WPA worker.

During the Great Depression, most computing laboratories had at least one staff member with enough mechanical expertise to repair or modify their desk calculators. Even the smallest laboratory had to follow a maintenance regimen for its machines, oiling gears and adjusting levers. Observatories relied on the same technicians that kept telescopes in working order. The larger organizations, such as the Iowa State Statistical Laboratory, kept a mechanic on their staffs. The Iowa State laboratory, formerly known as George Snedecor's Mathematical and Statistical Service, had to maintain an unusually large number of machines, as the laboratory acted as a general computing facility for the campus. It not only did calculations for any college faculty member but also provided adding machines and calculators to campus laboratories and offices. Laboratory staff kept the machines in good repair and trained the people who would use them.[4] With such a responsibility, it is perhaps not surprising that the laboratory director in 1938, Alva E. Brandt (1898–1975?), was a trained mechanical engineer and had once served as a professor of farm machinery at Oregon State University.[5]

The computing office at Bell Telephone Laboratories probably had the greatest access to trained engineers and inventors. Regularly, laboratory scientists turned their gaze on the computing staff of Clara Froelich and saw them as a model for some new calculating device. "In these laboratories," observed the staff scientist George Stibitz (1904–1995), "we have 10 or more girls, including at least one Hunter graduate, who spend most of their time dealing with [complex numbers]." Stibitz, a physicist, was a new addition to the mathematical division. In spite of his patronizing language toward the women, who were often a decade or more older than himself, he was one of those rare scientists who treated the computers as individuals. He seems to have been friendly with them, pausing over their desks to share a bit of conversation, comment on the upcoming weekend, and learn how they handled their calculations. He rarely described his early research without mentioning them.

By 1937, Stibitz had begun to think about building a machine that could perform complex arithmetic. Unlike L. J. Comrie in England, he was not inclined to adopt existing computing machines. "Although there are well-known rules for the use of ordinary computing machines to handle complex numbers," he observed, "the work is tedious and likely to lead to errors on the part of the operator."[6] Others at Bell Telephone Laboratories had already designed calculators for complex arithmetic, but they had met with mixed success. One laboratory engineer had created a special slide rule that could multiply complex numbers. Unlike an ordinary rule, this device could account for the two parts of a complex number, the value identified as the real part and the value identified as the imaginary part. In spite of its ingenuity, the new slide rule was never adopted by the human computers. In all likelihood, they found the device too cumbersome and judged that it was faster to do the work by hand.[7]

A second computing machine for complex arithmetic had been created by Thornton Fry, but this machine was a specialized instrument. Fry's machine, which he called an isograph, could be used to find the zeros of a polynomial. The zero of a polynomial is a value that makes the expression equal zero. These values are often called "roots," a term that suggests the part of a plant that is underground; hence roots are the values for which an expression vanishes to nothing. For the polynomial $x^2 - 5x + 6$, one of the roots is the value 3, because for $x = 3$, the value of the polynomial $x^2 - 5x + 6$ equals $3^2 - 5 \times 3 + 6$ or $9 - 15 + 6$, which is 0. All polynomials have roots, but they can be complex numbers and are often quite hard to find.

The isograph is best understood as an oracle for polynomials. It was not a true calculator but a device that could provide computers with information about their work. The idea had come from an analysis of gears that had been done jointly by Fry and Stibitz. The two had shown that the action of a certain combination of gears was best modeled by complex numbers. Fry took a similar set of gears and used it as the basis for the isograph. The device was big, about twelve feet in length, and was driven by an electric motor. A wheel on the left side of the machine could be used to advance the machine should the motor jam. Froelich, or one of the other computers, operated the isograph by setting a row of knobs that protruded from the base of the machine. They started their work by guessing a value that might make their polynomial equal zero. The isograph would process this guess, grinding its gears and producing a graph of circles on a plotting table. The computer would have to interpret these circles and determine whether the guess was too little or too big. The computer would then adjust the guess and ask for a second response from the machine. Continuing in this manner, with a little bit of strategy and a fair amount of patience, the computer could eventually find the values that made a polynomial equal zero.[8]

The isograph was clever, but in practice it proved no more useful than the complex slide rule. It offered assistance for only a narrow class of problems and required the computers to master a fairly esoteric set of controls. From his observations of the computing staff, Stibitz wanted to build a more intuitive machine, a general-purpose complex calculator that could add, subtract, multiply, and divide. Instead of using gears for this machine, he employed electrical circuits and binary arithmetic. Binary arithmetic was novel to computing machinery, even though it had been explored by the Harvard mathematician Benjamin Peirce some seventy years before. Peirce had noted that binary arithmetic could simplify calculation because the only symbols involved were 0s and 1s. Stibitz built upon this idea and developed an electrical circuit that could perform additions. Representing 1 as a positive voltage and 0 as no voltage, he demonstrated these ideas with a simple prototype. Borrowing a few "relays from a junk pile that Bell Labs maintained," he assembled a circuit that would add two single-digit binary numbers. "With a scrap of board, some snips of metal from a tobacco can, two relays, two flashlight bulbs, and a couple of dry cells," he recalled, "I assembled an adder on the kitchen table at our home."[9]

His prototype, which he labeled "Model K" for "kitchen," has since become a staple of elementary computer science classes. It could complete three ordinary summations: $0 + 0$, $0 + 1$, and $1 + 1$. By pressing on the scraps of metal, he would complete a circuit, which would activate a relay and light the combination of bulbs that represented the sum. By the time he completed Model K in November 1937, he had already moved ahead to more sophisticated circuits that could deal with larger numbers and all four arithmetic operations. He worked on several ideas that fall, including a binary version of Fry's isograph for finding polynomial roots.[10] These machines were rough drafts, attempts to master the problems of binary design. None of them was built. The design for his first operational binary machine required six months of careful effort. It was a partial calculator, a machine that could multiply and divide complex numbers. Stibitz added circuits for addition and subtraction only after the original units were operational.[11]

The complex calculator was built by laboratory technicians using standard telephone parts.[12] The machine cost $20,000 to build, about thirteen times the $1,500 salary of Clara Froelich, thirty-five times the pay of a Mathematical Tables Project computer, and fifty times the price of a traditional mechanical calculator. The Bell Laboratories computing staff seems to have kept the machine fairly busy, at least during the working day. Much of the demand on the device came from three divisions of the laboratory that dealt with telephone circuit design. The calculator was kept out of sight in a central equipment closet. The computers dealt only with a special keyboard that was connected to the calculator with ordinary telephone lines. This keyboard resembled an ordinary desk calculator. The laboratory built three of these keyboards, though only one could be used at a time, and placed them in the offices that made heavy use of complex numbers.[13] On special occasions, the laboratory gave outside researchers access to the calculator through long-distance phone lines. Stibitz demonstrated the machine at a meeting of the American Mathematical Society in New Hampshire. The mathematicians were invited to test the calculator themselves. One of the Aberdeen veterans,

33. The Bell Telephone Laboratories complex calculator

Norbert Wiener, spent several hours typing at the keyboard and watching the results appear on a roll of paper.[14]

The complex calculator of George Stibitz was the first of three machines that had its origins in a computing office of 1937. Taken as a whole, the three machines show how inventors adapted new technologies and new devices to the operations of computing laboratories. Stibitz developed binary arithmetic and relay circuits in order to give Clara Froelich a simple calculator for complex numbers. The second computing machine, which had its origins in the Iowa State Statistical Laboratory, attempted to meet a similar goal for a different problem, the calculations of least squares. By 1937, the laboratory had become a substantial contractor to the United States government. It received $35,000 a year to do tabulations and analyses for the WPA and for the U.S. Department of Agriculture, where the laboratory's former patron, Henry A. Wallace, served as secretary. Its staff processed data on farm production, analyzed crop experiments, and identified the trends in agricultural markets. In spite of their best efforts, the computers had been unable to mechanize the central calculations of least squares computation. They had acquired an IBM 601 multiplying punch, the same model that could be found at the Columbia University Astronomical Computing Bureau, but even this machine could assist only with the first and last steps of least squares computation. The remaining work was done with adding machines by a staff of seven.[15]

The Iowa State calculating machine was built by an outsider to the computing lab, a professor named John Vincent Atanasoff (1904–1995).[16] Atanasoff had been at Iowa State College for a little more than a decade. He was originally trained as an electrical engineer but had become interested in physics and had taken two years at the University of Wisconsin in order to complete a doctorate in the subject. Sometime in the early 1930s, he drifted into the statistical laboratory, intrigued with the stories of the mechanical tabulators. He did not seem to have a clear idea of how he might use the machines, for he later reported that he went "looking for a problem in theoretical physics that could be solved by IBM equipment."[17] The problem that he chose was a broader version of the least squares calculations that were regularly handled by laboratory computers, a problem called a simultaneous equation calculation.

In a simultaneous equation problem, a researcher has a certain number of unknown values that are defined by an equal number of equations. They are often taught with word problems. "When Caroline was born, Rose was twice as old as Ginny. Last year Rose was 10 percent older than Ginny. If Caroline is 28 this year, how old are Rose and Ginny?" This problem has three unknown values, the ages of the three sisters. These three values are defined by three equations, one for each sentence in the problem. Computers would find the three values by manipulating the three equations. Like complex arithmetic, such manipulations are detailed and time-consuming. The work becomes more difficult as the scale of the problem increases. A calculation with six unknown values is only twice the size of a problem with three unknown values, but it requires eight times the effort. A problem with twenty-four unknown values, the size of the least squares calculation that Alfred Cowles had proposed to Harold T. Davis, requires five hundred times the effort of a problem with three unknown values.[18]

John Atanasoff began his research into computing machines by modifying the laboratory's IBM tabulator. This work paralleled the development of the isograph at Bell Telephone Laboratories. Atanasoff experimented with the punched card technology to see what he might do with it. His results were no more successful than the isograph but for slightly different reasons. In this experiment, he worked closely with laboratory director A. E. Brandt. The two of them found a common bond in the work and seemed to enjoy tinkering with machinery. Before tackling the problem of simultaneous equations, they addressed a simpler calculation, one that came from the study of the light spectrum. This calculation was difficult, if not impossible, to perform on an unmodified tabulator. Atanasoff and Brandt found a way to handle the calculation with the tabulator, but their solution involved a new circuit and a special set of punched cards. The circuit plugged into the tabulator control board and took charge of the machine whenever it encountered one of the special cards. The modifications gave "trouble-free operation over long periods of time," according to Atanasoff and Brandt, but they handled only the intermediate problem, not the more difficult problem of simultaneous equations.[19]

Atanasoff designed a similar modification that would allow the tabulators to solve simultaneous equation problems, but before he could implement his idea, he lost his access to the IBM equipment. Later in life, he would suggest that IBM itself had barred him from the machines, but the reason was probably nothing more than the departure of his partner, A. E. Brandt.[20] In the spring of 1937, Brandt left the laboratory and took a job at the Iowa Agriculture and Home Economics Experiment Station. Brandt's successor was more interested in statistical work than in computing machinery and may have felt that Atanasoff's experiments interfered with the obligations of the laboratory.[21] The reason that removed Atanasoff from the statistical laboratory is less important than the consequences of that removal. No longer able to modify the IBM equipment, he turned to the idea of building an entirely new computing machine. This machine was more technologically daring, and yet it was a better match for the computers of the statistical laboratory.

As Atanasoff would tell his story, the basic principles of his new machine came in a late-night epiphany. Through the summer and fall of 1937, he considered several different ways of building a computing machine, but none of them would accomplish what he wanted to do. "I had outlined my objectives," he later recalled, "but nothing was happening and as the winter deepened, my despair grew." One evening, he left for his office, hoping "to resolve some of these questions." Instead of working at his desk, he got into his car and "started driving over the good highways of Iowa at a high rate of speed."[22] Moving away from the Iowa State campus and its statistical laboratory, he crossed half the state before reaching a roadside bar in Illinois. There, relaxed by the drive and perhaps by a drink, he identified the key elements for a computing machine that would solve least squares and simultaneous equation problems. This machine would be an entirely new device instead of a modified tabulator. It would use the binary number system, like Stibitz's machine, so that the arithmetic could be handled by electrical circuits.[23]

John Atanasoff would admit that he was "somewhat off the beaten track of computing machine gossip,"[24] though there was little such gossip to be found outside of Bell Telephone Laboratories or International Business Machines. With no access to the statistical laboratory and lacking an organization to support him, Atanasoff had to find a place to build his machine, as well as funds to pay for supplies and assistants. He spent about eighteen months building a simple demonstration model, his own "Model K." This machine was more sophisticated than the machine that Stibitz had demonstrated, but it did approximately the same thing: it added two binary numbers together. It gained him a grant of $650 from Iowa State College, enough to hire an assistant and start work on the full machine.[25]

To raise more money, Atanasoff went to the Rockefeller Foundation in New York City. The foundation was one of the larger financers of scientific research, and one of the foundation officers, Warren Weaver (1898–1978), had taught at the University of Wisconsin when Atanasoff was studying for his doctorate. The meeting did not proceed quite as Atanasoff might have hoped. Weaver had recently been hospitalized and had to be propped up with pillows.[26] He was not predisposed to Atanasoff, as he remembered the former Wisconsin graduate student as "rather bright but queer and opinionated."[27] He patiently listened to a description of the proposed machine and firmly stated that the Rockefeller Foundation did not support such research.[28] Yet, in the exchange, Weaver must have seen something of value, for he mentioned that a private foundation, the Research Corporation, supported engineering projects and might be willing to provide some money for Atanasoff's machine. He offered to let Atanasoff use his name in correspondence to the foundation and agreed to review Atanasoff's proposal.[29] The dismissal may have been disheartening, but it proved to be good advice, for the Research Corporation gave Atanasoff $5,330 for his computing project. This was a substantial grant for the time, even though it was about one quarter of the money American Telephone and Telegraph had spent on the complex calculator.[30]

Atanasoff constructed his machine in the basement of the Iowa State physics building. The device was about the size of a large desk. It had a reader for IBM punched cards and two rotating drums that held the numbers. When it was running, the two drums made a clicking noise, like a

34. Computing machine of John Atanasoff with operator

piece of cardboard slapping against the spokes of a bicycle. The operator stood in front of the machine and loaded the simultaneous equations onto these drums, one value at a time. Once an entire equation had been given to the machine, a special card punch would place all of the values on a card. Atanasoff had designed this punch so that it created holes with high-voltage sparks, rather than with a mechanical die. It recorded the numbers with a flash of blue light and a puff of smoke, an operation that would regularly singe the cards and occasionally set one alight. Each equation required one of these special cards, and each step of the calculation required that all special cards be repunched. The operator would stand in front of the machine, shuffling the cards back and forth, while the drums turned and snapped. Fresh cards would be drawn from a pile, and old cards would be discarded on a table or dropped to the floor.

Atanasoff's calculator was never a finished production machine like the complex calculator at Bell Telephone Laboratories, but by the most generous accounts, it did what it was intended to do. "It was good enough so that we were able to solve small systems of equations," wrote Atanasoff, though he acknowledged that the high-voltage card punch was troublesome. "We made substantial efforts to solve this flaw, including changes in the card material and careful changes in the voltage use for each material."[31] After working on the machine for two years, he left Iowa State College in order to take a job at the Naval Ordnance Laboratory in Washington, D.C. He apparently left no one at the college who was interested in preserving his machine or in keeping it operational. When the school's physics department decided to reclaim Atanasoff's office space, they disassembled the device, salvaged what they could for scrap, and disposed of the rest.[32]

Some thirty years later, Atanasoff's machine would acquire notoriety as a central exhibit in a court case that contested the patent on the electronic digital computer. In the trial, it would be named the Atanasoff-Berry Computer, and its builder would be identified as the inventor of the modern computer. The controversy that followed the verdict would last for two decades more and would create partisans who claimed "that the first electronic digital computer was constructed at Iowa State by J. V. Atanasoff and [his assistant] Cliff Berry"[33] and opponents who believed with equal fervor that the machine was not a computer, as it was "premature in its engineering conception and limited in its logical one."[34] This debate, with its implications for the reputations and fortunes of the participants, cannot be easily dismissed. Yet, in commanding both scholarly and public attention, it has obscured the position of the human computer in the late 1930s. Both Atanasoff's computing machine, be it a computer or no, and George Stibitz's complex calculator would have fit nicely into existing computing laboratories, just as the adding machines of the 1890s moved easily into the Coast Survey Office, the Nautical Almanac Office, and the Harvard Observatory.

The last computing machine of 1937 moves one step further from the offices of human computers, though it remained tied to the kinds of calculations that were being done by human computers. It was conceived at Harvard University by a graduate student in the school's electrical engineering program. The student, Howard Aiken (1900–1973), was studying the actions of electrons in vacuum tubes. The mathematical expressions that described the electrical forces inside a vacuum tube were a messy set of differential equations. Like all other problems driving the development of computation, they could not be solved in a simple, symbolic fashion. In common with the equations of Richardson's weather model, they described a phenomenon in three dimensions and would have required a substantial computing staff to calculate the solution. Harvard had access to funds from the National Youth Administration to pay the salaries of human computers, but such assistance was not sufficient for Aiken. "At the present time," he wrote, "there exist problems beyond our ability to solve, not because of theoretical difficulties but because of insufficient means of mechanical computation."[35]

Like Atanasoff, Aiken turned from the problems of physics to the problems of calculation. He designed a machine that used gears and wheels but had a special control mechanism that provided "automatic sequencing."[36] This mechanism would read a series of instructions from a paper tape and would direct the machine to perform those instructions. These instructions were almost a program, as we now use the term. By changing instruction tapes, the operator could make the machine perform complex arithmetic, solve simultaneous equations, compute orbits and trajectories, and reduce data.[37] In many ways, Aiken's idea was similar to Charles Babbage's second computing machine, the one he had called the Analytical Engine. Aiken discovered the work of Babbage while he was preparing the basic outline of his machine. He was even able to inspect a partial adding mechanism that had been built according to Babbage's specifications by his son. This connection between the nineteenth-century mathematician and the emerging computing machines is superficial, according to Aiken's biographer, I. Bernard Cohen. "At that time [Aiken] did not have a detailed and accurate knowledge of the purposes and principles of operation of Babbage's two proposed machines."[38]

Aiken accomplished what Babbage could not: he built a working relationship between a commercial business and a scientific computing laboratory. In 1937, Aiken was older than most graduate students. Cohen has characterized him as "tall, intelligent, somewhat arrogant [and] assertive." Aiken had supported his family since the age of fourteen, when he and his mother were abandoned by his father. As a high school student, he had taken night jobs while attending classes during the day. When he was an undergraduate at the University of Wisconsin, he had worked from four to midnight at the local electric and gas utility.[39] His position at Harvard freed him from the need to seek outside employment and allowed him to search for someone who might be able to sponsor his computing research.

He first presented his ideas to an engineer at the Monroe Calculator Company, "a very, very scholarly gentleman," Aiken recalled. The engineer quickly recognized what Aiken was attempting to do and "foresaw what I did not . . . the application to accounting." The engineer gave a favorable review of the machine, but the management of Monroe decided that they were not interested in the project.[40] Following this rejection, Aiken then turned to the computing staff of the Harvard Observatory. The observatory computing room operated much as it had in 1880 under Edward Pickering. A staff of computers and assistant astronomers, many of them women, measured photographs, interpreted data, and reduced the values recorded by the telescopes and sensors. The office had at least a few touches of modernity, such as mechanical adding machines, but it was more concerned with astronomy than with general methods of computation.[41]

35. Mark I mechanical computer at Harvard

The observatory director, through an indirect path, helped Aiken gain the attention of the senior managers at IBM. On a trip to New York, Aiken presented the IBM managers with a plan for a machine that would "be fully automatic in its operation once a process [was] established."[42] He visited the Columbia University Astronomical Computing Bureau, met Wallace Eckert, and studied the Orange Book. By the time his visit with IBM ended, Aiken had gained the attention of company president Thomas J. Watson. Watson was impressed with the proposal and offered to finance the project and build the machine in an IBM factory. Aiken would provide the general design and work with IBM engineers to develop the appropriate technology. Harvard would provide the computer center and operate the device.[43] In a move that suggested that the two groups would not long cooperate on the project, IBM decided to call the machine the Automatic Sequence Controlled Calculator, while Harvard would name it the Mark I. By the time the project was finished, IBM had invested $100,000 in the construction of the machine and donated another $100,000 to cover operational costs, a combined sum that approached the annual budget for the Mathematical Tables Project.

Viewing the new computing machines, George Stibitz prophesied that "Human agents will [soon] be referred to as 'operators' to distinguish them from 'computers' (Machines)."[44] Neither his machine nor that of John Atanasoff would take the title "computer" from human beings. The computers at Bell Telephone Laboratories may have operated the complex calculator, but they were more concerned with mathematics than with machinery. Atanasoff's machine handled only one modestly complex step of a large process. Both inventions were intermediate devices that did not quite reach the era of stored programs and still looked back at the age of oil cans and orangewood sticks. Even Howard Aiken's Mark I, the most sophisticated of the three machines, looked over its shoulder toward older technologies. One of Aiken's assistants captured the traditional nature of the Harvard computing laboratory when he described the Mark I as emitting "a distinct sound, not unlike the clatter of steel-shod horse's hooves clanging along a paved street."[45] Aiken generally employed his computing machine in work that could have been handled by the computing floor of the Mathematical Tables Project or the First World War computers of Aberdeen or even the *Nautical Almanac* computers of Charles Henry Davis. Shortly after the machine began operations, Aiken produced a set of mathematical tables. His volumes covered a different set of expressions from those being prepared by the computers of the Mathematical Tables Project, but the real difference between the two sets of computations was the difference between Harvard and the WPA, not the difference between machine calculation and handwork. The WPA reproduced its tables from mimeographed stencils. The Mark I tables were typeset and printed on fine paper. The WPA books were bound in a rough tan cloth and avoided references to work relief. Aiken used a fine blue cover and printed the university seal on the title page.[46]

CHAPTER FIFTEEN

Professional Ambition

> I couldn't find no job
> So I went to the WPA.
> WPA man told me:
> You got to live here a year and a day.
>
> Langston Hughes, "Out of Work" (1940)

LIKE THE ADDING MACHINES of the 1880s, the calculators of Stibitz, Atanasoff, and Aiken coincided with a world's fair. This fair, which opened in the spring of 1939, was hosted by New York City. "It is arranged," wrote the author H. G. Wells, "to assemble before us what can be done with human life today and what we shall almost certainly do with it . . . in the near future." After pausing for a digression, he added, "It is a promotion show."[1] Like the World's Columbian Exposition, now almost half a century in the past, the Long Island fair displayed the technologies that would be embraced by American culture. Visitors could examine television receivers, FM radios, prototypes of divided highways, and primitive fax machines. They could inspect the products of both Bell Telephone Laboratories and International Business Machines. IBM president Thomas Watson hosted a company conference at the fair and delivered a rousing speech on the future of punched card technology.[2]

In the midst of all the symbols of material progress stood the WPA hall with its proud and slightly self-contradictory inscription, "This building shows the wealth created by the skill and artistry of America's unemployed." The presence of the WPA had been controversial and a little embarrassing to government leaders. The opening of the exhibit had been delayed by labor troubles, an ironic touch that delighted opponents of the New Deal.[3] When the WPA finally allowed visitors into the building, more than three weeks had passed since the start of the fair. A signature book by the front door recorded the opinions of those who came into the exhibit during those first days. "The WPA must go," signed former presidential candidate Alf Landon, but his comment was altered by a WPA supporter so that it read, "The WPA must go on."[4]

"The exhibits cover every aspect of WPA activity," wrote one reporter, "from art, music, and drama to the manufacture of clothing for the poor." Above the displays, WPA employees had written slogans that portrayed the agency in a heroic light. "Work is the Right of every American," read one panel, and "Work Builds Better Communities," claimed another.[5] Over the science projects was written, "Work Increases Knowledge," a phrase that Gertrude Blanch and Arnold Lowan would have liked to claim for the Mathematical Tables Project. As far as we know, there were no calculations from the project on display.[6] Only the powers of integers had been officially published. The second book, the volume on the exponential function, was still being printed.

During that first summer of the fair, Gertrude Blanch was still attempting to accumulate the information that she needed for her computing plans. Like others before her, she was learning that there was no single literature of computation. When she needed some mathematical theorem or technical analysis, she would send a junior member of the planning committee to the New York Engineering Societies Library with instructions to scan through some collection of journals, page by page if necessary. The fact that many on the planning committee were immigrants or the children of immigrants simplified such searches, as much useful material could be found in foreign language publications, such as the *Archiv der Mathematik und Physik*, the *Mémoires couronnés et autres mémoires publiés par l'Académie Royale des Sciences, des Lettres et des Beaux-Arts de Belgique*, and the *Wissenschaftliche Schriften des Donetz-Tecknikums des Genossen Artjem zu Stalin*.

The one organization that might have been able to assist the Mathematical Tables Project with its library searches, the Subcommittee on the Bibliography of Mathematical Tables and Other Aids to Computation of the National Research Council, was still moribund and unfocused. By the spring of 1939, the leaders of the National Research Council had lost all hope that A. A. Bennett would be an effective leader of the group. After the initial flurry of activity in 1935 and 1936, Bennett had all but abandoned MTAC. His communications with the National Research Council had became a litany of excuses. "Unexpected and extended interruptions have retarded the work in a way that was not anticipated," he reported that April.[7] Most of these "unexpected interruptions" had come from the Aberdeen Proving Ground, where Bennett served as a consultant. The proving ground had expanded its computing facility with a differential analyzer, the machine that had been invented to solve differential equations. "During the last part of the year," wrote the base commander, "the analyzer has been used in the computation of two firing tables with gratifying results."[8] As gratifying as such results may have been, they did not suggest that the new computing machine would replace the human computers or eliminate all work for A. A. Bennett. The analyzer "saves a great amount of labor when a group of related trajectories are to be computed," reported one of Bennett's colleagues, but the device was sensitive and suffered from "mechanical inaccuracies." The adjustment of the analyzer was "a delicate matter, requiring so much time that for a single trajectory [it was] more economical to compute in the usual way."[9]

Sometime that spring, the leaders of the National Research Council quietly asked Bennett to resign his chairmanship. In his stead, they appointed Raymond Claire Archibald, who was, like Bennett, a professor of mathematics at Brown University. Archibald was a tall, imposing figure, filled with energy and topped by a head of hair that had retained its red color. He was a Canadian by birth and a distant cousin of Simon Newcomb, a connection that gave him great pride.[10] He was also unmarried, a fact that he prominently displayed in his biographies and resumes.[11] The council had twice passed over Archibald when it had sought a chair for MTAC, but in 1939, it was willing to accept anyone who might actually complete a bibliography. Archibald had already proven that he could be a leader of mathematicians, though not quite a leader with the stature of Veblen or even A. A. Bennett. During the First World

36. R. C. Archibald in his office at Brown University

War, Archibald had edited the *American Mathematical Monthly,* a prime job in ordinary times but one that seemed small compared to the experience of the Aberdeen veterans. Following the war, he had held several minor positions in the American Mathematical Society and, in the process, acquired the reputation of being difficult. He had once drawn a quarrel from Oswald Veblen over plans for financing the American Mathematical Society. "I gather from your letter," Veblen had scolded, "that you have not understood my position in the matter. But on re-reading my [note to you], I don't see how I can make it clearer. So I fear all I can do is to ask you to reread the last paragraph of that letter."[12]

When R. C. Archibald accepted the chair of MTAC, he put all of his considerable personality into the job. Within days of his appointment, he had written to the committee, asking members for "reactions, advice, comments and suggestions on various matters." He explained that the committee would act as a "clearing-house of information about Tables," that it would cooperate with England's Mathematical Tables Committee in order "to avoid duplication of effort," and that it might "take steps toward initiating the development of other tables which were thought to be desirable."[13] He proposed to establish a broad and inclusive committee that could direct any kind of computational work. Before the summer had ended, he had reorganized the group along lines that had been developed by the Mathematical Tables Committee of the British Association for the Advancement of Science, a structure that split the broad literature of mathematical tables into twenty-one different classes.

TYPOLOGY OF MATHEMATICAL TABLES
BRITISH ASSOCIATION FOR THE ADVANCEMENT OF SCIENCE

A. Arithmetical Tables
B. Tables of Powers
C. Logarithms
D. Circular Functions
E. Hyperbolic and Exponential Functions
F. Theory of Numbers
G. Higher Algebra
H. Tables for Numerical Solution of Equations
I. Tables Connected with Finite Differences
J. Summation of Series
K. Statistical Tables
L. Higher Mathematical Functions
M. Integral Tables
N. Interest and Investment
O. Actuarial Tables
P. Engineering Tables
Q. Astronomical Tables

R. Geodetic Tables
S. Physical Tables
T. Critical Tables of Chemistry
U. Navigation Tables

Many of these categories had little in common. Integral tables (M) involved no calculation at all. Interest tables and actuarial tables (N and O) were prepared for businesses, not laboratories. Many engineering tables (P) contained collections of data that had been gathered from experiments. In addition to these categories, Archibald proposed one final division, Z, that would not deal with tables at all but would prepare the bibliography of calculating machines. For each division on the list, he intended to create a small section of four or five members. The chairs of the sections plus Archibald would form an executive group.

The plan would create an unusually large committee for the National Research Council, and it did not take long for the organization's staff to object to Archibald's plans. "There is nothing in the by-laws or procedures of the Council to contravene what Professor Archibald proposes," lamented the council secretary. "However, I cannot help feeling that making a difference between classes of membership in this way introduces a rather invidious distinction."[14] Grateful for any action within the committee, the council was willing to let Archibald proceed, even though one scientist on the council compared the new MTAC structure to the myriad agencies of Roosevelt's New Deal and quipped that Archibald "apparently has caught the alphabet fever from the Government."[15]

Undeterred by the objections of the council, Archibald began making the appointments to the committee. He assigned H. T. Davis the chair of Section E, which dealt with material covered in Davis's encyclopedia of functions. L. J. Comrie was given the leadership of the computing machine group, Section Z. To that section, Archibald added George Stibitz of Bell Telephone Laboratories. For Section K, which dealt with statistical tables, he offered the chair to W. W. Edwards Deming (1900–1993), a U.S. Department of Agriculture statistician and a former student at Karl Pearson's Biometrics Laboratory. To catalog the literature of actuarial tables, Archibald turned to the New York insurance industry.[16] He also attempted to expand the international presence on the committee by recruiting Tadahiko Kubota of Japan's Tohoku University.[17]

For the section on astronomical tables, Section Q, Archibald turned to the obvious choice, Wallace Eckert of the Columbia University Astronomical Computing Bureau. "Many thanks for considering me in connection with this Committee," wrote Eckert in reply to Archibald. "I believe the work important, and am willing to do all I can to make the [committee] a success."[18] Other than L. J. Comrie, Eckert was the only section leader drawn from the astronomical community, a discipline that had once defined scientific computation. Eckert's little section also contained the only female member of MTAC, a Naval Observatory computer named Charlotte Krampe (1904–1969).[19]

By the fall of 1939, Archibald had appointed the core members of his committee and had raised a $15,000 publication fund from the Rockefeller Foundation.[20] At first, this money seemed to be a sign of Archibald's success, but it brought him into conflict with the leadership of the National Research Council. The council leadership believed that they had ultimate control over the funds, as MTAC was a subcommittee of their organization. Archibald claimed that he alone could decide how the money was spent. The disagreement was entirely academic until Archibald published his first bibliography and sent copies of it to the council. In the box, he included an invoice for $61.73.[21]

"I cannot understand the item at the bottom of your second page," wrote one leader of the council, Luther Eisenhart (1875–1965), "unless you mean that since the N.R.C. will use these copies for distribution the N.R.C. should pay the fund for them. This would raise an issue which I have never heard of before."[22] Whether or not the council had heard of the issue was of little matter to Archibald. He argued that the Rockefeller Foundation funds were his responsibility and that he was required to recover every dime he spent on publication, including the value of the books sent to the National Research Council.[23] Fundamentally, Eisenhart was sympathetic to Archibald. He, too, was an applied mathematician and understood that $15,000 publication funds were rarely given for mathematical research.[24] Yet Archibald's claims verged on insubordination, if not outright mutiny. He considered ousting his committee chair, but he confided to a colleague that "I do not know whom I could possibly secure to replace him and who would be willing and able to do the work in the way in which he has been doing it."[25]

The conflict waxed and waned for nearly two months as Archibald pushed his claim on the council and Eisenhart deflected his demands. Archibald seemed to grasp, though perhaps for misguided reasons, that he had a secure hold on the MTAC chair. For his part, Eisenhart wanted to resolve the conflict without showing any emotion and without weakening the discipline of the other subcommittees. The resolution came when Archibald, frustrated and impatient with the exchange, accused Eisenhart of using the "insidious political methods of the Roosevelt administration."[26] Eisenhart may actually have chuckled at receiving this overblown piece of rhetoric, but to Archibald, they were clearly fighting words. The MTAC chair held his ground until he was moved by the still, small voice of conscience. "I beg," he wrote to Eisenhart three days later, "that you regard the last five lines of the fourth paragraph of my letter to you of July 29 as never written, that we may be 'as before.'"[27] Eisenhart accepted the apology, confirmed the council's authority over MTAC, but spared Archibald's ego by paying the $61.73.[28]

It is easy to view the fight between Archibald and the National Research Council as a clash among strong personalities or the attempt of a maverick scientist to establish a small empire of his own. In the right hands, it might even have been turned into an entertaining evening of musical satire. However, the event captures the state of computing in 1939 and suggests that the discipline had reached a critical point in its history. With only a few exceptions, computing laboratories had always been under the control of scientists whose interests lay elsewhere. The directors of computing groups had been astronomers, surveyors, electrical engineers, ballistics engineers, physicists, statisticians, meteorologists, and economists. Only people like H. T. Davis and L. J. Comrie could claim to have computing as their primary interest, and of these two, Comrie had been trained as an astronomer.

With the emergence of Archibald and MTAC, Comrie and the Scientific Computing Service, Davis and his books on mathematical functions, Blanch and Lowan and the Mathematical Tables Project, human computers began to claim that they had an independent discipline, that they commanded a body of knowledge that stood apart from astronomy, mathematics, and statistics. This was not a philosophical question but a practical issue with practical consequences. If computing was an independent body of knowledge, then it should have a standard way of training new computers, a publication to disseminate new ideas, and an organization that could coordinate research. "If it be proposed to advance some truth, or to foster some feeling by the encouragement of a great example," wrote Alexis de Tocqueville, "[Americans] form a society."[29] In 1939, MTAC appeared to be the germ of a new computing society, but before it could claim such a role, Archibald would have to gain the support and confidence of the major computing organizations. His work was complicated by the emergence of the new computing machines and by the start of a new world war, which promised far more work for human computers than had ever been found by Oswald Veblen and his friends in the summer of 1918.

September 1939 brought a new urgency to America's scientific and industrial research. On the third day of the month, Germany ended the armistice of 1918 by invading Poland. Great Britain rushed to the aid of the Poles, and France followed. In increasing numbers, Americans believed that they would be drawn into the war, but this time they were not rushing into the fray. They had seen enough of battle during those brief months in 1918, when the American Expeditionary Force had fought on the western front.[30] Following the invasion, President Franklin Roosevelt had to remind the country that "though we may desire detachment, we are forced to realize that every word that comes through the air, every ship that sails the sea, every battle that is fought, does affect the American future."[31]

The scientific veterans of the First World War volunteered for a second round of military research. Having watched the army take control of the National Research Council in 1917, they formed a new coordinating committee that included the interests of both the army and the navy but kept the balance of power in the hands of civilian scientists. This organization, the National Defense Research Committee, divided its work among broad alphabetic classes of military needs. The committee had a Division A for armor and ordnance, B for bombs and explosives, C for communications, D for detection devices and controls. To these four sections was added a fifth, Division E, that dealt with the rights of the inventors. There was no division M for mathematics and computation, as such research did not seem to be a high priority, even though both military services moved to strengthen their existing computing laboratories. The army invested $800,000 in the Aberdeen Ballistics Research Laboratory, using part of those funds to expand the computing laboratory.[32] The navy, which had just retired the director of its Nautical Almanac Office, went in search of a leader who understood computing machinery.

The almanac had seen little change since the retirement of Simon Newcomb, some forty years before. In Simon Newcomb's last years, the navy had closed the almanac office in Foggy Bottom and moved the computers to the new Naval Observatory, which overlooked the monuments and government offices from atop St. Alban's Hill. Since that time, the group had been overseen by a sequence of dependable but unimaginative directors. A review by the naval officers remarked that almanac directors of the early twentieth century "did see to it that the Almanacs were produced on time," but they had done nothing to advance the capabilities of the office.[33] Using blunter language, L. J. Comrie concluded that the computing laboratory was "stagnant."[34] Under either judgment, the almanac was unprepared to handle the work that came with the start of hostilities. For nearly thirty years, the American almanac staff had shared the burden of ephemeris computations with the almanac offices of England, France, Germany, Russia, and Spain.[35] This collaboration had begun in 1912, when the United States had officially abandoned the prime meridian in Washington. It had been suspended for the First World War and been resumed in the 1920s. Now, for the second time in twenty years, the almanac staff found that they were computing the tables for the *Nautical Almanac* without the full assistance of its European collaborators. Germany withdrew from the agreement to share almanac computations, and France found itself unable to contribute its part. As the staff struggled with additional work, naval officers were designing a new almanac for airplane navigators. This almanac would be exceptionally detailed, as airplane navigators needed astronomical positions at fifteen-minute intervals. Without this kind of detail, navigational computations would forever lag behind the actual position of the plane by a distance of 10 or 50 or even 200 miles.[36]

There were many senior astronomers and mathematicians who were qualified to lead the almanac office and at least a few who were ambitious to fill the role, but the navy wanted only one individual, Wallace Eckert of the Thomas J. Watson Astronomical Computing Bureau, as the Columbia University computing facility was now called. Eckert was flattered by the initial overture but was not especially interested in the position. He was happy in New York and saw no reason to leave. The superintendent of the Naval Observatory, unwilling to accept Eckert's refusal, persisted in his efforts to entice the Columbia astronomer to Washington. He appealed to Eckert's patriotism, suggested that Eckert would have more influence in the national capital than in New York, and openly flattered Eckert with the notion that he "would go down in history as a second Newcomb." In a moment of emotional vulnerability that seemed out of place for a naval officer, the superintendent confessed, "I have never wished for anything so hard in my life as that you will accept the position and come to Washington."[37]

Eckert finally accepted the position in the spring of 1940. He left the

37. Wallace Eckert

Columbia Astronomical Computing Bureau in the hands of an assistant and moved to Washington. He arrived at the almanac intending to install IBM equipment in the office and teach the computers the methods that he had developed in New York, but as he settled into his new office, Eckert discovered that he would not have the same kind of resources Columbia University received from Thomas Watson and IBM. With dismay, he reported, "The funds available for punched card equipment were sufficient to form the nucleus of a scientific computing laboratory for part of the year only." But Eckert was not a blind partisan of punched card technology, and he knew that there were other ways of producing an almanac, ways that might even prove to be less expensive than the leases for IBM equipment. Borrowing an idea from L. J. Comrie, Eckert adapted a Burroughs accounting machine to act as a difference engine. Once the staff had prepared the initial values for an astronomical table, the accounting machine could compute the rest. Though the Burroughs machine was not intended for scientific calculations, it proved sufficient to the task at hand. By the end of Eckert's first year in Washington, he could report that "the work of the office as a whole has been advanced two or three months," even though he did not have the equipment or the staff that he desired.[38]

In June 1940, the Mathematical Tables Project stood as the largest scientific computing organization in the United States, dwarfing the combined staff of the Aberdeen Proving Ground, the American *Nautical Almanac,* the Thomas J. Watson Astronomical Computing Bureau, and the mathematics division of Bell Telephone Laboratories. The WPA computing floor was home to three hundred computers. Half of this group still computed with paper and pencil; the remainder pulled the cranks of old adding machines.[39] Another hundred workers checked results, did preparatory calculations, and edited the finished numbers. The group had published six volumes of tables and an equal number of scholarly articles. The planning committee was working on ten volumes more.[40] Yet the project remained on the fringe of the scientific community. Arnold Lowan could not begin his day, could not open his mail, could not talk with his workers without being reminded that he was running a work relief effort. The WPA dictated the hours he worked, the problems he could undertake, and the laborers he could hire. The label of work relief stuck to him like the dirt of the streets, and no matter how hard he stamped his feet, he could not shake the dust from his boots. "The first requisite of a satisfactory organization of science for war is that it must attract first-rate scientists," wrote the historian James Phinney Baxter. "One outstanding man will succeed where ten mediocrities will simply fumble."[41] As the country prepared for war, as R. C. Archibald brought direction and dis-

38. Mathematical Tables Project computers with adding machines

cipline to the Subcommittee on Bibliography of Mathematical Tables and Other Aids to Computation, and as Wallace Eckert acquired difference engines and punched card equipment for the *Nautical Almanac* staff, Lowan felt a constant pressure to show that his project held at least a few first-rate minds and was not a collection of three hundred unemployable mediocrities.

Nothing symbolized professionalism more than the presence of computing equipment, be that equipment punched card tabulators, electric desk calculators, or even the weary Sunstrands. WPA officials were unwilling to grant more machines to the project, even though Arnold Lowan argued that machine-driven calculators would allow him to hire more handicapped workers. A "one armed operator using the new Frieden calculator was able to produce 40% more work than an unimpaired worker using a calculator which is not fully automatic," read one of his proposals.[42] WPA officials responded by asking, "Which important tables will have to be abandoned if it is impossible to secure machines?" and then summarily declining the request.[43]

Lowan was more successful in raising money and acquiring machines when he discovered that he could look beyond the WPA office. Sometime in late 1940 or early 1941, he realized that the Mathematical Tables Project was legally an independent organization that could solicit contracts from any institution, including the project sponsor, the National Bureau of Standards, and the other offices of the United States government. Sensing that there might be an opportunity to do some work for the military, Lowan offered the services of the Mathematical Tables Project to both the army and the navy. He was at first surprised, but ultimately pleased, to discover that both branches of the military employed large computing organizations to support surveyors and the construction of maps.[44] In the spring of 1941, the army offered Lowan a contract to handle the most repetitious of survey calculations, the preparation of map grids. Map grids reconciled the local geometry of left and right, north and south, with the curvature of the earth. They were particularly useful for field commanders when they looked for sites to place artillery batteries or sought ways to march troops across unmarked land. The army needed such grids for areas that were vulnerable to attack—Alaska, Puerto Rico, the Philippines, and the Caribbean basin—and they were willing to make the Mathematical Tables Project a generous offer if it could do the work quickly. The contract included funds for the purchase of new electrical adding machines and the lease of IBM tabulators. "We greatly appreciate your cooperation in providing this equipment," wrote Lyman Briggs, who had to approve the final contract. He added that the calculations for the map grids would be "undertaken as soon as the necessary IBM equipment can be obtained."[45]

The army contract represented a moment of promise at a time when the future of the Mathematical Tables Project had started to grow dim. The WPA, in response to the shift of government funds toward military preparation, was shrinking week by week. In the first six months of 1941, the WPA liquidated half of its assets and released four hundred thousand workers.[46] Just after Lowan consummated the contract with the army, he received the worrisome news that the WPA was considering closing the Mathematical Tables Project.[47] No decision had been made, but Lowan tried to stall a closure by requesting that the project be certified as essential for national defense, a status that could be granted only by the secretary of war or the secretary of the navy. The prospects for such certification were not good, as most certified WPA projects were either military construction efforts or salvage teams that were gathering scrap materials from city dumps.[48]

As he had in times past, Lowan turned to Philip Morse of the Massachusetts Institute of Technology for assistance. Lowan asked Morse to use whatever influence he might have with the office of the secretary of war. "I wish to point out that if the desired classification as a Certified Defense Project is granted by the War Department," he wrote, "many of the difficulties under which we are laboring would be obviated." As an incentive, he pointed out that if they were certified, the computers could finish a large table of interest to Morse and do any other work that he would need for defense projects. Lowan mailed this letter from his home in Brooklyn, not from the Mathematical Tables Project office on Manhattan, and ended the correspondence with the warning, "For obvious reasons, it would be advisable not to mention this letter in your recommendation to the Secretary of War."[49]

For about a year, Lowan had been conducting a clandestine correspondence with Morse, posting the letters from Brooklyn and asking for replies to his home address. Often these letters were paralleled by official messages that were logged in the WPA correspondence file. In such cases, Lowan would tell Morse, "An office letter of somewhat similar contents will reach you . . . in due course."[50] There was nothing illegal about what they were doing. There was no promise of funds, no exchange of favors, no hints of special treatment. Lowan was merely trying to orchestrate support for his project, though he was trying to do so in a way that appeared free from the very actions he undertook.

About a half dozen scientists received unofficial letters from Lowan, including his former mentor at the Institute for Advanced Study, John von Neumann. Lowan asked von Neumann for a letter of recommendation for the project. He liked to publish such letters as a way of attracting new work. Von Neumann was quick to oblige, but he sent a short and tepid letter, the kind of recommendation that one might pen for a distant and slightly disreputable nephew. Lowan responded immediately to von Neumann, firmly rejecting the letter and asking that it be rewritten. "In its present form, . . . the phrase 'To Whom it May Concern' and the general tone of the statement stamp it as being solicited and, therefore, unsuitable for the purpose for which it was intended, as I stated in my previous letter." Not trusting von Neumann to write a proper reply, he dictated the form of the new recommendation. "Include in it a personal letter, addressed to me, say, commenting on the program outlined in the April circular, . . . and perhaps making further suggestions for the computation of new functions of importance in applied mathematics."[51] Without further comment, von Neumann did as he was bidden.

Lowan never saw any contradictions in his approach to promoting the Mathematical Tables Project. He had been trained as a scientist, a searcher for truth, or at least his best understanding of truth, and yet believed that truth was rarely to be found in the political sphere of this world. At the WPA, he had struggled with contradictory demands: the need to produce accurate mathematical tables and the requirement to employ the greatest number of workers, the goal of keeping a large computing floor busy and a managerial structure that made it difficult to prepare new computing plans for that floor. Five separate offices reviewed Lowan's computing plans.[52] Each office could reject plans and require

Lowan to rewrite them. During the two and a half years of operation, Lowan had written emergency telegrams to Washington, begging WPA officials to approve his computing plans. On one occasion he had to file a late telegram, warning, "We have only 10 days work for manual computing group."[53]

In the fall of 1941, the Mathematical Tables Project moved in a twilight world, without enough support to become a permanent operation and yet not facing the kind of opposition that would result in termination. Most scientists, including those engaged in calculation, simply held no strong opinions about the group. When R. C. Archibald tried to rally the MTAC committee to recommend certification for the Mathematical Tables Project, he discovered that the majority of his committee members would not respond to his calls and letters.[54] The only strong voice came from Wallace Eckert at the *Nautical Almanac*. "I am not prepared to certify that the organization should be looked upon as a war necessity whose resources should not be in any wise curtailed," Eckert wrote. He noted that "many worthy programs are being curtailed and even abandoned."[55]

The December 7 attack on Pearl Harbor went unremarked in the records of the Mathematical Tables Project. Monday, December 8th, was much like Friday the 5th. There were worksheets to collect, computing plans to check, agreements to make with suppliers. In three quick days, the country was pulled into the war. "There is a lot of fanfare and excitement about the dramatic occasion," complained the writer Anne Morrow Lindbergh, "and yet I feel chiefly a desperate lack of dignity, lack of seriousness, lack of humility about the whole." Lowan may not have shared Lindbergh's assessment of public attitudes, but in some way, he shared the plight of her husband, the aviator Charles Lindbergh. In December 1941, Charles Lindbergh had no obvious role in the war effort. Until the attack, he had been an ardent opponent of American intervention, a leader of the "America First" movement. Charles "wants terribly to get to work," observed his wife, "constructive work—almost with his hands, work in his craft again, have his contribution count—his experience, his technique, his training."[56]

Like the famous aviator, Lowan desperately wanted to work at his craft, to have his contribution count. Each day he went to work hoping that the notice of defense certification would be lying on the top of his wire in-basket and fearing that the termination letter would be there in its place. Every night, he returned home anticipating a letter of guidance from Phil Morse. A month after the Pearl Harbor attack, Morse urged Lowan to expand his efforts to contact military research laboratories and contractors. There "will have to be built up a technique whereby you can be in more continuous touch with some of the Defense activities," he wrote, "so that your project can be shifted from time to time as the computation needs change." He told Lowan to put operations in the hands of his assistant, Gertrude Blanch, and spend his time meeting with the scientists who needed computing services. This kind of work could not be easily done with an "exchange of letters," Morse explained to Lowan, "but probably should be done by frequent visits to various defense research centers."[57]

Though Lowan was a vigilant publicist, he was not comfortable with the kind of personal contact that Morse suggested. He preferred to work from his Manhattan office, or his Brooklyn home, and promote the project through letters and fliers. He let Morse serve as the public representative of the project, even though he would have benefited from more contact with the members of the National Defense Research Committee.[58] In March of 1942, Morse brought the fate of the Mathematical Tables Project to the attention of the National Defense Research Committee and urged the group to place the project under the control of its Division D, the division whose activities included exterior ballistics, bombing trajectories, and the aiming of guns. This last subject, known as "Fire Control," was developing into a substantial field of mathematical research.[59] The National Defense Research Committee was willing to take responsibility for the Mathematical Tables Project, but they were thwarted by the project's relief status. "Congress is exceedingly jealous of its prerogative in laying down the rules according to which government funds may be spent," reported the committee attorney, and hence "it would be pretty dangerous to attempt [to take control of the Mathematical Tables Project] pending further study of the situation."[60] Accepting this recommendation, the National Defense Research Committee left Lowan with no secure future.

Defense certification came quite unexpectedly to the Mathematical Tables Project in April 1942, but it offered no special status, as almost all of the surviving WPA projects were considered vital to the war.[61] The New York office of the WPA informed Lowan that they were "interested in the work proposed for the coming year" and that they could only support the Mathematical Tables Project "provided workers are available."[62] The deserving poor were increasingly scarce as unemployment fell and WPA workers abandoned the relief projects for better-paying jobs in war industries. The Mathematical Tables Project lost several junior members of its planning committee to local manufacturers. Lowan was able to staunch the flight of these workers by transferring the entire planning committee to the payroll of the National Bureau of Standards and offering them competitive salaries.[63] He was less successful on another problem, the departure of his IBM equipment. The army notified Lowan that they were canceling their contract for map computations and reclaiming the tabulator, the card punches, and the sorters. They could no longer provide resources to a relief project, they explained, for "only our Civil Service employees will be used to operate these machines."[64]

The defense certification did little to strengthen the position of the Mathematical Tables Project, and it raised a new issue that threatened the organization's survival. Shortly after the secretary of the army signed the certification order, he instructed Arnold Lowan and Gertrude Blanch to apply for Federal Bureau of Investigation (FBI) security clearances.[65] In theory, the entire staff would have to be cleared in order to accept defense projects, but the first step was to ensure that the leaders were approved. Lowan and Blanch dutifully complied with the request only to have their applications placed in a security limbo, neither accepted nor rejected. We do not know the doubts that the FBI harbored about Lowan, but we do know that they believed that Blanch was excessively liberal and sympathetic to the Soviet Union. The evidence was circumstantial, but according to FBI analysts, it accumulated into a compelling case. The FBI reported that Blanch had been seen purchasing the *American Worker*, a communist newspaper; that she was a registered member of the American Labor Party, which openly sympathized with communists; and that she had signed campaign petitions for communist political candidates. Unknown to her, the WPA had identified her as "potentially disloyal," a claim of hearsay that was as impossible to refute as it was to prove. The most damning evidence against Blanch was her sister, Elizabeth, with whom she shared an apartment. Elizabeth freely acknowledged that she was a registered member of the Communist Party and actively recruited new members for the organization.[66]

At the time, the Soviet Union was an ally of the United States against Germany, but that was of little matter. The Russian revolution and the Soviet claims of world domination had frightened many in the West. Liberal thinkers argued that the communists would eventually move into the comfortable ranks of the middle class and drop their aggressive ways, but there was little evidence of this in 1941. The Soviet Union was actively spying on its ally, spiriting scientific information back to Moscow in diplomatic pouches and air flights over Alaska. At one point, a Soviet engineer had approached the Columbia Astronomical Computing Bureau and asked for "permission to visit your laboratory in order to receive some information as regards organization and method [*sic*] used there." The request came through Amtorg, the notorious trading company that served as a cover for Soviet spies. The new director of the laboratory forwarded the request to his predecessor, Wallace Eckert. Eckert told the Amtorg official that the Columbia laboratory was open to all but then recounted the story of Boris Numerov, who had visited the laboratory four years before and had subsequently disappeared. "I sincerely hope that his interest in machines was not construed by his government as treason," Eckert wrote. "I am shivering a little bit," wrote the new director after the exchange. "I would not be surprised if I wouldn't hear from them at all, and frankly I just as soon would not."[67]

The Mathematical Tables Project never had any visit from Amtorg or any other known Soviet agency, but the label of "untrustworthy" joined that of "work relief" in the public image of the project. Gertrude Blanch and Arnold Lowan tried to compensate for their lack of security clearances by taking a public loyalty oath and soliciting testimonial letters from friends and supporters, including the faithful Philip Morse.[68] These actions did nothing to improve the reputation of the Mathematical Tables Project, as Blanch and Lowan discovered when they were asked to prepare tables for the LORAN project.

LORAN, an acronym for Long-Range navigation, was a joint project of the Massachusetts Institute of Technology and Bell Telephone Laboratories. It used pairs of special radio stations to guide ships and planes. The two stations would be separated by tens or hundreds of miles, yet they would exchange a special synchronized signal, a little blip that flew back and forth from one station to the other like an electronic tennis ball flying through the ether. Navigators would determine their position by timing the arrival of these electronic blips. From one pair of times, they could determine that their plane or ship was located somewhere on a curved path. If they conducted this operation twice, using two different pairs of stations, they would be able to construct two curves that intersected at the exact point of their position.

MIT engineers demonstrated the LORAN system in June 1942. They used a single pair of radio stations, one located on the tip of Long Island and the other housed in an old Coast Guard Station on the Delaware shore. They placed a LORAN receiver on a navy airship, taught the navigator how to perform the special calculations, and released the craft at Atlantic City. The airship meandered up and down the coastline, going nowhere in particular but giving the navigator a chance to do the arithmetic.[69] When the craft returned to its base, the scientific staff declared the experiment a success, even though the navigator was able to do only one-half of the calculations. "We are ready to begin the computations necessary for the production of [navigation] charts," wrote one of the engineers, but he noted that the navy "cannot now take on the routine but voluminous calculations required." The Bureau of Navigation, which included the Nautical Almanac Office, did not have enough staff to prepare the tables, and there was no other large computing office within the service. Through Phil Morse, the LORAN engineers learned about the

Mathematical Tables Project. "This group would gladly undertake the calculations," commented one engineer, "but they already have a number of other jobs progressing."[70]

Arnold Lowan was pleased to put aside other jobs for the LORAN work. He assigned the navigation calculations to a small team of computers led by Milton Abramowitz (1913–1958), the longest-serving staff member at the Mathematical Tables Project. Lowan had recruited Abramowitz in 1937, when he was a young graduate student at Brooklyn College. When Blanch had come for her initial inspection of the Mathematical Tables Project office, Abramowitz had been working alone at an old table in the big, empty computing room. In the first years of operation, he was the staff sergeant, the floor leader of the group. As he gained experience and completed his graduate studies, he advanced through the project ranks. He led the special computing group, became a key member of the planning committee, and, during the year of army map grid calculations, took command of the project's punched card equipment.

The MIT engineers estimated that "the services of about a dozen of the better computers would be satisfactory" to prepare the tables, but that was not enough.[71] When Abramowitz prepared a sample table, he would regularly appear at Lowan's desk, inquire after the director's health, and politely request an additional computer. Lowan, who spoke with a heavy Romanian accent, would complain that he did not have enough computers to fulfill important requests from the army and navy, but in the end he would give Abramowitz the additional worker. By the early fall, Lowan had lost so many computers to LORAN, or to other projects, that he was no longer able to take new assignments. Looking for a way to strengthen his computing staff, he requested that ten of his computers be allowed to work forty hours per week instead of the thirty-two hours mandated by WPA regulations.[72] The New York City WPA office denied the petition, stating that the "Emergency Relief act is intended to furnish temporary work for needy people. It is supposed to encourage workers to gain private employment."[73]

With a shrinking staff and no dispensation to work a full forty hours, the Mathematical Tables Project struggled ahead as best it could. The members of the planning committee accepted some of the extra work, spending their evenings on the computing floor, finishing the calculations, going over results, trying to get the most from their staff. Their efforts were complicated when Lowan received an urgent letter from Phil Morse that raised the issue of security. "I hope," he wrote, "you have arrangements whereby you are certain that the results of the calculations do not leak out to the outside world without your control."[74] Lowan assured him that human computers had access only to intermediate calculations and that the senior staff had control of the final results. The reason for Morse's concern soon appeared. In early November, the Mathematical Tables Project completed the sample navigation table for the LORAN project after devoting 5,000 hours of computation to the task.[75] The table was reviewed by MIT engineers, who judged the result "very satisfactory" and recommended that the project be given a contract to prepare all of the tables. The navy acknowledged the recommendation but after considering the issues involved, concluded that they would look for some other group to do the calculations. "This decision was based on the security aspects of our work," wrote one engineer, "and was made by Naval Operations."[76]

On December 3, less than two weeks after the LORAN decision, President Franklin Roosevelt announced that the WPA would be terminated and all the projects liquidated. Under the termination plan, the Mathematical Tables Project would operate through March 1943 in order to consolidate its accomplishments and end its activity gracefully. For Arnold Lowan, the news was a blow to the chest, a painful way to end four years of dedicated work. The phrase used by Roosevelt to describe the closure, "honorably discharged," stung with irony, for Lowan wanted nothing more than to contribute his bit to military research. He was not ready to concede that his computing lab would have to die the death of a relief project, so one more time he returned home to Brooklyn, placed a sheet of paper on his table, and wrote to Philip Morse, "I would welcome any suggestions you may have."[77]

When the other members of the Mathematical Tables Project reflected upon the difficult days in late 1942, they liked to believe that L. J. Comrie had saved their jobs. According to Ida Rhodes, L. J. Comrie was listening to the radio on the night of December 3 and heard the news that the WPA would be terminated. "Not bothering to affix his wooden leg, he hopped to the nearest telegraph office and sent a 'hot wire' to President Roosevelt, stating that under no circumstances must the Math Tables Project be allowed to perish—adding some choice language about boondoggling in other projects." Upon receiving this telegram, as Rhodes told the story, Roosevelt was supposed to have ordered an investigation. A government inspector, a young woman in a uniform, was immediately dispatched to New York. She called upon the project, reviewed its accomplishments, and recommended that the National Defense Research Council take responsibility for the group. Her opinion carried weight in the highest offices of government, and so the project was saved.[78]

There is some evidence to support Rhodes's story. Comrie did hear a radio broadcast on December 3 and, a few days later, sent a telegram to President Roosevelt. The message was free of "choice language about boondoogling" but said simply: "British scientists engaged on war work hope you will provide for continued activity of New York Work Projects Administration Mathematical Tables Project."[79] There is no comment on the artificial leg and no record that the telegram made it from Roosevelt to the office of the WPA. There was too large a gap between the responsibilities of the president and the little project that Lowan oversaw. Furthermore, Comrie had no standing with the government of the United States. He had only limited affiliation with American scientific institutions and commanded no American votes. Even if the telegram had reached the WPA office and been entered into its communication log, the program officers would have recognized that Comrie had written several times to the agency and that his last letter had not been so complimentary. Only six month before, Comrie had complained that the computation of the Mathematical Tables Project "seems to me extravagant, and to savour of computing gone amok."[80] Comrie's December telegram reached the Mathematical Tables Project only because Comrie sent a copy of it to Arnold Lowan and Lowan distributed copies of the message to all of his supporters.[81]

In fact, when Comrie sent his urgent telegram, two efforts were under way to commandeer the project for military work. The first came from the navy, which had discovered that there were no other large computing laboratories in the United States that were prepared to handle LORAN computations. The woman Rhodes recalls visiting the project was likely Regina Schlachter, a lieutenant (junior grade) attached to the navy's Hydrographic Office. Schlachter visited the Mathematical Tables Project one month before Comrie sent his telegram to Roosevelt. Schlachter examined the operations of the project, determined that the calculations could be secured, and recommended that the navy claim the services of one mathematician, Milton Abramowitz, sixty calculating machines, ten typists, and forty-nine computers.[82]

With the resources of the Mathematical Tables Project, Lt. Schlachter organized a new computing office, named the New York Hydrographic Project. She found space for the group at the Hudson Terminal Building, a complex of offices in Lower Manhattan that was later redeveloped into the World Trade Center. Workers from the nearby Brooklyn Navy Yard cleared the rooms, reinforced the doors, and installed a safe. At the request of the navy, Lt. Schlachter installed a twenty-four-hour guard at the facility. She deferred all mathematical questions to Abramowitz, who was assisted by an MIT professor with a reserve commission.[83]

The hydrographic computers were a small cross section of New York's population. There were twenty-six men and twenty-three women. The group included a Weinberg, a Sinclair, a Nabokov, an O'Brien, a Dalrimple, and a Cordova. Some of the computers came from Brooklyn, some from the Bronx, and a few from the neighborhood of Harlem. The navy made no attempt to investigate the loyalty of these workers. Instead, it tried to isolate the computers and prevent them from having access to the final tables. The computing sheets made no reference to LORAN or navigation. There was no indication of time, of radio frequencies, or even of longitude and latitude. When computers resigned, the navy tended to replace them with the wives of servicemen, reasoning that women with a personal stake in the success of the military would be unlikely to betray the office.[84]

The actions of the navy were matched by a decision from the National Defense Research Committee. In the fall of 1942, the committee was in the middle of a major reorganization, which expanded the number of divisions. During this effort, the committee created a new division called the Applied Mathematics Panel and put this group under the leadership of Rockefeller Foundation mathematician Warren Weaver.[85] This panel was the offspring of the division devoted to fire control, a division that was also led by Weaver. Weaver had argued for the Applied Mathematics Panel because "the demands to carry out analytical studies kept increasing rapidly"[86] and because, without it, new devices could not be designed, tested, manufactured, and deployed "in time to affect the conduct of the war."[87] In general, the "analytical studies" were expansions of ballistics work, "the mathematical analysis of certain fundamental problems."[88] These studies developed mathematical models for bombing runs, antiaircraft fire, shock wave propagation, and other aspects of weapons operations.

Weaver recruited Thornton Fry of Bell Telephone Laboratories to be the vice-chair of the committee and asked Oswald Veblen to bring his experience from the First World War. In all, about ten mathematicians served on the Applied Mathematics Panel, including Princeton professors S. S. Wilks (1906–1964) and Marston Morse (1892–1977) and New York University mathematician Richard Courant (1888–1972). For a time, the brother of Oswald Veblen's First World War colleague Forest Ray Moulton served as staff to the committee.[89] At the first meeting of the panel in January 1943, Weaver opened the discussions by identifying computing as "a large and important need" and argued that the work of the panel would "doubtless involve several broad contracts with groups such as the Lowan WPA Computing Group, the Thomas J. Watson Astronomical Computing Bureau, the Computing Center at MIT, etc."[90]

The Thomas J. Watson Astronomical Computing Bureau was the new name for Wallace Eckert's old laboratory at Columbia. The Computing Center at MIT was a small group of computers working directly with Phil Morse.[91] As both of these organizations were university research labs, they were obvious candidates for Applied Mathematics Panel contracts. The Mathematical Tables Project was more problematic. Weaver told the first meeting of the panel that he had met with Arnold Lowan and Gertrude Blanch in mid-November, three full weeks before the WPA announced its liquidation. He reported that it was a good discussion and that he came away from the meeting with a better understanding of how the group operated and the kind of work it was able to do. He wanted the panel to take control of the project, though he admitted that no "significant fraction of the group could be cleared" and that the computers would have to be limited to "work of such a general character that could be unclassified."[92]

Among the members of the Applied Mathematics Panel, only Warren Weaver was interested in taking responsibility for the Mathematical Tables Project. Weaver went looking for more support by polling the leaders of the National Defense Research Committee divisions to see if they might make use of a large computing office. On January 14, he sent a mimeographed letter to the nineteen leaders and waited for the replies. In less than two weeks, he had four votes in favor of the Mathematical Tables Project and fifteen votes against. As he reviewed the comments from the different divisions, Weaver decided to ignore the poll and his own concerns about security. "Of my own knowledge," he told the Applied Mathematics Panel, "I can say that if Dr. Lowan's group were disbanded, another group of several dozen computers (at least) would have to be set up in NDRC somewhere to take care of the calculations which Dr. Lowan is at present carrying on for various sections of the NDRC."[93]

For a time, Weaver went looking for an alternative to the Mathematical Tables Project, a computing office that had no history of work relief and no baggage of security problems, but there was no organization with a similar expertise in scientific computation. "I hear that you have considerable computing machines at Vassar [College] and that you have some experience with them," he wrote to a mathematics professor named Grace Hopper (1906–1992). "Are you interested in doing work for the Applied Mathematics Panel?"[94] Hopper was available for the work, but she was hoping to win a commission with the navy. The navy was resisting her overtures, claiming that she was too old, at the age of thirty-six, and was underweight. Undaunted, Hopper persisted and finally received her commission. The navy assigned her to Howard Aiken's computing facility at Harvard, which handled a variety of calculations for weapons and communications research.[95]

Unable to find any alternative to the Mathematical Tables Project, Weaver began to bring the group under the control of the Applied Mathematics Panel and was surprised to discover that Arnold Lowan was circulating his own plan for the project. Lowan proposed two options to the Applied Mathematics Panel. The first would establish a group of six mathematicians and fifty computers; the second would keep the same number of mathematicians but retain only twenty-five computers. When asked about the plan, Weaver confessed that he was "just a little embarrassed by this whole situation." He had no direction from the National Defense Research Committee, and "my only information concerning the budgetary possibilities is contained in an estimate which I believe Dr. Lowan prepared . . . and a copy reached me by an indirect route." After studying the document, Weaver decided to accept the smaller of the two plans. With a single letter, he swept away all the restrictions of the WPA and most of the stigma of work relief. The Mathematical Tables Project would be a contractor to the Applied Mathematics Panel. It would operate as an office of the National Bureau of Standards, though Lowan would take his orders from the panel. The agenda for the group was no longer set by a New Deal agency in Washington but by a committee of well-respected mathematicians.[96]

Even though Arnold Lowan would have to sacrifice seventy-five computers, he was pleased with the offer from the Applied Mathematics Panel and told Philip Morse that this "very satisfactory arrangement is unquestionably due to a great extent to your constant efforts on our behalf, for which please accept the expression of our warm gratitude."[97] Lowan had about two weeks to finish his obligations to the WPA and prepare for his new assignment. March 15 was the last day of operation as a relief project. The computers packed their equipment, burned old computing sheets, and disposed of the posters that had guided them. As the Mathematical Tables Project shed the seventy-five computers, it also shed the manual computing division. When the project opened its new office, all computers would use adding machines or mechanical calculators.[98] Ida Rhodes would identify this move as the time that "life began,"[99] the moment when the Mathematical Tables Project finally dropped the trappings of work relief and become a professional computing organization.

CHAPTER SIXTEEN

The Midtown New York Glide Bomb Club

> I am asked to think out an abstract problem when I am very
> tired out with a multitude of infinitesimal concrete and
> immediate problems. . . .
> Anne Morrow Lindbergh, *War Within and Without* (1943)

THE WINTER OF 1943 marked the start of the imperial age of the human computer, the era of great growth for scientific computing laboratories. It seemed as if all the combatants discovered a need for organized computing that winter. A German group started preparing mathematical tables at the Technische Hochschule in Darmstadt.[1] Japan, which had received material from the Mathematical Tables Project through 1942, formed a computing group in Tokyo.[2] The British government operated computing groups in Bath, Wynton, Cambridge, and London.[3] Within the United States, there were at least twenty computing organizations at work that winter, including laboratories in Washington, Hampton Roads, Aberdeen, Philadelphia, Providence, Princeton, Pasadena, Ames, Lynn, Los Alamos, Dahlgren, Chicago, Oak Ridge, and New York City. Most of these calculating staffs were small, consisting of five to ten computers. Langley Field, a major aeronautical research center in Virginia, employed about a dozen such groups, each assigned to a specific research division. "Some [groups] have as many as ten computers," explained a history of the center, "while others have one computer who often devotes a part of her time to typing and secretarial duties."[4] Only a few computing laboratories were as large as the New York Hydrographic Project with its forty-nine veterans of the Mathematical Tables Project or the thirty-person computing office of the Naval Weapons Laboratory at Dahlgren, Virginia.[5]

Amidst this growth of computing offices, the MTAC committee finally came to life and began to chronicle the literature of calculation. Nearly eighteen months after his confrontation with Luther Eisenhart and the National Research Council forced him to retreat, R. C. Archibald had returned to his post in the summer of 1942 and announced a new goal for the committee. "Our *Guides* are very slow in appearing," he wrote. "Hence I have been led to the conviction that it would be very desirable to establish a quarterly publication called *Mathematical Tables*" in order to circulate the committee's bibliographies and reports.[6] The proposal surprised the members of the National Research Council. It would "be like issuing a professional journal," complained the permanent secretary.[7] After the argument over the $61.73, the council members were uncomfortable with the idea and tentatively tried to check Archibald. They approached Warren Weaver, in his role with the Rockefeller Foundation, and asked if the money granted to Archibald by the foundation could be used to finance a publication. Weaver confessed that he had a "certain horror" at being associated with Archibald's idea, but he also stated that the Rockefeller Foundation would not stop the new periodical.[8] The members of the MTAC committee, from L. J. Comrie to Charlotte Krampe, were slower to respond, but they generally liked the proposal. Wallace Eckert wrote that the periodical "would serve a very useful purpose" but warned that it "would probably become a financial headache" and that the "present is not the most auspicious time to start it."[9]

Never one to wait for favorable times, Archibald pressed ahead, leaving even the most sympathetic members of his committee behind. "With the load I have to carry," he wrote to the MTAC committee, "I can not possibly undertake either to discuss everything with you before hand or send all copy to you before publication."[10] He completed the first issue of the journal, entitled *Mathematical Tables and Other Aids to Computation,* in February 1943. To his credit, he recognized that the journal could not flourish if he was the sole contributor and apologized to his readers, "R. C. A. greatly regrets the apparent necessity for numerous personal contributions in this issue, as well as in the second."[11] The issue contained a great deal of useful information, including lists of tables, errata, book reviews, and articles on methods of calculation. The only thing that seemed out of place was a piece devoted to the computing machines of the seventeenth century, a favorite subject of Archibald's.

Mathematical Tables and Other Aids to Computation provided American computers with the first systematic reports on computing activities. Before the journal reached a wide audience, many computers did not know what organizations existed and what work was being done. In early February 1943, the members of a new computing group at the University of Pennsylvania did not know how they might contact the Mathematical Tables Project. One of the group's leaders, John Brainerd (1904–1988), sent a letter to the project sponsor, Lyman Briggs at the National Bureau of Standards. Brainerd explained that he was undertaking a large computing effort for the Aberdeen Proving Ground and was searching for human computers and computing expertise. He hoped that the Mathematical Tables Project was still operating and that it might provide him with human computers or handle some of his calculations or provide him with training materials.[12] Brainerd needed especially sophisticated com-

puters, computers with a good background in mathematics. The Mathematical Tables Project might have seemed an unlikely source of such
computers, but Gertrude Blanch had initiated an extensive training program in 1941. She and other members of the planning committee developed a series of eight mathematics courses, which they offered over the
lunch hour. The first course discussed the properties of elementary arithmetic; the intermediate ones covered standard high school algebra,
trigonometry, and college calculus; the final course presented the methods of the planning committee: matrix calculations, the theory of differences, and special functions. The teachers treated the courses as a formal
school, requiring the students to attend every session and asking them to
"do a reasonable amount of 'home work' on their own time."[13]

Lyman Briggs replied to Brainerd's letter just as the Mathematical Tables Project was preparing to move from its old WPA office to the rooms
rented by the Applied Mathematics Panel. He explained to Brainerd that
the project had found a home for the duration of the war and was able to
accept outside assignments. "I think you will be glad," Brainerd told his
colleagues, "to note the action which is being taken in connection with
the computation project."[14] The enthusiasm of this initial contact quickly
faded as the leaders of the two computing organizations employed different strategies in their work. Arnold Lowan, of the Mathematical Tables Project, was a classical physicist who understood the rules of divided
labor. For him, computing machinery was an aid that "facilitated and
abridged" the efforts of his staff. Brainerd was a professor of electrical
engineering at the University of Pennsylvania. He organized his office
around a large computing machine, a differential analyzer, and used
human labor to compensate for the machine's shortcomings. Brainerd's
computers were machine operators, as George Stibitz had prophesied,
but these operators were not mere drudges, for they needed a thorough
mathematical education in order to do their work.

In 1937, the University of Pennsylvania had acquired a differential analyzer in conjunction with the Aberdeen Proving Ground. The proving
ground had financed the differential analyzer under an agreement that allowed ballistics researchers to use this machine in times of war. Until the
spring of 1942, the analyzer had been used by engineering professors and
graduate students. Like most university research equipment, this machine
received regular but intermittent use. Four or five times a term, it would
calculate a curve associated with some electrical component or circuit.
Occasionally, it would serve as the object of an experiment by a graduate
student interested in electromechanical controls. Once or twice a year,
the university was able to rent the device to a local company. For other
periods, the machine stood idle, gathering dust and dripping oil.[15]

In June 1942, proving ground officials notified the University of Pennsylvania that they needed to use the differential analyzer for ballistics research and offered to reimburse the school $3.00 an hour for operational costs: electricity, the wages of mechanics, supplies, and the salaries of any staff that were needed to oversee the calculations. A small group of Aberdeen researchers took the train north from the proving ground to inspect the machine. The analyzer was housed in a nondescript brick building just a few blocks from the railroad station. Their first test of the machine, a trajectory for 4.7" antiaircraft shells, was disappointing. "Upon arrival," wrote a member of the Aberdeen staff, "it was apparent that a desirable rate of analyzer output had not been achieved." The output from the machine substantially deviated from a hand-calculated trajectory. "The Philadelphia analyzer . . . has not been under the compulsion of the great accuracy demanded at Aberdeen," observed a proving ground researcher, "and therefore has not been as assiduously cared for as the Aberdeen analyzer." To "attempt to maintain [high accuracy] with the Philadelphia analyzer," concluded the army, "required an exorbitantly high number of adjustments and test runs."[16]

At a hastily called meeting between university officials and army officers, John Brainerd presented a plan that would produce results within 0.5 percent of hand-computed values. This plan called for a few modifications to the machine, strict operational standards, and a staff of human computers to oversee every step of the calculation.[17] An early test of the new procedure achieved the specified accuracy but at the cost of substantial hand calculation. It is "desirable to expand the Philadelphia unit somewhat at once," concluded the army, in order "to train and prepare its personnel for handling the contemplated output of the analyzer." The calm words of the military report camouflaged the problem facing the Pennsylvania faculty. The university did not have enough college-educated computers for its analyzer staff. They had hoped to find twenty to thirty women with bachelor's degrees in mathematics or physics, but after scouring the school's alumna lists, they had identified only eight who held the appropriate degree. Brainerd had offered each of them a position as an assistant computer with a salary of $1,620 per annum, but he believed that no more than three or four would accept these positions. With no other obvious options, Brainerd concluded that the university would have to prepare a curriculum for human computers and operate training classes.[18] He found money for this endeavor at the government's Engineering, Science, and Management War Training Program and borrowed course materials from Aberdeen veteran Gilbert Bliss, who had taught ballistics classes to civilians at the University of Chicago.[19]

Brainerd's plan had serious problems, as he freely admitted. The university lacked enough instructors qualified to teach mathematical ballistics. The only faculty willing to train the women were three retired professors, whom many judged "no longer up to the strain of teaching day long courses."[20] Nothing improved the prospects for the training courses until Adele Goldstine (1920–1964) walked into Brainerd's office in September. Goldstine was the wife of the officer that the army had assigned to monitor the computing work at the university. She was a slight woman but poised and filled with energy. She had the education that Brainerd needed, a bachelor's degree in mathematics from Hunter College for Women in New York City, a master's degree from the University of Michigan, and a connection to mathematical ballistics. Her husband had studied with Gilbert Bliss and had helped Bliss prepare a textbook on mathematical ballistics.[21] Within a few weeks of her arrival, Adele Goldstine had taken command of the training program. According to her husband, she immediately "got rid of the deadwood," the three retired professors, replaced them with two younger instructors, and helped teach the first group of students, twenty-one in number. These students completed the training that fall, swore the required oath of allegiance, and started work as computers.[22]

From the start, the University of Pennsylvania recruited only "women college graduates." The sign "Women Only" marked the door of the computing office, which was a converted fraternity house.[23] This decision was not based on any dictate from the Ballistics Research Laboratory, for the Aberdeen computing staff included both men and women.[24] In all likelihood, it was motivated by common stereotypes concerning office work and gender: that men were difficult to recruit for office work in wartime, that single-gender office staffs were easier to manage then mixed-gender staffs, that women were somehow specially suited for calculating.

Between classes, Goldstine spent much of her time recruiting potential computers. By the winter of 1943, John Brainerd had concluded that the university needed a staff two or three times larger than his initial estimate. They might require seventy or even eighty computers to keep the differential analyzer fully occupied and have a sufficient number of workers in reserve. In the winter of 1943, the school had less than half that number.[25] Brainerd returned to the University of Pennsylvania alumna lists, sent circulars to the American Mathematics Society and the American Association of University Women, and wrote to university faculty to ask the professors to volunteer their daughters or their daughters' friends.[26] As a last resort, Goldstine took to the road, visiting Bryn Mawr and Swathmore Colleges in suburban Philadelphia, Goucher College in Baltimore, Douglass in New Jersey, and her own Hunter College.[27] "I've arranged to be at Queens College Tues[day]," she wrote to Brainerd from a hotel in New York, but she confessed that she did not expect much, as "next week is exam week. Also I was not able to arrange for any very effective means of advertising the job." Even when she was able to notify students of the opportunities at the University of Pennsylvania, she found few interested applicants. At one college, she found "only 25 or so women seniors all of whom have good prospects in their own fields and so probably could not be enticed by our offer."[28]

Goldstine returned from her travels just as R. C. Archibald was printing the second issue of *Mathematical Tables and Other Aids to Computation*. Though much of the publication was devoted to traditional mathematical tables, a few articles at the back dealt with computing machinery. The first, by L. J. Comrie, explained how traditional business machines could be adapted to scientific computation. The second, by Bell Telephone Laboratories mathematician Claude Shannon, discussed the operations of the differential analyzer.[29] With human computers hard to find and an old analyzer struggling to meet the precision requirements, the computing staff was looking to build an improved computing machine, an electronic version of the differential analyzer. This new machine, tentatively called the Electronic Numerical Integrator and Computer, or ENIAC, would have no mechanical parts that could slip or jam or in some other way induce inaccuracy.[30]

Even though he had an embarrassing departure from the British Nautical Almanac Office, L. J. Comrie remained the single most important source of computing information for English scientists. His company, Scientific Computing Service Ltd., was one of four major organizations that were handling ballistics, ordnance, and navigation calculations for the British government. The second group was the British Nautical Almanac Office. Like their American counterparts, these computers no longer shared the burden of producing an almanac with the French and Germans and hence had an extra burden of calculation. The third and fourth groups were the computing laboratories at the University of Manchester and Cambridge University. These two schools owned and operated differential analyzers, just like the University of Pennsylvania.[31]

In the winter of 1943, the British government formed a fifth computing office, one that could undertake general-purpose calculations for both the military and the war industries. The group, called the Admiralty Computing Service, was the creation of Donald Sadler (1908–1987), Comrie's replacement at the *Nautical Almanac,* and John Todd (1915–), a professor at King's College. Todd had been educated at Cambridge under the watchful eye of John Littlewood, the mathematician who had developed ballistics theories in the First World War.[32] Todd had taken a modest interest in computing problems as a student, but he did not become fully involved with computational mathematics until 1938, when he met L. J. Comrie at a meeting of the British Association for the Advancement of Science. Comrie befriended the young mathematician, in-

39. John Todd of the Admiralty Computing Service

troduced him to the association's Mathematical Tables Committee, and eventually taught him the operation of the Brunsviga calculator, repeating the lessons that he had learned from Karl Pearson.[33]

Todd had been drafted at the start of the war and assigned to a naval office that was studying ways of protecting ships from German mines. "I found the work boring," he recalled, "and it was not very effective."[34] Todd's wife, the mathematician Olga Taussky (1906–1995), analyzed the vibration in aircraft structures for a government ministry. Her work produced large systems of equations with unknown values. "A large group of young girls," she related, "drafted into war work, did the calculations on hand-operated machines."[35] From observing his wife and reflecting on his own experiences, Todd concluded that the war effort would benefit from a general-purpose computing organization. "I realized that pure mathematicians, such as I," he later wrote, "could be more useful in dealing with computational matters and relieve those with applied training and interests from what they considered as chores."[36] In private, he was a little more pointed. "After a year of working for the navy, I decided that mathematicians could make tables better than physicists."[37] He joined forces with the almanac director, Donald Sadler, and created the Admiralty Computing Service.

Unlike most other computing offices, the Admiralty Computing Service had two divisions, a staff of ten computers and a small group of mathematical analysts. The computers were managed by Sadler and were located in Bath, the eighteenth-century resort town where the almanac had been evacuated for the duration of the war. They occupied a prefabricated military building situated on an old Georgian estate. The computing staff consisted of students and young teachers, most of them male, who were unable or unwilling to serve in the military. Sadler described them as "nurtured by comprehensive special training," as the skill of computation "cannot adequately be 'picked up' in the course of day-to-day work."[38] They worked three to a room in their military hut, sharing tired desks and improvised tables. Their equipment, worn but serviceable, came from the Greenwich office of the almanac and included L. J. Comrie's old National Accounting Machine. Work began at eight in the morning, ended at five in the evening, and continued for a half day on Saturday. The weekly schedule included time for instruction, discussion, and a meeting for the review of results. "Sadler was a real martinet for getting rid of errors," one computer recalled. "If you made a mistake on some work and if it went out, he'd give you such a dressing down that the whole office would know."[39]

The second division of the Admiralty Computing Service, the mathematical analysts, worked in London and were overseen by Todd. London was a dangerous place, but it was also the home of the major scientific and engineering offices. "[John and I] moved 18 times during the war," Olga Taussky later explained to a friend, the First World War computer Frances Cave-Browne-Cave, "and our belongings were hit by a flying bomb."[40] Moving past the damaged buildings and the rubble in the street, Todd traveled from office to office, talking with engineers, listening to government officials, reviewing military plans. "Often we could not help them with the problems they first presented to us," he recalled, "but I usually found a different problem that we could do."[41] For all that his clients knew, the computations were done somewhere beyond Paddington train station, where Todd began his journeys to Bath. Once or twice a week he would pass through the station, carrying requests for calculations and returning with finished results.

In the spring of 1943, Todd made the trip to Bath with John von Neumann, who had come to England in order to inspect British scientific efforts. Von Neumann then was working for the Ballistics Research Laboratory at Aberdeen and other American research projects. He had requested an opportunity to see the computing facility and L. J. Comrie's famous accounting machine. Todd and von Neumann spent a day in Bath, talking with the computers and observing the operation of the office. On the trip back to London, the two of them discussed a new way of doing interpolation with the accounting machine. The train windows were blackened to avoid drawing the attention of German aircraft, so the two mathematicians had no distractions in the passing scenery. Taking out a piece of the "rather poor quality paper issued to government scientists at that time," they began to prepare a computing plan. They worked as the train passed the royal castle at Windsor, the munitions plants at Slough, and the shuttered shops of Ealing. By the time the dark coaches reached the London station, they had completed their work. "It was a fixed program," Todd wrote, but it did not quite eliminate the need for computers, as "it involved a lot of human intervention."[42] The experience intrigued von Neumann in a way that five years of circulars from the Mathematical Tables Project had not. Von Neumann had kept his distance from the WPA computing floor in Lower Manhattan even though he had promised to respond to Lowan's letters. With one trip to Bath, his views changed. "It is not necessary for me to tell you what [our visits] meant to me," he wrote to Todd after the war, "and that, in particular, I received at that period, a decisive impulse which determined my interest in computing machines."[43]

The two parts of the Admiralty Computing Service had obvious counterparts in the contractors of the Applied Mathematics Panel, but the American effort was far more complicated than John Todd's organization. In England, Todd was free to make most of the key decisions, but in the United States, all requests for mathematical and computational assistance were reviewed by a committee of mathematicians. This executive committee met weekly in the conference room of the Rockefeller Foundation, a sumptuous private suite on the 64th floor of Rockefeller Center's RCA Building.[44] Meetings would begin with a luncheon, which gave the members an opportunity to chat about the issues of the day and discuss new developments in mathematics. At an early meeting, while Rockefeller Center waiters poured drinks and brought the plates of food, one member complained of "American indifference to the German 60 ton rocket," which he described as a false faith that the Atlantic Ocean would protect the country.[45] The mathematicians generally agreed that the German missile program was worrisome, yet they had also concluded that "the bombing of New York would be futile since an explosion outside of a building would break windows but not damage the structure itself, except very old brick types of structure."[46]

For the most part, the luncheon conversations were an opportunity to return to the summer of 1918, the season at Aberdeen that Norbert Weiner had likened to a term at an English college. In that conference room, they would not think about the carrier battles of the South Pacific, the soldiers described by Ernie Pyle in his dispatches from Europe, or the Willies and Joes that cartoonist Bill Mauldin drew for Stars and Stripes. Instead, the executive committee would turn their attention to their fa-

40. Warren Weaver being decorated for his service on the Applied Mathematics Panel

vorite topic, mathematics. "At lunch, there was an interesting discussion of the character of 'probability,'" reported the Applied Mathematics Panel chair, Warren Weaver, after a meeting in the spring of 1943. Probability had become important to several projects before the panel, but practical applications were not the subject of conversation. The mathematicians were interested in the philosophical foundation of chance. Some at the meeting argued that there was no such thing as a random event and that probability was nothing more than a clever use of set theory. "As frequently happens," Weaver observed, "the argument settled down to the question of the most useful definition or connotation of words."[47]

Once the meal was finished and the dessert dishes cleared from the table, the mathematicians turned to their business. The first items on the agenda were reports from the panel's major contractors. Most of these organizations were located in or near New York City. One member of the panel mockingly referred to his research group as the "Mid Town New York Glide Bomb Club."[48] The largest contractor was Columbia University, which was home to four different research centers: an applied mathematics group, a statistical group, a bombing studies group, and the Thomas J. Watson Astronomical Computing Bureau. It alone accounted for half of the Applied Mathematics Panel budget. In 1943, the remaining budget, save 5 percent, was spent within a one-hundred-and-fifty-mile radius of Manhattan.[49] Recipients of the panel's largesse included the Mathematical Tables Project, New York University, Brown University, Princeton University, and the Institute for Advanced Study. There were no contracts with the University of Chicago, none with Iowa State College, and none with the University of Michigan. There was no contract for the Harvard mathematics department, until one of its faculty began raising a public fuss about the panel and Warren Weaver responded by offering the Massachusetts school a token assignment.[50]

After reviewing the contractor reports, the mathematicians of the Applied Mathematics Panel turned to new requests for mathematical work. Some of the requests came directly from the military, but most originated at the war research laboratories. Each member of the panel was responsible for working with a division of the National Defense Research Committee and identifying potential projects for the Applied Mathematics Panel. As they discussed the new requests, the panel members would sketch a rough solution to the problem. Most of these solutions required little mathematical skill beyond that taught to undergraduates, but they usually demanded attention to details and the careful consideration of special situations. By the time each discussion ended, the panel members would have a sense of the effort required for the problem, the kind of individual who might handle the work, and the value of the result. They declined several projects on the grounds that they were not worth the effort.[51]

After they had accepted a request, the panel would assign it to one of their contractors, such as the Applied Mathematics Group at Columbia University or the Bombing Analysis Group at Princeton University. The research staff of the contractors were generally young professors or graduate students. Most were mathematicians, though several were economists, such as Milton Friedman, or engineers, like Julian Bigelow. Weaver characterized the contractor staff as "high grade persons who may admittedly not be geniuses, but who have unfailing energy, curiosity and imagination, and a reasonable set of technical tools." The Applied Mathematics Panel rarely quibbled about the specific training of its research staff, though it did note that the most important quality for success was "the unselfish willingness to work at someone else's problem."[52]

Each of the contractors maintained some kind of computing staff. The three mathematics groups at Columbia employed twenty human computers plus the tabulating machine operators at the Thomas J. Watson Astronomical Computing Bureau.[53] New York University built its computing staff around a young graduate student, Eugene Isaacson (1919–). Isaacson was first exposed to the methods of computation at the Mathematical Tables Project. "I learned about using the mechanical calculators and about computing from this fine team," he would later assert. At first, he worked by himself, but he was soon assisted by Nerina Runge Courant. Courant took up the work not as a student but as a wife and a daughter. She was married to the head researcher at the university, Richard Courant, and was the daughter of Carl Runge (1856–1927), a mathematician who had refined the methods of solving differential equations. Her connections were a tie to the past, a reminder that fathers and husbands had once provided opportunities for women at the American *Nautical Almanac* and the Harvard Observatory. In this war, the opportunity came from the scale of the mobilization, not from family relations. When the New York University computing office expanded, adding six more computers, the school appointed Isaacson to lead the group, not Nerina Courant.[54]

The Mathematical Tables Project served as the reserve computing unit for the Applied Mathematics Panel and was the second-largest item on the panel's budget.[55] Normally, the panel would review all computing requests and forward the ones that they approved to Arnold Lowan. For some large problems, they would occasionally solicit competitive bids. In the summer of 1943, they asked three computing groups to estimate the amount of time required to produce a table of complex numbers. Responding for the Mathematical Tables Project, Lowan wrote that it would take about twelve weeks to prepare the table and check the results. At the Thomas J. Watson Astronomical Computing Bureau, the bid was prepared by Jan Schilt (1896–1982), the astronomer who had replaced Wallace Eckert as director. Schilt's analysis suggested that the bureau's mechanical tabulators could prepare the tables in eight and a half weeks, though the time would double if the Applied Mathematics Panel wanted the calculations checked for machine errors. The last bid came from the IBM Corporation. A company engineer estimated that IBM could complete the calculations in only seven and a half weeks and argued that there was no need to duplicate the calculations to check the results. The Applied Mathematics Panel did not see an obvious choice among the three proposals. At length, they decided that the table was not worth the expense and abandoned it.[56]

Through the middle of the war, the Applied Mathematics Panel found that the expense of human computers was close to the cost of machine calculation. In a competitive bid between the Mathematical Tables Project and Bell Telephone Laboratories, Arnold Lowan estimated that it would cost $1,000 for his staff to do the work, while the computers under Thornton Fry stated that the calculations would require $3,000. The panel rejected both bids, judging that they exceeded the value of the computation. An engineer at Bell Telephone Laboratories decided to produce a machine that would "automatically grind out and record results, using third order differences." When the news of the machine reached Warren Weaver at the Applied Mathematics Panel, he ruefully noted that the "estimated cost of [the] Gadget [was] about $3,000," three times the bid from the Mathematical Tables Project.[57]

Computing machines were more efficient than human computers only when they could operate continuously, when they could do repeated calculations without special preparations. A punched card tabulator could work much faster than a human being, but this advantage was lost if an operator had to spend days preparing the machine. The differential analyzer was proving to be a good way of preparing ballistics tables only when it could compute trajectory after trajectory with little change to the machine. The problem of solving linear equations offered the same kind of opportunity for mechanical computation, as the rules for solving such equations did not change from problem to problem. In the fall of 1943, the Applied Mathematics Panel received a request from the Army Signal Corps to compute twenty-six values from twenty-six equations. Warren Weaver noted that the scale of this problem was remarkably close to the capacity of a machine proposed by his former student, the Iowa State College professor John Atanasoff. "We have recently run into problems which necessitate the rapid solution of systems of linear algebraic equations," he wrote to the dean of Iowa State College. "Could you inform me concerning the status of the electrical machine which Atanasoff designed for this purpose?"[58] The dean replied that the request had come too late. All that remained of the machine was a pile of scrap metal, a box of salvaged circuit parts, and the two drums that had once served as the machine's memory.[59]

Some accounts of the Applied Mathematics Panel describe the work as if it were accomplished under battlefield deadlines with late-night mathematical analyses and forced marches at computing machines.[60] In fact, much of the work, mathematical and computational, was done under strikingly ordinary conditions. For most of the war, the computers of the Mathematical Tables Project were able to work a standard shift, beginning their days at eight in the morning, ending at five, and taking an hour for lunch. Except at a few moments of crisis, the computers spent about 30 percent of their time finishing tables that they had begun under the

WPA. "Gertrude Blanch abhorred a vacuum," recalled one computer,[61] so she used the old projects to keep the computers busy.[62] The project also acted as the reserve staff for LORAN. By the summer of 1943, it provided the New York Hydrographic Office with a couple of computers each day, as well as typists, secretaries, and proofreaders.[63] Arnold Lowan kept a running tally of the debt, which eventually amounted to 3,150 days of labor.[64]

In the late fall of 1943, the Mathematical Tables Project experienced a brief season of double-shift work, a period when the computers began calculating at 8:00 AM, finished at midnight, and went home through empty streets and cold night air. It was an experience that pulled them together and made them feel connected to the soldiers who were training for the invasion of France. They took pride in the knowledge that the calculations were intended for planners of Operation Overlord, the code name for the D-day invasion of Europe. This assignment had its origin in a bombing sortie that had failed to reach its target in France. Before heading back across the English Channel, one plane had lightened its load by dropping its bombs over the beaches of Normandy. The crew reported that their actions "set off a strange series of explosions" in the area, indicating that the beaches were probably mined. This news would have been unremarkable except for the fact that Normandy was the planned site for the D-day invasion. When the Overlord planners received this news, they decided to prepare a bombing mission to clear the defenses. The planes would drop high explosives on the beach and rely on the shock waves to detonate the mines.[65] To prepare this operation, the planners requested tables that would estimate the number of mines that could be cleared by a squadron of planes.[66]

The Applied Mathematics Panel approved the request for beach-clearing tables in the fall of 1943 and assigned it to Jerzy Neyman (1894–1981), a statistician at the University of California. In many ways, he was a poor choice for the Applied Mathematics Panel, as more than one historian has noted. Neyman disliked working with the military and had "a tendency to postpone the computational chores assigned him by the panel" and instead pursue "highly general theoretical studies of great interest to statisticians but little use to practical-minded generals."[67] He even had trouble working with other mathematicians. Neyman had originally been a subcontractor to the Princeton University research effort, but in the fall of 1943, he had called upon Warren Weaver to ask if he could be treated as an independent researcher. Weaver recorded that Neyman engaged in "considerable hemming and hawing, considerable artificial emphasis on the fact that [the Princeton mathematicians] are 'good fellows,'" as he found the courage to explain that he had "no affinity" for the Princetonians. Weaver's assistant, who knew that the relations between Neyman and the Princeton group had caused considerable problems for her boss, recorded that Weaver "keeps his face reasonably straight, and expresses the opinion that it may barely be possible to work out some sort of a divorce."[68]

For the mine-clearing problem, Neyman used a statistical model for "train bombing," the practice of dropping bombs from a plane at regular intervals. He treated the train of bombs as a problem of geometric probability. The bombs became circles, which fell to their target like a handful of coins dropped on a tile floor. Some of the circles fell to the left, some to the right; some grew large, others shrank to a dot. Neyman's analysis estimated the number of handfuls that would be required in order to cover the floor.[69] The analysis required a substantial amount of computation to move from coins on the floor to bombing tables, more than Neyman could handle by himself. He had a small computing staff in California, six students and an assistant, who shared five computing machines.[70] These students could handle small projects, but like Neyman, they were more interested in the mathematics than in the calculation and tended to defer their numerical work until the late evening hours.[71]

Weaver had first tried to find a punched card facility to do the mine-clearing calculations. He talked with three different groups, the University of California business office, the laboratory of chemist Linus Pauling (1901–1994) at the California Institute of Technology, and the Thomas J. Watson Astronomical Computing Bureau in New York. The University of California was unable to take the work, but the other two offices welcomed the task.[72] "We would be very glad to team up with Neyman on any project that seems worthwhile to you," Pauling told Weaver, adding, "The men here . . . have had now a great deal of experience with the use of punched card machines for mathematical calculations."[73] The Watson Laboratory reported that they were doing some work for Wallace Eckert at the *Nautical Almanac*, "but they seem to think that this could be put to one side." Weaver urged Neyman to send his analysis to the Watson Astronomical Computing Bureau, as the "costs are exceedingly moderate due to the fact that the IBM company furnishes all equipment, etc. so that we would need to pay only stipends of the people involved and consumed supplies."[74]

In the end, neither Pauling's lab nor the Watson Astronomical Computing Bureau handled the computation. Warren Weaver assigned the job to the Mathematical Tables Project, and Gertrude Blanch prepared the computing plan.[75] At first, Blanch believed the work could be accomplished by a handful of her workers. Following the progress from California, Neyman soon realized that Blanch's plan did not capture his intent. "Soon after the computations were started, it appeared necessary to alter the program," he reported to Weaver, "which means in fact to extend it." The new plan required more effort from the Mathematical Tables Project computers. Before long, the entire staff was spending two full shifts working on nothing but Neyman's calculations. "I am sorry for underestimating the amount of computations done by Dr. Lowan," Neyman apologized. In all, the calculations had consumed twenty-three times the labor that he had anticipated.[76] The final report was completed, after three full weeks of labor, on December 17, 1943.[77]

As with many of the war computations produced by the Mathematical Tables Project, Blanch and Lowan sent their results to the Applied Mathematics Panel and had only the vaguest idea how they would be used. It was like sending offspring into the world and never knowing what these children would accomplish, what trials they would face, where they would make their home. At times, Lowan would comfort himself, thinking that this work was a humble but key part of the war effort. It was like the proverbial nail which, if lost, would cause the loss of a horseshoe and set in motion a chain of disasters that would precipitate the loss of a horse, a rider, and ultimately the battle itself. Lowan desperately wanted to connect the Mathematical Tables Project to the successes of the war, and so he avoided the moments of sober contemplation, which would have reminded a more secure leader that a horseshoe is generally affixed to the hoof not with a single nail but with six. Neyman's tables represented but one way of preparing the landing site at Normandy. After surveying the beaches more closely, the planners of Operation Overlord concluded that there were no mines blocking the invasion. The bombers that would have been assigned to mine clearing were deployed against artillery batteries.[78] The computations were filed away and never used.

Just before the turn of the new year, the Applied Mathematics Panel was approached by a commander from the navy's Bureau of Ordnance. The officer reported that the bureau wished to purchase a computing machine to handle exterior ballistics calculations, but they had "absolutely no one who can survey the machines available." Their scientists were "inclined to favor one that uses digital computation," but they knew little about such devices. The commander asked the panel members to prepare a report on computing machines and make a recommendation to the navy. The commander's superiors indicated that the Bureau of Ordnance would need "the backing of an Applied Mathematics Panel recommendation in order to secure a satisfactory machine."[79]

This request was awkward for Weaver. As much as he wanted to prepare a survey of computing machines, he believed that none of the Applied Mathematics Panel scientists could produce such a report without bias. George Stibitz was the panel member best prepared to write such a report, but he was predisposed to electric machines built from relays, such as his complex calculator. That winter, he was designing a second machine with the technology. This device was an interpolator, a machine that could compute intermediate values of a function.[80] After weighing the virtues of expertise against the problems of conflicted interests, Warren Weaver asked Stibitz to prepare the review. In an attempt to ensure that the report was balanced, he asked the Bell Telephone Laboratories researcher to work with a committee that included a naval officer and an MIT professor, whom he characterized as being "familiar with the electronic type of computer and with the IBM equipment."[81]

Stibitz's committee restricted their attention to large machines, such as the calculator that Howard Aiken had begun in 1938. This machine, which had been under construction for much of the war, was nearing completion at an IBM factory. IBM engineers had tested large parts of the device and were preparing to ship it to Harvard. The committee also considered the differential analyzers that were operating at MIT, Aberdeen, and the University of Pennsylvania. The Stibitz committee ignored the ENIAC, the digital differential analyzer under construction at Pennsylvania.[82] The project was far from finished, and hence there was not much to report. It was still classified by the army, but those outsiders who knew about it had doubts about its future. Its lead designers, J. Presper Eckert (1919–1995) and John Mauchly (1907–1980), did not have much of a pedigree. Mauchly was a former teacher at a small religious college outside of Philadelphia. He had been introduced to computational problems during the Depression, when he had organized a statistical laboratory with National Youth Administration funds.[83] Eckert was a recent graduate of the university's electrical engineering program. He had been known as a clever student, but he had not been at the top of his class, nor had he ever built a large machine.[84]

The report did not have much influence over the navy's computing plans. As Stibitz was preparing the report, the Bureau of Ordnance was making arrangements to assume authority over Aiken's machine at Harvard and was considering a more advanced version of the device.[85] Still, the navy was satisfied with the paper and circulated it to their officers.[86] Stibitz followed this review with studies of punched card equipment, relay computers, and interpolating machines. Gertrude Blanch contributed a small part to one report on computing machinery. She was asked by Stibitz to "determine which iterative [computational] methods lend themselves best to the instrumentation of a modern computing device."[87] Given the limitations of standard punched card equipment, it was not entirely clear that any of the computing machinery would be as flexible as a staff of human computers. Blanch studied the details of the Bell Telephone Laboratories computing machines, including the new interpolator and the design for a more sophisticated calculator that was still under construction. After she grasped that these machines could perform lists of instructions, she reported that most of her "techniques should work well on relay computers."[88]

In its first year as a contractor to the Applied Mathematics Panel, the Mathematical Tables Project had drawn few signs of respect from the panel's senior mathematicians. They seemed to view the group as a secondary research unit, an organization much inferior to Columbia and Princeton. None of the panel members had even visited the offices of their second-largest contractor, preferring instead to send Warren Weaver's administrative assistant, Mina Rees (1902–1997), to communicate with Arnold Lowan. In correspondence they tended to call the project director "Mr. Lowan" rather than "Doctor Lowan," the honorific they reserved for scientists that they did not know, or the unadorned "Lowan," the form they reserved for themselves.[89] Their attitude toward the group began to change when Cornelius Lanczos (1893–1974) joined the planning committee. Lanczos was a well-respected applied mathematician and had served for a year as a research assistant to Albert Einstein. He was one of the many Jewish mathematicians who had fled Eastern Europe in the 1930s and settled in the United States. For a time, he had held a position at Purdue University, but he was a poor match for the school. "I am trying desperately to get away from here," he had written to Einstein.[90] He was so desperate that he was willing to forgo a regular university appointment and take a position at a former relief project.

Lanczos never served as a traditional planner, never prepared a computing plan, never oversaw the computing staff. Instead, he acted like a visiting scholar, an expert on the methods of calculation who could teach new techniques to Gertrude Blanch, Ida Rhodes, and the other members of the planning committee. Starting in the winter of 1944, he offered seminars on numerical methods, advertising them through the Applied Mathematics Panel and nearby New York University. "His lectures attracted a wide audience, not only from the Project, but from mathematicians at local universities," recalled Ida Rhodes.[91] These lectures brought a small glimmer of respect from the Applied Mathematics Panel. By March, they were starting to address Lowan in more informal terms and to refer to the project as "Lowan's Group."[92] More important, they were pointing to the Mathematical Tables Project as a successful computing organization. They encouraged prospective computers to visit the organization and copy its operating procedures. Among the visitors that winter was a group of scientists that was preparing to build a computing laboratory for the Manhattan Project in Los Alamos, New Mexico.

Computing laboratories were familiar institutions to the atomic scientists, as most of the major university physics departments had some kind of computing staff. Yet these academic laboratories were far smaller than the scale demanded by the effort to build the bomb. The computing office at the University of Chicago, one of the larger contractors to the Manhattan Project, consisted of just one faculty wife and a few graduate students.[93] The senior leaders of Los Alamos wanted to model their organization on the largest computing offices of the Applied Mathematics Panel, the Thomas J. Watson Bureau and the Mathematical Tables Project.

The Watson Laboratory received the first visitors from the Los Alamos staff, a couple named Mary and Stanley Frankel. Stanley Frankel had managed the computing bureau at the University of California that had handled the calculations for isotope separation, the problem of extracting the type of uranium that could be used in a chain reaction.[94] It was a small group with none of the equipment that could be found at Columbia. The Frankels spent about three days at the bureau, working with director Jan Schilt and a young graduate student named Everett Yowell (1920–).[95] Yowell had the rare distinction of being a second-generation computer. His father, also named Everett Yowell (1870–1959), had computed for the Naval Observatory from 1901 to 1906. The elder Yowell was part of the generation that had known Simon Newcomb, Myrrick Doolittle, and the computers of 1918 Aberdeen.[96] After his service as a computer, the senior Yowell had become a mathematics instructor at the U.S. Naval Academy and then had returned to the family home in Ohio to become the head of the Cincinnati Observatory. The younger Everett Yowell spent his youth playing in the halls and chambers of the observatory. His father taught him how to use a telescope, how to record the position of an object, how to reduce astronomical data. His texts were the classic books: Crelle's *Tables,* Newcomb's *Positional Astronomy.* At the age of twelve, Yowell assisted his father on an expedition to study a solar eclipse. He entered college with a firm understanding of traditional astronomy and arrived at Columbia knowing the methods of hand computers.[97]

During his first year at the school, Yowell had little contact with the Watson Bureau. "I was sort of drafted as an operator during the summer of '42," he recalled. The facility was beginning to do calculations for war research and had lost much of its skilled staff. Eckert was in Washington, and many of the younger workers had left for the military. Yowell learned the techniques of punched card computation by studying Eckert's Orange Book and by experimenting with the machines. Over the course of a year, he became an expert on wiring plugboards, the mechanisms that controlled the tabulators. Plugboards were flat panels, about the size of a large notebook, that were filled with holes that represented the different operations of the tabulator. By connecting the holes with short cables, Yowell could direct the flow of data through the machines and implement the methods of the Orange Book.[98]

The Mathematical Tables Project received its Los Alamos visitor a few weeks later, a researcher named Donald Flanders. Knowing the limitations of punched card tabulators, Flanders was organizing a hand computing group that would be known within the laboratory as T-5. The T-5 group was a typical wartime computing office with about twenty computers.[99] It earned a certain distinction because of its association with the physicist Richard Feynman (1918–1988). Feynman was a junior staff member at Los Alamos, and he worked with Stan Frankel to prepare computing plans for T-5. One of his plans recalled the work of de Prony or the early computing floor of the Mathematical Tables Project. Feynman divided the computation into specific tasks, such as additions, square roots, and divisions, and then assigned each task to a specific computer. Like de Prony's computers, one T-5 computer did nothing but add. A second took square roots, using a mechanical calculator. A third only multiplied.

Instead of creating computing sheets, Feynman used standard index cards to hold the results of the computations. These cards passed from computer to computer as the calculation progressed. "We went through our cycle this way until we got all the bugs out," recalled Feynman, and it "turned out that the speed at which we were able to do it was a hell of a lot faster than the other way, where every single person did all the steps. We got speed with this system that was the predicted speed for the IBM machine."[100] This claim, the notion that the T-5 computers could equal the speed of a punched card office, was tested late in the war when Feynman organized a contest between the human computers and the Los Alamos IBM facility. He arranged for both groups to do a calculation for the plutonium bomb, the "Fat Man." For two days, the human computers kept pace with the machines. "But on the third day," reported an observer, "the punched-card machine operation began to move decisively ahead, as the people performing the hand computing could not sustain their initial fast pace, while the machines did not tire and continued at their steady pace."[101]

The competition between the T-5 computers and the punched card equipment is generally reported as a scientific version of the tortoise and hare fable, a story that predicted the triumph of computing machinery and a sign that human computers would soon be replaced by the electronic computer. The result can also be interpreted the other way, as suggesting that, through much of the war, human computers were closely matched to their mechanical counterparts. Since human computers did not demand the kind of preparation required by punched card machines, they outperformed the tabulators on many military calculations. The Los Alamos scientists relied on human computers to check large calculations. The plutonium bomb calculations were compared to a similar set of numbers that had been prepared on Howard Aiken's Mark I at Harvard.[102] As the war entered its last year, human computers might still be considered the equals of automatic computing machinery.

CHAPTER SEVENTEEN

The Victor's Share

> We cannot retrace our steps.
> Going forward may be the same as going backward.
> We cannot retrace our steps retrace our steps.
> Gertrude Stein, *The Mother of Us All* (1947)

SOMETIME IN 1944, computers became "girls." The University of Pennsylvania hired "girl computers"; Warren Weaver started calling Applied Mathematics Panel computers "girls"; Oswald Veblen, who had once led a team of computing men, used the term "girls"; George Stibitz began ranking calculating projects in "girl-years" of effort.[1] One member of the Applied Mathematics Panel defined the unit "kilogirl," a term that presumably referred to a thousand hours of computing labor, though in at least one letter it suggested an Amazonian team of computers.[2] L. J. Comrie, in an article entitled "Careers for Girls," stated that girls "can be made proficient and give good service [as scientific computers] in the year before they (or many of them) graduate to married life and become experts with the housekeeping accounts."[3] Even at this date, computing was not the sole domain of women. It was really the job of the dispossessed, the opportunity granted to those who lacked the financial or societal standing to pursue a scientific career. Women probably constituted the largest number of computers, but they were joined by African Americans, Jews, the Irish, the handicapped, and the merely poor. The Mathematical Tables Project employed several polio victims as computers, while the Langley research center kept an office of twelve African American computers carefully segregated from the rest of the staff.[4]

For all of the Applied Mathematics Panel computers, female and male, the end of the war was first glimpsed on September 25, 1944, when Warren Weaver received a letter reminding him that the National Defense Research Committee was "a war time agency which will go out of existence at the end of the war." The letter informed the panel that they needed to prepare their plans for demobilizing the organization.[5] There was scant military news to support the idea that the war was nearly over and that research was no longer needed. Germany, though in retreat, still commanded a strong military that could inflict substantial injury on Allied forces. Japan promised a long and bloody fight for the control of its home islands. Still, the senior leadership of American science felt that the last months of the war "would involve almost no research and very little development work."[6] They viewed the National Defense Research Committee as "an emergency organization" that "deserved to die a dignified death."[7]

"For planning purposes," Weaver told the members of the Applied Mathematics Panel, "we are assuming that Germany will fall by November 15, 1944."[8] Shortly after the end of the European war, the panel would start terminating research programs and liquidating contracts. The panel would have to provide guidelines on how project records should be preserved, which materials could be published, and when computing groups could be disbanded.[9] As the panel developed its demobilization plans, Warren Weaver recorded that the members were "becoming deeply interested in the post-war possibility of the important new computing techniques, which are being developed during this war." They discussed computing machines over an eight-month period, sometimes reviewing their progress as part of a formal meeting, at other times sharing speculations about Stibitz or Aiken or the Aberdeen Proving Ground over lunch. "Quite outside of the mere furnishing of labor-saving assistance," Weaver explained, these machines "may, in fact, have theoretical consequences of very great significance."[10]

In its discussions, the panel was beginning to recognize that there might be a substantial demand for computing services after the war. Weaver observed that the navy was using Aiken's Harvard machine for three shifts a day, seven days a week. From this, he concluded that "two machines could be kept busy on [naval] Ordnance work alone" and that the navy should consider building "a central machine . . . under a broad agency in Washington."[11] He argued that the navy should make this last machine available to both military and civilian researchers, as it might "take care of a variety of problems."[12]

All of the major computing laboratories had a full load of requests that winter. The New York Hydrographic Office was completing LORAN tables for the Pacific Ocean. The group's leader, Milton Abramowitz, had lost almost half of his staff, yet he had reduced the computing time from six weeks per table to four. He had received some help from the Mathematical Tables Project staff, to be sure, but he had achieved much of this improvement by developing a new method of preparing the tables.[13]

The Mathematical Tables Project found that it could add only a few numbers each week to the unfinished WPA volumes. It was handling about two dozen calculations at once. A computer might work on two or three or four assignments, advancing each a little each day. That winter, the big calculation was a rather gruesome optimization problem. It attempted to determine the "combination of [high explosive] and [incendi-

ary bombs] that should be used for specific Japanese targets." High explosives would shatter Japanese buildings but not set them on fire. Incen-
diaries would light fires, but the damage would be limited if the buildings
remained intact. A well-planned combination of the two kinds of bomb
would create firestorms that would destroy Japanese cities.[14] The mathematical analysis of this problem had been assigned to Jerzy Neyman,
who, again, had failed to meet his deadlines and keep his expenditures on
budget. The Mathematical Tables Project computers had to work at a
stiff pace in order to complete the work before the start of the Pacific
bombing campaign in March 1945.[15]

With two years of experience, the Applied Mathematics Panel was becoming reluctant to label every computing problem as essential to the
war effort. When General Electric requested thirty computers, the panel
demurred. General Electric wanted these computers to prepare tables for
the antiaircraft systems of the B-29 bombers, the planes that would lead
the bombing campaign against Japan and would carry the atomic bomb.
The work was important, but Warren Weaver informed General Electric
that the Applied Mathematics Panel had no computers to spare and "suggested that [the company] use the computation facilities of some insurance company."[16]

General Electric was not alone in its rejection. The Applied Mathematics Panel accepted few new problems that winter. Those that went
from the panel to the Mathematical Tables Project or the Thomas J. Watson Astronomical Computing Bureau tended to be expansions of old calculations. The work was becoming routine, and hence it is probably not
surprising that Warren Weaver became interested in an unusual problem
that had no obvious mathematical solution. In March, he told the panel
of an issue that was "being held very secret." The secret would not last
long, he informed them, because "all Police Departments in the US have
been informed about it." Nevertheless, the panel should treat the request
as confidential, for any public discussion of the problem might incite
panic among American citizens. Large balloons had been discovered
across the country. They were constructed out of rice paper and were
about thirty feet in diameter. Most had landed in Oregon and California,
but several had made it over the Rocky Mountains and landed as far east
as Michigan. They apparently came from Japan, but no one knew where
they were launched or what they were designed to accomplish. Some
members of the National Defense Research Committee believed that the
Japanese planned "to use bombs in dry weather to start large scale forest
or crop fires." Others suggested that the balloons might be intended to
carry poisonous gas or deadly viruses. Weaver could offer no theory of
his own, though he argued that they must be weapons of last resort, as

any damage to the United States mainland would be "a poor bargain in view of our demonstrated ability to put bombs on Japan."[17]

Weaver asked the panel members to estimate the number of balloons that had reached North America, given "the number of recoveries which have been made and of the tolerable assumptions concerning the probability of recovery." It seemed likely that most of the balloons were lost at sea and that many of the remainder had landed in remote sections of the coastal mountains or in the deserts of the western states. He also asked the mathematicians to consider "what other questions related to the problem can have rough numerical answers."[18] With hindsight, it is difficult to understand how Weaver's announcement could have been anything more than a reaction to an unsettling piece of news. Given how little the military knew about the balloons, the Applied Mathematics Panel would have found it difficult to prepare a mathematical analysis of the attacks. Weaver's response may have also been shaped by a growing anxiety among the members of the panel that "mathematics is not adequately represented" in the country's plans for postwar science.[19] President Roosevelt had recently announced a new organization, the Research Board for National Security, to coordinate scientific research after the war. Only a single mathematician, Oswald Veblen, had been appointed to this committee.

As a whole, the members of the Applied Mathematics Panel had no coherent vision for postwar mathematics. Few of their number showed any interest in building a new institution for postwar mathematical research. John von Neumann argued that mathematics should have no organizational hierarchy, as "decentralization would be the more efficient manner of organization."[20] Richard Courant, from New York University, wanted a small executive body to coordinate research but warned that "the enthusiasm among capable scientists for war research will abate after a short time." Organized mathematics research would have "no attraction whatsoever for scientists of high quality unless research can be conducted with as few strings attached as possible."[21] Only Warren Weaver seemed interested in promoting a government office for mathematical research, yet he described such an office only in terms of the way that it would operate. The American government should "set up laboratories in a place where scientific people like to be," he stated. It would also have to "pay high salaries" and would need to "provide better materials and greater leisure than are commonly provided in government service."[22]

The death of Franklin Roosevelt in April introduced new uncertainty into the discussions of the Applied Mathematics Panel. "To many," wrote historian David McCullough, "it was not just that the greatest of men had fallen, but that the least of men—or at any rate the least likely of men—had assumed his place."[23] The mathematicians knew little about his successor, Harry Truman. No one knew how the new president felt about government sponsorship of science or whether he would be able to get the approval of Congress for any ambitious scientific program. The members of the Applied Mathematics Panel were somber when they convened their first meeting after Roosevelt's death. Warren Weaver was ill, and Thorton Fry held the meeting at Bell Telephone Laboratories. They heard reports from several projects, approved a few minor actions, and accepted a request for calculations from New York University. They assigned the work to the Mathematical Tables Project but asked that Arnold Lowan "secure some time estimate" before starting the work. Just as the meeting broke up, the panel agreed that "meetings in the future [will] be held at intervals of 3 to 4 weeks until further notice."[24]

The mathematical work for the Second World War was coming to an end, but the future of the computing staffs remained undecided. The Applied Mathematics Panel did not even raise the issue until June, when Oswald Veblen suggested that the staff of the Mathematical Tables Project might be useful to the Research Board for National Security. Thornton Fry, still acting as chair, showed no enthusiasm for the idea, noting that there were many new computing machines, and "presumably after the war [they will] not be fully used." He acknowledged "that if this group is broken up, it cannot be put together again," but the only employment that he foresaw for Arnold Lowan, Gertrude Blanch, and the others was the preparation of "out of hour uses for these machines," computations on the second or third shift.[25] Fundamentally, the panel did not give the computers the same professional status that they accorded mathematicians. When appointing a mathematician "to keep track of the types of analytical problems arising at Aberdeen; of the computing problems which are involved in these; and of their relations to modern computation devices," they wanted the appointment to have a government rank of at least P-6, more than halfway up the professional appointment scale, and even a P-7 or "a P-8 is not absolutely excluded."[26] When they dealt with the senior computers, who did the same kind of work, they had no such scruples. At the Mathematical Tables Project, Arnold Lowan held the highest rank at a P-5. Gertrude Blanch, who had already done the tasks required by the Aberdeen position, was a P-3. The rest of the Mathematical Tables Project planning committee held a P-2 or even a lowly P-1 ranking.[27]

The first months of summer brought the initial skirmishes for the spoils of war, and when the fights were over, the results remained inconclusive for the Applied Mathematics Panel computers. The first trophies of the conflict were visits to the German scientific institutions, the rocket development center at Peenemünde, the large industrial laboratories along the Rhine, Werner Heisenberg's nuclear research facilities. A select cadre of British, French, and American scientists followed the advancing Allied troops, seizing scientific documents, confiscating equipment, and interrogating research personnel. Often, these visits were an opportunity to greet old friends, to become reacquainted with a familiar scientific colleague who had been isolated from Allied researchers by the war. It was also a chance to look into the enemy's lair and to evaluate the science that had been used against the conquering troops, the convoys of the North Atlantic, and the civilians of Europe.

The Allied generals did not consider computational laboratories to be of much importance until the British army mistakenly seized the German mathematician Alwin Walther. The British had confused Walther with a senior rocket engineer who had the nearly identical name of Helmuth Walter. When they took Alwin Walther to London for interrogation, the mathematician protested that he had nothing to do with the rocket program but actually operated a computing facility at the Technische Hochschule in the town of Darmstadt. To prove his claim, he produced a battered photograph, which he had been carrying for just such a situation. The picture showed him walking arm in arm with another man, who he claimed was the mathematician Richard Courant of the Applied Mathematics Panel.[28] The British military, looking for a mathematician to confirm or deny the story, turned to Olga Taussky, the consultant working with the Ministry of Aircraft Production.

Taussky interviewed Walther and concluded that he was telling the truth. Before Walther was released, she returned with her husband, John Todd of the Admiralty Computing Service, for a second interview. Walther told his interrogators that the Darmstadt facility was the largest computing office in Germany and that there was a second mathematical laboratory, the Mathematisches Reichsinstitut, or National Mathematics Laboratory, in Oberwolfach. He explained that the Oberwolfach offices had done mathematical analyses for aircraft design, ballistics trajectories, and the guidance systems of the V-2 missile. The Technische Hochschule at Darmstadt had served as a "calculating workshop" for the German military. Its researchers had developed a number of computing machines, including a differential analyzer, extensions to punched card tabulators, and a device that could solve systems of linear equations, though it was a very different machine from the one created by John Atanasoff. The Technische Hochschule also had a substantial computer staff and had produced a number of tables.[29] These stories intrigued Taussky and her husband. "Business was not as brisk at the Admiralty then as it had been earlier," Todd reported, and hence he was free to travel. When he returned to the Admiralty offices, he discussed the German institute with his colleagues. Together, they "conceived an intelligence mission to investigate mathematics in Germany."[30]

From his work with the Admiralty Computing Service, Todd was familiar enough with the British war ministries that he could identify the individuals who might support a trip to the defeated Germany and could present the idea in a way that would gain their support. "Before the week was out we were officers in the Royal Navy Volunteer Reserve," he reported, "with open orders and maps of all Germany." They were instructed to visit "targets of opportunity," which included the Mathematisches Reichsinstitut, the Technische Hochschule, and the University at Göttingen, long the premier training facility for mathematicians. Outfitted with naval commando uniforms and given a fresh round of inoculations, Todd and five colleagues from the Admiralty flew to Brussels and went in search of German mathematics.[31]

"Before the group had spent a week in Germany, its numbers began to dwindle. One member fell ill and returned to London. A second was reassigned to an expedition that was preparing to visit the north magnetic pole. A third mathematician accepted an invitation to view a solar eclipse and was killed in a plane crash. The remaining three, Todd, a second mathematician, and a translator, crossed a devastated landscape as they continued without their colleagues. They had to be alert for unexploded ordnance, German deserters, booby traps, and the final defenders, the Volksstrum, or provincial volunteers. A member of the Mathematisches Reichsinstitut stated that "a feeling of anxious suspense hung in the air" during the first weeks after the surrender. To protect themselves, they destroyed items that would connect the mathematics institute with the Nazi regime. "The meadow behind the house saw dozens of copies of the Hitler bible, 'Mein Kampf,' go up in flames and vanish," she wrote. Though they gladly sacrificed the most common totem of the Nazi regime, they spared copies of another volume of propaganda, *The National Socialist View of History*. In these volumes, "a wide margin had been left for notes," she explained, and the staff was willing to risk incrimination in order to have the scrap paper. In the end, the gamble was rewarded, for the books were never used as evidence against the institute, and the pages "proved to be most advantageous in many ways during the coming period of extreme need."[32]

By the time Todd arrived at Darmstadt, Alwin Walther had returned to the organization and resumed his work. The Technische Hochschule had a staff of about ninety. Twelve of them were PhD mathematicians; the rest were young women, "some quite young," according to Todd. A staff member at the facility identified the computers as "female university entrants with a flair for mathematics whom the state compulsorily assigned" to the Technische Hochschule and who were known as "Walther's harem."[33] The largest section of the institute, consisting of about thirty computers, worked for the V-2 missile program calculating ballistics trajectories. Two smaller groups of computers processed wind tunnel data and calculated the propagation of electromagnetic waves, the kind of work that was needed for radar research. The institute had two other computing offices. One operated a differential analyzer; the other used adding machines and punched card tabulators to do miscellaneous calculations, the sort of work that was handled by the Mathematical Tables Project in the United States.[34] The tabulators had been seized by the German army from the occupied states. The Darmstadt staff claimed that it accepted the machines reluctantly, "hoping it might benefit from the machines for its research activities—a hope that proved to be false."[35]

During June and July 1945, Todd spent six weeks in Germany and visited eight centers of mathematical research. He moved methodically from institute to institute, collecting papers, seizing books, interviewing mathematicians, and occasionally finding old friends. From the experience, he concluded that "German mathematical research was not centralized until the very end of the war."[36] From what he saw, he felt that the Admiralty Computing Service had done a better job of organizing mathematicians and solving problems for the military. Still, he was impressed with the quality of German mathematical research and returned home with a substantial collection of texts, tables, and reports of computation. Traveling with restrictions on military luggage, he stuffed his trophies into pockets and jackets, shirt sleeves and trouser legs. When he stepped on a scale to be weighed for the flight home, the diminutive mathematician registered 300 pounds.[37]

Before Todd returned to London, a British scientific officer notified the Applied Mathematics Panel of the mathematical mission to Germany and asked whether the Americans wanted to send a representative. The request reached the desk of Thornton Fry, who was still running the panel in the absence of Warren Weaver. Fry knew that it would be easy to add an American to the party, as several Applied Mathematics Panel researchers were already working in London on field projects. However, he concluded that Todd would do an adequate job and declined the offer.[38] Arnold Lowan reached a different conclusion when he learned of Todd's trip. "Imperative that a representative of MTP should go to Darmstadt,"[39] he wrote to the Applied Mathematics Panel. Lowan had read a preliminary report from the Technische Hochschule and had concluded, correctly, that the group was similar to his own. He hoped that a visit to the enemy's computing center might give added importance to his group. It might show that Germany profited from organized calculation and that the United States would strengthen its defenses by expanding the Mathematical Tables Project. Thornton Fry received this request coolly. "I can find nothing in the report of the Darmstadt activities to justify sending anyone to Darmstadt on AMP account," he concluded. In what Todd had found there were not "any strong indications that they were much ahead of us. In the matter of methods of computation and especially in the development of computing machines, they seem to be far behind us." Even if the Germans had achieved some interesting results, he was not prepared to let Arnold Lowan investigate. "I do not know what future Dr. Briggs has in mind for the Lowan group," Fry wrote, "and therefore cannot guess how he would feel about sending Lowan over as a Bureau of Standards representative."[40]

Lowan undoubtably appreciated that the denial from Fry suggested an uncertain future for the Mathematical Tables Project. However, this news was accompanied by a special request for calculation, a request that must have reminded Lowan how far the group had come. This request came from Richard Courant at New York University for some unnamed third party and was marked as top secret. Even though no one in the group had received a security clearance, they had occasionally handled sensitive calculations. The calculations would be passed to Warren Weaver at the Applied Mathematics Panel, who would strip the mathematics of every reference to the physical setting, including the units of all quantities involved in the work. Though the members of the planning committee could usually grasp the general nature of the application, they were rarely able to piece together the full problem. During the winter and spring of 1945, the Mathematical Tables Project had handled classified computations for microwave radar and the targeting of depth charges.[41] "I did not completely understand what we were doing," said one committee member, "until I read [the official history] after the war."[42]

This new request was an explosion problem. It traced shock waves through different materials. Superficial evidence suggests that the request came from the Manhattan Project and may have been a duplicate or more refined calculation for the Fat Man bomb. The memo was passed through the explosives division of the National Defense Research Committee, which was helping the scientists at Los Alamos design the mechanism for detonating the plutonium bomb. The Los Alamos scientists requested duplicate calculations of many key elements in order to verify their work. The first test of the plutonium bomb was only a few weeks away, and engineers were already assembling the test equipment at Alamogordo. Even if the calculation did not come from the Manhattan Project, it was still important evidence that the group, once judged impossible to secure, had gained some measure of trust and prestige among the wartime scientists, for it came with a stern warning:

You will not show this paper to any member of your group.
You will make no reference to this paper in any of your own work.
No copies of the paper will be made.
The paper will be returned as soon as it has served its purpose.[43]

With the surrender of Japan on August 16, 1945, the contractors of the Applied Mathematics Panel dropped their tools in the field and returned to their homes. The final demobilization began quickly, as the mathematical contractors released workers, declassified results, published significant discoveries, and recorded the history of their organizations. By the end of September, the bulk of the projects were liquidated, including the statistical studies at Columbia, the remaining work of Jerzy Neyman, the explosion analyses at New York University, and the bombing mathematics at the Institute for Advanced Study. The rest of the contractors, save the Mathematical Tables Project, had scheduled their final dates of operation. Most would finish their work in October. Only the mathematicians of Brown University would operate through the month of November.[44]

In this final flurry of activity, the Applied Mathematics Panel had arranged for the navy to fund the Mathematical Tables Project as a special-purpose computing laboratory. This arrangement would reunite the two divisions, joining the New York Hydrographic Office computers, who were already under navy authority, to those who had been directed by the panel. The navy, viewing the project as the seed of a larger computing laboratory, agreed to provide Arnold Lowan with IBM punched card tabulators. Naval officers would take no role in the daily operation of the center and stated that they would support "special computations for various navy bureaus, [and] also the work on basic tables."[45] The consolidation required one final move. The computers of the Mathematical Tables Project had to pull their mathematical books off the shelves, pack their notes, box their adding machines, and join the New York Hydrographic Office staff in the Hudson Terminal Building. The combined facility was the best office that had ever been given to the group. It had separate space for the punched card equipment, a room for the planning committee, and a view of New York Harbor.[46]

As the computers prepared to move to their new offices, Arnold Lowan, accompanied by Milton Abramowitz, traveled north to attend a conference on computation. The conference was sponsored by the Subcommittee on the Bibliography of Mathematical Tables and Other Aids to Computation and was the evidence of R. C. Archibald's faith and efforts. After a dozen years of uncertain fortune, the committee was able to organize a weekend discussion of computation for eighty-four researchers, "those chiefly active in connection with mechanical computa-

41. Hudson Terminal Building, last home of the Mathematical Tables Project

tion on both sides of the Atlantic," according to Archibald.[47] Much of this new authority came from the success of the committee's journal, *Mathematical Tables and Other Aids to Computation*. In a little more than three years, the journal's subscription list had grown to nearly three hundred. The periodical could be found with every contractor for the Applied Mathematics Panel and was generally accepted as the scholarly record of human computers.[48]

The meeting, held at the Massachusetts Institute of Technology, was a mixture of the old and the new, Depression-era methods and wartime accomplishments. L. J. Comrie mingled with John von Neumann. Arnold Lowan watched Howard Aiken operate his Mark I. From the same stage, speakers talked about difference engines and differential analyzers, the punched card machines of the *Nautical Almanac* and the relay computers of Bell Laboratories. "The conference was most notably successful," Archibald reported, "and one heard on every side expressions of the hope that such a conference might become an annual event."[49] Yet this meeting was the only time that the war computers gathered as equals, the one moment when mathematicians and human computers, punched card clerks and differential analyzer operators, electrical and mechanical engineers came together and talked about their experiences with equations and numbers. A second meeting, held just three months later, gave clear signs that machine designers were starting to outpace human computers. This meeting was a press conference, held at the University of Pennsylvania, that announced the age of electronic computation. In the building that held the old differential analyzer, university officials unveiled its replacement, the ENIAC. The machine had been proposed in 1943, but it had not been finished by the end of the war. It had done its first complete calculations in November, just at the time of the MIT conference.[50]

Like the meeting that had been organized by MTAC, the ENIAC announcement pointed toward the future while not letting go of the past. The machine was fast because it was entirely electronic. The calculations did not have to pause for a gear to turn or a telephone relay to click. It was more precise than older machines because it computed digitally. Each number was represented as a series of electrical pulses rather than as the turn of a wheel or the level of a voltage. The output was more useful than that of the differential analyzer because it came as numbers, not as an ambiguous graph. Yet for all of its benefits, the ENIAC was not quite a modern computer. It was really a collection of electronic calculators. Most of these devices were nothing more than adding machines, a few did multiplication, and one took square roots. The engineers prepared for a calculation by connecting these units with large, black cables. Following the computing plan, they arranged the cables so that they took a number from a punched card, passed it to an adder, sent it to the square root unit, returned it to the adder, and finally left it on the multiplying unit. The staff called this cabling process "programming," but only the word, not the actions, would be the legacy of the machine.[51]

The press conference was a time of great pride and accomplishment, but it was also marked with a bit of impatience. As the ENIAC became operational, the University of Pennsylvania researchers had come to appreciate the limitations of their machine. "We designed [the ENIAC]," recounted Adele Goldstine's husband, Herman, "and immediately lost interest in it."[52] In building this machine, they had recognized that they could design a more flexible machine that was controlled by a special list of electronic instructions rather than by the bulky cables. These lists would inherit the name "program." The design team had conceived a way to modify the ENIAC so that they could control it with a primitive program, but they were more interested in building a new machine that would be entirely programmable.[53]

The new, programmable machines were at least two years away, and in the interim, the University of Pennsylvania had to live with the ENIAC. The press conference generated a tremendous interest in computing devices. In response to requests to study and use their machine, the school organized a seven-week course on computing machinery, a conference that would be called the Moore School Lectures. When the course convened in July 1946, the list of lecturers constituted "a Who's Who of computing of the day." Most of the speakers came from the ENIAC project, though George Stibitz, Howard Aiken, and John von Neumann also conducted sessions. The talks had little in common with the discussions at Archibald's conference three months before. The Moore lectures dealt with circuit design and the preparation of problems for machine computation. The students were not human computers but "a select group of seasoned professional engineers and mathematicians."[54] The discussions made little reference to human computing groups, and few in attendance had any experience with organized calculation. No one from the Mathematical Tables Project was invited to attend. L. J. Comrie was not chosen as a representative of Great Britain. There were no computers from any Nautical Almanac Office, the Manhattan Project, or any of the projects of the Applied Mathematics Panel.[55]

Had any human computers been at the Moore School Lectures, they would have heard a somewhat fanciful history of calculating devices that ignored the contributions of workers like themselves. This history, the first of the lectures, also overlooked the punched card machines of Herman Hollerith and described the difference engine of Charles Babbage as "a special purpose [device] developed for the satisfaction of personal curiosity or as an intellectual stunt."[56] Historians of the conference claim that the talk was meant to "entertain and inspire,"[57] but a close examination of the text suggests that it was an attempt to build a distinguished lineage for the electronic computing machine, a pedigree that ignored the influence of commerce and the hard labor of human computers. To many at the talk, the human computer was already starting to fade from memory. Most of the wartime computing groups had been shut down, reduced to a small remnant, or replaced by punched card equipment. The major employer of human computers, the Applied Mathematics Panel, had finally ceased operations after one last meeting. Led by Warren

Weaver, the mathematicians had gathered in their old conference room to celebrate their accomplishments. There would be those who would criticize the panel and argue that it missed an opportunity to advance the cause of mathematics, but this was not the time for such discussions.[58] The members of the panel spoke their praise to mathematics, to science, and to their accomplishments, though no one felt it necessary to express gratitude toward the workers who had undertaken the calculations.[59]

Even though Arnold Lowan and Gertrude Blanch were not invited to the Moore School Lectures in the summer of 1946 or to the final meeting of the Applied Mathematics Panel, they were confident about the future. The navy and the Army Air Corps had guaranteed funds to sustain the operations of the Mathematical Tables Project for two more years, and the military services seemed likely to continue providing that money for a subsequent period, at least until they were in possession of a comprehensive facility with electronic computers. Even then, they seemed likely to retain the members of the Mathematical Tables Project as the support staff for the new computing machines.[60] For the moment, the military provided the project with new and interesting computations, work beyond the traditional labor of preparing mathematical function and LORAN tables. The services asked for guided missile trajectories, radar wave deflections, and shock wave propagations. In addition to this work, the Mathematical Tables Project was now receiving requests for calculations relating to the development of atomic energy, such as problems that described the interactions of particles or analyzed reactor designs.[61]

The only worrisome development for the project was a change of leadership at the National Bureau of Standards. The bureau had been the fixed point of the Mathematical Tables Project through the fluctuations of the Depression and the war. Lyman Briggs, director of the bureau, had been the champion of the project, sponsoring it for the WPA, serving as the executive manager under the Applied Mathematics Panel, and helping to obtain the postwar funding from the military.[62] The planning committee of the project viewed him as the "Beloved Boss" who had supported Arnold Lowan, corrected his mistakes, both gently and not so gently, and spoken for him in the high councils of science.[63] Briggs had retired in the fall of 1945 and had been succeeded by the physicist Edward Condon (1902–1974). When he first arrived at the Bureau of Standards, Condon had questioned the value of the group. He had changed his mind after a visit to the project's offices, when he met Lowan, talked with the planning committee, and observed the computing staff. Condon concluded that the Mathematical Tables Project might be able to contribute to the National Bureau of Standards, though he also observed that there was no reason for Lowan to report directly to the bureau director. After returning to New York, he informed Lowan that the project would be guided by his new assistant, John Curtiss.[64]

Curtiss was familiar with the work of the project and sympathetic to the group. Back in 1940, he had reviewed the first volumes from the Mathematical Tables Project in the *American Mathematical Monthly* and found them quite acceptable. By training, he was a statistician, but he came from a mathematical family and was well connected in the mathematical community. His father had been a professor at Northwestern University and had served as president of the Mathematical Association of America, the professional society devoted to mathematical instruction. Curtiss had studied at the University of Iowa and at Harvard. His rise to prominence had come during the war, when he had taken a naval commission and served as a statistician for the Bureau of Ships. Most of his work concerned quality control, the statistical methods that tested batches of manufactured products to ensure that all of them operated properly or met basic standards. At the end of the war, he was recommended to Condon as a mathematician who understood the nature of organizational politics.[65]

Perhaps remembering Curtiss's review, one of the few favorable signs in the early days of the Mathematical Tables Project, Arnold Lowan welcomed the arrival of John Curtiss and invited the statistician to visit New York. Curtiss accepted the offer and, like Condon before him, came away from the project offices with a favorable impression. He decided that the group should be a key part of a new mathematical research organization that he hoped to create within the National Bureau of Standards. For the moment, he was using the name "National Applied Mathematics Laboratories" to describe his idea. In his plan, the laboratories would have four distinct units. The first would research the methods of applied mathematics, the second would consider problems of applied statistics, and the third would develop new computing machines. The Mathematical Tables Project would be the fourth and final laboratory within the group. Renamed the Computation Laboratory, it would provide "a general computing service of high quality and large capacity, for use by private industry, Government agencies, educational institutions, etc."[66]

In Curtiss's plan, the Computation Laboratory would be larger and better equipped than the old Mathematical Tables Project. It would have desk calculators, difference engines, punched card tabulators, "special analogue equipment for the solution of algebraic equations," and finally "two general purpose automatic electronic digital computing machines of large capacity."[67] Arnold Lowan was delighted with this idea and immediately started referring to the Mathematical Tables Project as the Computation Laboratory, even though the new title would not become official for another eighteen months and the old name would stick to the group as if there was no other way to describe the former WPA project. There were many details to complete before Curtiss's plan could be implemented. Curtiss and his boss, Edward Condon, had to obtain funding from Congress and explain how the National Applied Mathematics Laboratories would work with other government agencies. One minor point to resolve was the location of the new Computation Laboratory. Curtiss had stated that the laboratory might be located in either New York or Washington, D.C., but Lowan wanted the laboratory to remain in New York. New York was the nation's financial capital, headquarters to much of its industry, close to the bulk of its major universities, and the home of the American Mathematical Society. It was also the city in which Arnold Lowan lived and the place where he held a second job, his professorship at Yeshiva University.

In the summer of 1947, Curtiss announced that the Mathematical Tables Project would have to move to Washington, D.C., as it was "an integral part of the National Bureau of Standards, rather than a field office."[68] Arnold Lowan immediately protested the decision and claimed that it would cause "complete demoralization of our personnel."[69] Indeed, outside observers could detect that something was wrong that summer, even though the staff kept to their tasks. The computers were "so upset that they didn't know what to do," reported Everett Yowell of the Thomas J. Watson Computing Bureau.[70] Yowell's presence at the project offices was the first cooperation between the two New York computing institutions, and it should have been another sign that the Mathematical Tables Project was finally becoming part of the scientific community. With so many staff members preoccupied with the potential move, it was an opportunity lost. Yowell largely interacted only with Milton Abramowitz, teaching him some punched card techniques to earn "a little extra money."[71]

Believing that he was fighting one more time for the survival of the Mathematical Tables Project, Arnold Lowan again turned to his scientific allies: Philip Morse, John von Neumann, Julius Stratton, and R. C. Archibald. Letters came from Lowan's home in Brooklyn, marked "Personal" and bearing the now familiar tag line, "For obvious reasons I would appreciate your keeping this letter in strict confidence."[72] With Phil Morse, he explored the possibility of bringing the Mathematical Tables Project under the authority of the Atomic Energy Commission and moving the group to the new Brookhaven Laboratory on Long Island. After discussing the idea with friends on the Atomic Energy Commission, Morse reported that Brookhaven would not accept the computing group. He never mentioned whether the Atomic Energy Commission was at all interested in the Mathematical Tables Project but instead reported that the new laboratory, which was located in an area then considered quite isolated, could not provide housing for the computers.[73] Turning next to John von Neumann, Lowan suggested that the staff of the Mathematical Tables Project might work as a computing office for the Institute for Advanced Study, though of course they would remain in New York. Von Neumann, who was already planning to build his own electronic computer, never raised the issue with his colleagues at the institute, writing that "my own impression is that the Institute is not a suitable vehicle for such a function."[74]

As rejection followed rejection, Lowan's letters became more frequent, more urgent, more desperate. Through the winter of 1947–48, Morse was the recipient of four letters, each more anxious than the last. Lowan described his problems yet again, asked why he had received no reply, requested a meeting with his supporter, and finally begged for any contact.[75] "I have reason to believe that Dr. Curtiss intends to proceed with his plan of either transferring the [Computation Laboratory] to Washington even at the risk of wrecking it," he wrote, "or to curtail it considerably by even abolishing some of the jobs of the mathematicians."[76] His writing became tinged with self-pity, as the legacy of the Mathematical Tables Project became intertwined with every little slight that he had felt as its leader. He complained that he had been denied a promotion, that Curtiss had already appointed his successor, that his accomplishments were being ignored.[77]

Finding no aid from his traditional supporters, Lowan looked outside the scientific community for any assistance that might be offered. He turned first to the New York City congressional delegation and asked for help from Representatives Emmanuel Cellars of Brooklyn and Jacob Javits of Manhattan. The two congressmen wrote to the leaders of the National Bureau of Standards, presented Lowan's arguments, and asked for an explanation. Both congressmen got polite responses from John Curtiss that offered no compromise on his plan to move the Mathematical Tables Project to Washington.[78]

As the date for the move approached, Arnold Lowan asked for the assistance of the union that represented his human computers, the United Public Workers of America. The United Public Workers had organized the Mathematical Tables Project computers in 1938 as part of a broader effort to represent the clerical workers of the WPA. The union was on the more radical side of 1930s labor organizations. It was a member of the CIO, the Congress of Industrial Organizations, and kept its offices in the same building that housed the American Communist Party. It had mounted at least one strike against the WPA, in 1939, though the correspondence of Arnold Lowan suggests that the labor action spared the Mathematical Tables Project.[79]

The United Public Workers followed the lead of Arnold Lowan in

42. Phil Morse in his computing laboratory after
the war

pressing its case. They contacted Cellars and Javits in Congress, R. C.
Archibald and Phil Morse in the scientific community, John Curtiss and
the senior scientists of the National Bureau of Standards. The union's
first letters were polite and deferential. "The employees of the Computation Laboratory appreciate the interest you have shown in the problem of
its location," the union president wrote to Phil Morse. "May we take the
liberty of writing to you once more on the subject."[80] The union argued
that the planned move would destroy jobs and damage the legacy of the
Mathematical Tables Project. "It may never be possible," wrote the president, "to fully make up for the loss of skilled personnel that the move
would entail."[81]

Subsequent letters were not so deferential to the authority of the scientists. The union attacked John Curtiss and portrayed him as naive, op-
portunistic, and prejudiced. Quoting a memo that had been supplied by
Lowan, the union president charged that "twenty per cent of the present
employees of the [Mathematical Tables Project] who are Negroes are not
expected to go to Washington because of the Jim Crow conditions existing there." Furthermore, "fifty percent of the employees are Jewish and
fear increased discrimination in Washington."[82] The charges had more

than a grain of truth, for Washington was a segregated city. "White-collar work and employment in skilled trades dominated in Washington," wrote one historian of the city, "but access to such jobs was anything but equal racially."[83] The letter stirred only one scientist to action, the volatile R. C. Archibald. Archibald, who had praised the Mathematical Tables Project in the pages of *Mathematical Tables and Other Aids to Computation* and in *Science,* took up his pen to rail against "the attempted transfer of the [Mathematical Tables Project] from New York to Washington" and the "brutal treatment of Lowan" at the hands of John Curtiss, whom he characterized as "not even a second rate mathematician."[84]

Neither the United Public Workers nor R. C. Archibald could keep the Mathematical Tables Project in New York City. The death blow for the group was delivered by John von Neumann, who wrote one final letter at the request of Arnold Lowan. In the letter, von Neumann told John Curtiss that he had "always had a great admiration for the work of the Mathematical Tables Project–Computation Laboratory" and added, "I think that this organization constituted and still constitutes an ideal computing group on the non-automatic level." He felt that the project had done "very excellent and valuable work in the past, and is likely to do so in the future." The strong tone of the letter began to waver when von Neumann suggested "that a group of this type will become obsolescent when automatic devices become widely distributed," and it collapsed in the last paragraph, when he deferred to Curtiss's judgment. "If you are satisfied that [it is impossible to fund the group in New York], then I concur with you that a gradual transfer to Washington is the only possible solution."[85]

Arnold Lowan probably never read von Neumann's letter, but he knew that its effect was "extremely unfavorable." After receiving the letter and meeting with the members of the Mathematical Tables Project, John Curtiss announced that the move would begin immediately and that the punched card unit would be the first group transferred to Washington. He offered jobs to most members of the planning committee and to eighteen of the human computers, about half of a staff working in New York.[86] In order to limit Arnold Lowan's ability to thwart the move, Curtiss offered two alternatives to the project leader. Lowan could come to Washington, take charge of the new Computation Laboratory, and receive a promotion. If he did not wish to do that, Lowan could remain in New York as the director of a fifteen-person computing office. The Bureau of Standards would provide funds for exactly one year. After that, Lowan would have to finance the group himself.[87]

In the history of human computers, this is the one moment that might have provoked a labor action, a strike by those who worked with numbers, but the computers were unwilling to back Arnold Lowan. The United Public Workers made one final attempt to start a protest by charging that Curtiss's plan "would certainly lead to the Bureau's having two relatively inefficient computing groups," but they found that there was no support for their position among either the computers or the scientists.[88] By midsummer, they had abandoned their efforts on behalf of the Mathematical Tables Project. That July, the new Computation Laboratory opened its doors in downtown Washington, and simultaneously, Arnold Lowan changed the name of his office to "Computation Laboratory, New York."

During the ten-year history of the Mathematical Tables Project, John von Neumann was a distant figure whose letters offered nothing but vague encouragement and deferred hope. He never visited the project offices, though he made time to visit the Admiralty Computing Service in England, and he never offered an unconditional blessing to Arnold Lowan. Only in the final days of the Mathematical Tables Project, just as his letter was sealing the project's fate, did von Neuman ask Arnold Lowan for computational assistance. He asked whether the computers would test a new technique called "linear programming." Von Neumann was not really interested in the results of the test computation, just as he was not especially interested in the future of the Mathematical Tables Project. He requested the calculation because he wanted to use the human computers as surrogates for computing machines, as a means of projecting the operation of a programmable electronic computer.

John von Neumann may have initiated the request, but he did not invent the method of linear programming. The technique was developed by a student of Jerzy Neyman named George Dantzig (1914–). Dantzig had spent much of the war analyzing operational problems for the Army Air Corps. A typical problem looked for the best way of storing spare parts for aircraft. The idea was to find the number of storehouses that would provide the best access to parts yet at the same time minimize the expense of keeping a large inventory.[89] Such problems had posed serious difficulties for the air corps, according to one of Dantzig's contemporaries, and had "required the labors of hundreds of highly trained staff officers."[90]

Dantzig had been able to test his method of linear programming only on small, simple problems, as the work had the same kind of demands as least squares or simultaneous equations. A small problem, such as the task of locating two or three storehouses, was simple and straightforward. As problems grew, the amount of calculation expanded rapidly. If the problem was doubled to six storehouses, then the calculations would require eight times the effort. Dantzig would have been restricted to these simple problems had he not been given the opportunity to present his method to von Neumann. He visited von Neumann at the Institute for Advanced Study and encountered an impatient mathematician. "In under one minute, I slapped the geometric and the algebraic version of the problem on the [black]board," he recalled. Von Neumann quickly grasped the nature of the problem, took the chalk, and "then proceeded for the next hour and a half to lecture me on the mathematical theory of linear programs."[91] As it happened, Dantzig's work was related to research that von Neumann had already completed. Von Neumann wanted to see Dantzig's method tested on a large problem, one that had a classic standing in the economics literature. This problem attempted to identify the cheapest possible diet from among seventy-seven different foods. The list of foods began with wheat flour and ended with strawberry preserves. The diet had to provide 3,000 calories and minimum amounts of eight different nutrients.[92] The calculations for this problem included 29,856 additions, 15,315 multiplications, and 1,243 divisions.[93] Von Neumann had hoped that the Aberdeen Proving Ground might agree to do this calculation on the ENIAC, but the ballistics researchers refused the request, so he turned to Arnold Lowan and the Mathematical Tables Project.[94]

The calculations began in mid-April, just as Arnold Lowan was preparing for his meeting with John Curtiss. The work required twenty-five computers and was overseen by a junior member of the planning committee. It was a somber time in the office, but the work progressed steadily. Dantzig monitored the efforts of the computers and kept von Neumann informed of their progress. Von Neumann would take the letters from Dantzig, turn them over, and calculate the amount of time that the ENIAC would spend on the same work. Dantzig's last letter reported that twenty-five computers had completed the work in twenty-one days. In half a page of pencil scratching and little diagrams, von Neumann concluded that the ENIAC could do the same work in about nine hours.[95] This was the only number that he would ever use from the Mathematical Tables Project. He never saw the final calculations. "The setting up of a computational procedure for this type of problem is the main objective," acknowledged the final report on the work, "rather than the solution of this specific diet problem."[96]

The computers began to disperse almost as soon as Dantzig's calculations came to an end. Some went to Washington; a few, including Ida Rhodes, started to learn about the new electronic computers;[97] most started looking for jobs in New York City. By January 1949, Lowan had fewer than fifteen computers working for him, and they were all assigned to old problems that had been started under the authority of the WPA. Struggling to keep his facility operating, he still hoped that Philip Morse would be the savior of his New York office this final time, but he was coming to accept that there would be no future for his organization. "The fact that you have not replied to my [last letter]," he wrote to Morse, "would seem to substantiate my feeling that you are seemingly unable to do anything which would change the trend of events."[98] The return mail brought the news he had dreaded. "I very much fear you are right," Morse had written. "It seems that the situation you are up against is well nigh unbreakable. I don't like it but there it is."[99] Having served, in its last moment of public glory, as a proxy for the digital electronic computer, the Mathematical Tables Project closed its doors for the final time on Friday, September 30, 1949.

CHAPTER EIGHTEEN

I Alone Am Left to Tell Thee

> The real power, the power we have to fight for night and day, is not power over things, but over men. . . .
>
> George Orwell, *1984* (1949)

EVEN THOUGH THE FINAL DAYS of the Mathematical Tables Project were filled with drama and emotion, and even though they engaged an unusual cast of characters, they were nonetheless part of a conventional scientific decision. Two scientists, each seeing a different direction for a project, shared a common claim over a single pool of resources. One of the two, John Curtiss, ultimately prevailed and steered the course of the Mathematical Tables Project to his liking. At the end, no one questioned the credentials of the loser, Arnold Lowan, or thought that he was unfit to lead a computing group or believed that he was incapable of developing new methods of scientific calculation. Most of the American computing groups of the Second World War ended in such a manner. As the country returned to the peacetime economy, the government reduced the budget for scientific and engineering research. In response, the leaders of the research laboratories, with or without an argument, cut their staffs, including their human computers.

As a world war gave place to a cold war, the United States began to rebuild and expand its research laboratories. The navy supported science through its Office of Naval Research. The air force created a private research company in California, the RAND Corporation. A dozen universities created laboratories in order to provide research services to the military. Of these laboratories, the only facility to develop a large computing staff was the one devoted to the mathematical methods of computation, the Institute for Numerical Analysis at the University of California, Los Angeles (UCLA).[1] The institute, a division of the National Bureau of Standards, did much productive work during a period of controversy and criticism. This criticism was not concerned with the quality of work done at the Institute for Numerical Analysis or the need for standard methods of computation or even the proper allocation of the institute's budget. It touched on the right of the institute's staff to claim the title of scientist and to hold stewardship over the country's scientific legacy. This criticism came not from within the family but from without, not from the greater community of scientists but from the American public. At the center of this criticism was a quiet but ambitious human computer, the former technical leader of the Mathematical Tables Project, Gertrude Blanch.

In the spring of 1948, Gertrude Blanch was having no part of Arnold Lowan's confrontation with John Curtiss. While Lowan was rallying the supporters of the Mathematical Tables Project, Blanch was quietly going about her business. In earlier years, she might have avoided the controversy by disappearing into John von Neumann's test of linear programming, but she needed no such excuse this time, as she was shortly to depart for California. Curtiss had asked her to be the assistant director for computation at the Institute for Numerical Analysis, a position that would oversee a computing office and engage in some mathematical research. She viewed the new job as an opportunity to move away from Lowan's shadow and establish her own reputation. Already she was preparing articles for *Mathematical Tables and Other Aids to Computation* and talks for the Eastern Association for Computing Machinery, a new society for those interested in electronic computers.[2]

Blanch spent the month of April cleaning her desk, packing her books, and retrieving those few trinkets that reminded her of all that the Mathematical Tables Project had accomplished. On her last day, the computers took a break from their linear programming calculations and gave her a farewell party. Surrounded by adding machines and piles of paper, they shared one last plate of food with Blanch, offered her a parting gift, and left her a card expressing their best wishes for her future. The computers came from the working-class neighborhoods around New York City: Brooklyn, Yonkers, Jersey City, the Bronx, Harlem, and the Lower East Side.[3] A few of them would be offered positions in Washington, D.C., but most would soon be looking for work in New York. None of them would be following Blanch.

At the start of May, Blanch emptied her apartment and said good-bye to her sister and to her nephews and nieces. She had no family of her own to keep her in New York, no man to tie her heart to Brooklyn, at least none that had been discovered in a routine security check.[4] Her train took three days to make the trip to California, time that would allow her to think about the future as the scenery went past the windows. The train lingered by the stockyards of Chicago before passing through the farms of Iowa and climbing the Rocky Mountains. She had been west once before, a visit to relatives in Mexico, but this trip was a fundamental change in her life. She was leaving her friends and her home to pursue one of the new opportunities that had been created for American scientists. There was more than a little risk in the trip, as the United States Congress had become more willing to support scientific research but had not entirely determined how the institutions of democracy should interact with scientists and scientific laboratories. The Research Board for National Security had failed in its first year and had not been replaced. The National Bureau of Standards was acting as the interim coordinator of postwar research.[5]

Blanch was more concerned with personal issues than with national policy. She was nearing her fiftieth birthday, and she truly wanted to make a mark on the world. The war had brought many women into science, but only a few of them were finding a place in the time of peace. Ida Rhodes and Irene Stegun, both members of the Mathematical Tables Project planning committee, had jobs at the National Bureau of Standards. Mina Rees, the assistant to Warren Weaver, had found a good position with the Office of Naval Research. Grace Hopper, the Vassar professor who had "considerable experience" with computing machines, had received a commission in the navy and was making her career in the laboratory of Howard Aiken. Blanch believed that the Institute for Numerical Analysis would be her intellectual home. John Curtiss had told her that it would function like an academic research department, "a group of peers working together as a university,"[6] though Curtiss had clearly indicated that the institute would be "a computing service containing both standard equipment and high speed equipment" for the aircraft industry of Southern California.[7] As she approached Los Angeles, she must have pondered her new role. Would she be able to pursue her own research, or would she spend her days doing calculations for others? Would she have a place of her own, or would she always be subservient to someone else? At the end of her career, would she be able to leave a legacy for others to follow, or would she have been more productive if she had never left that well-appointed photographic equipment office and taken a job in a work relief agency?

Blanch found a new apartment across the street from the UCLA campus, a short walk from her office. The Institute for Numerical Analysis was housed in an old rehabilitation hospital, a temporary wooden structure that had been constructed during the war. The building, which was only a single story tall, was located in an undeveloped section of the campus. When she first entered the institute building, in May 1948, Blanch encountered a scene reminiscent of the first days of the Mathematical Tables Project. The building was empty except for a corner office that was occupied by the institute's administrator, Albert Cahn. The computing office was an open room furnished with large tables. The computers would be able to open the windows to catch the breezes from the ocean or work outside on a central veranda that looked across the scrub growth toward the west.

The institute administrator, Albert Cahn, was no Arnold Lowan and had no authority over the scientific work of the organization. He handled correspondence for the institute, managed the organization's money, and promoted the new computing service.[8] In later years, the staff of the institute would tell the story that Cahn had been hired for the new institute because he had been found "sleeping in a common room at Princeton [University] without a job."[9] He had spent the war at the Chicago office of the Manhattan Project. Holding only a master's degree in physics, he had handled the tasks that were too simple, too routine, or too boring for the senior scientists with doctoral degrees. After the war, he had gone in search of an academic position, hoping that the cachet of the atomic bomb would help him find a good job. Receiving no offers, or at least none that interested him, he drifted back to the universities of the East Coast and found his way to Princeton.

Only two other offices were occupied when Blanch arrived. John Todd and Olga Taussky had come from England to be the first visiting mathematicians at the institute. It was "a very welcome and perhaps deserved change," remarked Taussky. The war had been invigorating and exciting, but it had taken its toll. She had found the demands of maintaining a home to be "rather strong on top of the mathematical activities."[10] Neither of the two was pleased by the prospects for British mathematics. The Admiralty Computing Service had little to do after the war, and many of its computers were preparing to join a new British national mathematics laboratory. Taussky and Todd were invited to be part of this group, but they "didn't think much of [the] leadership."[11] Discovering that it would be difficult to start a new research program within the University of London, they accepted an invitation to come to the institute. Much of the institute's work would be conducted by visitors, such as Todd and Taussky. It would have only eight permanent mathematical staff. The rest would come and go as the institute considered different problems and ideas.[12]

By the end of the summer, Blanch had assembled a computing office that resembled the wartime Mathematical Tables Project. The group was "founded partly on the older hand-machine techniques," observed the institute director, "and partly on theories radically new in numerical analysis."[13] It had sixteen computers, seven men and nine women, who had been recruited from the students of UCLA and the residents of West Los Angeles.[14] They had electric desk calculators, machines that could multiply and divide as well as add and subtract. They also had an accounting machine that they could use as a difference engine. Their first project was a new variation on the classical calculations of comet orbits, which had been requested by the U.S. Army. The army was building upon the spoils it had seized at the German rocket research center of Peenemünde and was developing a new generation of high-performance rockets. The work was still in its preliminary stages, but army officers were already thinking

43. John Todd, Olga Taussky, and John Curtiss on excursion from Institute for Numerical Analysis

about the difficulties of boosting a payload beyond the atmosphere. To prepare for the time when they could put a satellite in outer space, they requested tables of "rocket and comet orbits" from Blanch's computers.[15]

Most members of the institute staff believed that the human computers would be temporary workers, quickly replaced by a new "automatic computing machine," as electronic computers were then called. Initially, the National Bureau of Standards intended to purchase a computing machine for the Institute for Numerical Analysis. Three American companies were offering to build such machines: Raytheon Corporation of Massachusetts, Electronic Research Associates of Minnesota, and UNIVAC Corporation of Pennsylvania, which had been formed by the ENIAC designers, J. Presper Eckert and John Mauchly. John Curtiss was not satisfied with any of these proposals and convinced the senior staff at the bureau that the Institute for Numerical Analysis should build its own machine. He argued that a machine could be built quickly and that it would develop improved computing technologies.[16]

Under the best of circumstances, a speedy construction would take two or even three years after the first designers arrived at UCLA. To provide an interim machine-computing service, the institute acquired a device called the Card-Programmed Calculator. The Card-Programmed Calculator stood halfway between the IBM punched card tabulators and the new electronic computers. It had been created by engineers at an IBM customer, Northrop Aircraft Company. The engineers had recognized that they could create a fairly powerful computing device by connecting a new IBM card tabulator to one of the company's accounting machines. IBM had not intended for the machines to be joined in this fashion, but they quickly realized the value of the combination and adopted it as an official product. They eventually leased about 700 of these machines, far more than any of their early electronic computers.[17]

By the fall of 1948, the institute was offering computing services with the Card-Programmed Calculator to both academic researchers and commercial firms in California, including aircraft manufacturers and oil companies. Within six months, Albert Cahn could report that "calls upon this service are such that the facility has been expanded to almost double the size contemplated when the institute was established."[18] Originally, the new calculating machine was overseen by one of Blanch's computers, a woman named Roselyn Seidel, but with the increased demands, the leaders of the National Bureau of Standards wanted the device to be managed by someone experienced with punched card machines.[19] After brief negotiations, John Curtiss convinced Everett Yowell to leave Columbia University and join the Institute for Numerical Analysis. The mathematical staff considered Yowell a "significant appointment" to the institute, but Yowell recalled that the human computers were less enthusiastic. "Ros [Seidel] was not happy to have me hired over her."[20]

The Card-Programmed Calculator was not a full electronic computer, though it was a considerable improvement over the old punched card tabulators. Using special punched cards and an IBM plugboard, an operator could instruct the machine to undertake a complicated series of operations. It also had an electronic memory that could store forty-eight numbers of ten digits each.[21] With this memory, the machine could handle small simultaneous equation problems, such as the least squares calculations that had stymied George Snedecor at Iowa State University twenty years before. However, the new machine could not handle large problems, including the one that had appeared in the Mathematical Tables Project test of linear programming.

Some simultaneous equation calculations posed unusual problems for Blanch's computers. On these calculations, the computers could follow every step of the computing plan, check the work with a desk calculator to ensure than every step was done properly, and still produce values that were wildly incorrect. Though some blamed the computing plan, Blanch discovered a difficulty that would eventually be called "ill conditioning." Ill-conditioned simultaneous equation problems are fundamentally unstable, just as a coin balanced on its edge is unstable. Rounding the values of an ill-conditioned problem, a simple and innocuous act, can cause the calculation to collapse into a meaningless mess of figures. The only way to fix this problem is to reorganize the computing plan, producing a plan that is algebraically equivalent to the original calculation but avoids certain combinations of the four arithmetic operations.[22]

During the construction of the new machine, the mathematicians at the institute started exploring computing techniques that were difficult to do by hand. They experimented with linear programming and invited John von Neumann to visit the institute. Another mathematician developed "relaxation techniques," computing methods that began with a rough guess of the final answer and then slowly adjusted that guess. Many at the institute became interested in "Monte Carlo" techniques, methods that used random numbers to calculate answers.[23] The institute computers worked on such calculations as best they could, though they could rarely handle a large problem. The staff never developed the kind of computing skill that was found at the Mathematical Tables Project, as few computers stayed at the institute for more than a year. The rapid departure of human computers never seemed to bother the mathematicians, as they were looking ahead to the new computing machine.[24]

The electronic computer took three years to complete and went through three changes of name. Employees at the institute called the machine "Zephyr," after the great western wind, while the staff of the National Bureau of Standards in Washington referred to it as "Sirocco," the hot air from the desert. Eventually, the two groups settled on the government acronym SWAC, which stood for Standards Western Automatic Computer.[25] Though the members of the institute occasionally felt isolated from the academic centers of the East Coast, they generally had no regrets about their distance from Washington. John Todd wrote that a "certain distance from Washington was certainly desirable, for some mathematicians are uncomfortable with strict dress code and regular hours."[26] Their location may have spared them the need to arrive at work at 8:00 in the morning or to wear formal business clothing, but it did not insulate them from the political turmoil of the late 1940s.

The political conflict of this era was rooted in the looming problems of the Cold War, the legacy of Franklin Roosevelt, and the frustrations of the Republican Party. The Cold War placed the United States in a contest for global dominance with a frightening and powerful enemy. Roosevelt had reordered the political landscape in a way that kept the Democratic Party in power for over sixteen years. The Republicans, frustrated by the Democratic hold on power, were "traumatized and bitterly divided," observed the journalist David Halberstam.[27] The Republicans had always been a minority party and had held power only by pulling support away from the Democrats. During the period of Republican dominance in the nineteenth century, conservative leaders had often found it useful to discredit Democratic opponents by calling them secessionists, politicians sympathetic to the old Confederate states. Seventy-five years later, a new generation of Republican leaders accused the Democrats of being communists, agents of the Bolshevik revolution, traitors. The fact that some of the Democrats had actually been communists, or at least had been sympathetic to the Soviet Union, bolstered such charges. By the fall of 1949, Republicans equated communism with treason and pointed to the Soviet atomic bomb program for their proof. The Soviet Union had detonated a nuclear weapon on August 29, 1949, an event that surprised American weapons experts.[28] Only a few weeks before, the Central Intelligence Agency had predicted "that [a Soviet] atomic bomb cannot be completed before mid-1951."[29]

The center of the American political maelstrom was the House of Representatives Un-American Activities Committee, often called HUAC. It was filled with disaffected Republicans who were either valiantly defending the United States from an insidious enemy or trying to make a name for themselves by pinning the label of communism on public figures, depending upon one's point of view. The committee investigated the senior members of the Truman administration, the writers, directors, and actors of Hollywood, and the scientists in government service. In 1947, the House Un-American Activities Committee had begun an investigation of Edward Condon, the director of the Bureau of Standards, an investigation that struck uncomfortably close to the Institute for Numerical Analysis.

Condon had liberal inclinations, though he had never been a member of the Communist Party, and he had worked on the atomic bomb during the war. He spent most of the war at the University of California, but he had briefly served at the Los Alamos laboratory. At Los Alamos, he been engaged in "several arguments about security regulations," according to historian Jessica Wang, and soon resigned his position. In his resignation, he stated his belief that military control of scientific information was impeding work on the bomb. "To his mind, intellectual freedom and international cooperation were intimately linked," wrote Wang. "Scientific progess required open communications, free from military requirements of secrecy."[30] When the House Un-American Activities Committee learned of Condon's record at Los Alamos and his support for the open dissemination of atomic research, the committee reviewed his case and declared that Condon was "one of the weakest links in our atomic security."[31]

In general, the House of Representatives investigated only public figures and senior members of the administration, such as Condon. Junior employees, such as those who worked for the Institute for Numerical Analysis, were investigated by administrative committees that had been established by the Truman administration as a way of deflecting Republican criticism. The committee that oversaw the National Bureau of Standards and the Institute for Numerical Analysis was Department of Commerce Loyalty Board Number Two. In December, this board announced that it would require "pre-appointment loyalty checks of all research associates and guest workers who are located at National Bureau of Standards for more than one week."[32] This order included everyone working at the Institute for Numerical Analysis, the administrative staff, the eight permanent researchers, and the twenty annual visitors.

The first member of the Institute for Numerical Analysis to be called before Department of Commerce Loyalty Board Number Two was the senior administrator, Albert Cahn. No record has been found of Cahn's case, and all we can do is speculate that Cahn belonged to the broad class of liberal scientists, a group that worried many loyalty investigators. One of the few concrete facts we know of Cahn is that he signed the July 17 Petition, a document drafted by Manhattan Project scientists after the first test of the plutonium bomb. The petition requested that President Truman delay the use of the new weapon.[33] Manhattan Project officials annotated one copy of the petition, indicating whether each signatory was important or unimportant to the project.[34] While this act may have brought Cahn to the Loyalty Board, it may also have been only one of several pieces of evidence against him. In all, seventy scientists signed the petition, and some of them were never questioned by an investigative panel.[35]

After Cahn was summoned before the Loyalty Board, fourteen months passed before his case was decided. The delay did not bode well for him. During this period, Klaus Fuchs confessed to passing atomic secrets to the Soviet Union; Alger Hiss was convicted of perjury; Mao Tse-tung and his followers established the People's Republic of China; Senator Joseph McCarthy gave a rambling but inflammatory speech in which he claimed to have evidence of communists in the State Department; North Korean troops invaded South Korea; and, finally, the FBI arrested Julius and Ethel Rosenberg on charges of spying for the Soviets. When the Loyalty Board finally reviewed Cahn's case, they concluded that the evidence against Cahn seemed to fit into a broader pattern of threats against the

United States. They judged that the institute administrator was a security risk and placed him on administrative leave.[36]

Following Albert Cahn came Gertrude Blanch. Blanch was especially vulnerable, as she had already failed not one but two security investigations. The first investigation had been conducted in 1942, when the Mathematical Tables Project was certified as an essential relief project. That review had denied her a security clearance. The second investigation had occurred in 1946, when Blanch had been invited to join the computing staff at Los Alamos. In that review, the FBI quickly uncovered the results of the 1942 investigation and declared Blanch untrustworthy. When the administrators of Los Alamos received this verdict, they quietly withdrew their invitation.[37] Both of these judgments were part of the record presented to Department of Commerce Loyalty Board Number Two in the spring of 1951. When the board considered her case, they concurred with the earlier decisions and judged her untrustworthy, but before she could be placed on administrative leave, Blanch appealed the ruling and requested a formal hearing before the board. The board accepted her request and scheduled her hearing for May 1952.[38]

"There are only a few times," wrote the 1950s sociologist William Whyte, when an individual "can wrench his destiny into his own hands— and if he does not fight then, he will make a surrender that will later mock him."[39] By nature, Gertrude Blanch avoided public confrontations. Her moments of strength were private moments. She had quietly postponed a college education in order to support her mother. She had gently guided the poverty-stricken computers of the Mathematical Tables Project. On at least one occasion, she had stood firm against the bluster of Arnold Lowan. In this last situation, Blanch had to confront the charges against her or surrender her place as a scientist. The case against her was based on the five points that had been identified in 1942. The first three were circumstantial. First, the FBI had an informant who claimed to have seen her purchase a copy of the *Daily Worker* sometime during the late 1930s. Next, at approximately the same time, the New York office of the WPA identified her as a "Red." Finally, she had shared an apartment with her sister and brother-in-law, who were open members of the Communist Party. The last two points were harder to dismiss. During the 1930s, Blanch had registered as a member of the Democratic Labor Party, an organization that the FBI claimed was "captured by the Communists." She had also signed petitions for political candidates who openly identified themselves as communists.[40]

Blanch approached her loyalty hearing with the same logical care that she brought to her computing plans. She knew that she was a minor figure in the drama and would gain nothing from grand statements or denunciations. She could not behave as the author Lillian Hellman had behaved in front of the House Un-American Activities Committee and declaim, "I cannot and will not cut my conscience to fit this year's fashions."[41] Her best strategy lay in arguing that the charges were false, in showing that she was a valuable government scientist, and in suggesting that there was nothing to gain from dismissing her. She gathered letters of support from her fellow scientists, asking those who knew her personally to attest to her loyalty and her value to the government. She also drafted and redrafted her statement for the board, refining the logic, honing the evidence, creating the kind of defense that only a mathematician could make.

"Let us assume, for a moment," Blanch began, "that both my sister and my brother-in-law could be called somewhat radically inclined. It does not follow that I, too, must share their views—in fact, the probability is not even high that there is a correlation between their views and mine." It was not an easy task to dispose of a communist who shared the same parents and who had once shared the same house. By starting with this issue, Blanch risked losing credibility with the board, but if she could make her case, then the rest would be easy to handle. She spent a few minutes discussing her relationship with her sister, commenting on how sisters could be close but have different political opinions, and then moved to the subject of communist periodicals. "I personally do not read the *Daily Worker*," she remarked; "the newspaper does not happen to be to my taste, nor does it reflect my political sympathies." She ignored the issue of her voting registration and would not deny that she had signed petitions, though she admitted only to signing papers that called for the "admittance of Jews to Palestine." She concluded her presentation by stating, "I think I may say that I am conservative in my tastes, and I have never leaned toward radical movements of any sort."[42]

Blanch understood herself well enough to know that she was not a revolutionary and that she preferred to work within existing social structures, even at times like these. If she were a communist, then she was the sort of communist that had been educated at the czar's expense, a communist who liked nothing better than Western art, music, and theater, a communist who deeply desired to purchase a home of her own. Her most radical idea was the notion that women could be mathematicians, that they could work outside the family, that they could have a role in public affairs. Her presentation was bolstered by interviews with neighbors and coworkers. A few suggested that she might have liberal inclinations, but none questioned her commitment to the United States. When the board passed judgment on her case, they declared that they had "no objection on grounds of Loyalty" to her continued employment at the institute.[43] The hearing resolved the charges against Blanch and allowed her to return to her job, but it did not end the threats to the National Bureau of Standards and the Institute for Numerical Analysis.

The Bureau of Standards was attacked in the winter of 1953, after Dwight D. Eisenhower was inaugurated as president and returned the Republican Party to power. The new cabinet member responsible for the National Bureau of Standards, Secretary of Commerce Sinclair Weeks, was highly critical of government workers. He frequently spoke of removing employees who were trying to "hamper, hoodwink and wreck the new administration." He complained about "the theories of foreign socialists" and "the notions of local egg-heads" and finally promised that he was "going to improve the situation by finding means to replace [disloyal employees]."[44]

In the first weeks of his administration, John Curtiss came within Weeks's sights. Curtiss was an unconventional man, a "bachelor who enjoyed fast cars and plenty of good food and drink," according to his friend John Todd.[45] He was known to hold large and boisterous parties in his small apartment. "He invited hundreds (it seemed) of guests," reported one member of the institute staff, "and one was very lucky if you managed to get inside."[46] Behind the flamboyant lifestyle was the suggestion, uncomfortable to the age, that John Curtiss might be a homosexual. Government agencies generally considered such individuals vulnerable to blackmail and hence poor security risks.[47] In early March, Curtiss was informed that he had been identified as a likely homosexual and that he must choose between a public dismissal and a quiet resignation. His colleagues encouraged him to appeal the ruling. After considering his situation, Curtiss decided to leave quietly. "I have a great desire *not* to be a cause celebre," he wrote, and added that he desired to have "a scientific career which will be a little more constructive than that of a professional victim."[48]

A month after Curtiss left, Secretary Weeks turned on the director of the National Bureau of Standards, a physicist named Allen Astin. Astin had replaced the embattled Edward Condon, who had resigned during his investigation by the House Un-American Activities Committee. Condon "cited his low government salary as his reason for leaving," observed one historian of Condon's career, "but more likely the continual burden of having to respond to [the House Un-American Activities Committee] accusations had grown to outweigh the appeal of public service."[49]

Astin conflicted with Secretary Weeks over an event that was interpreted one way by National Bureau of Standards scientists and quite another way by members of the business community. It concerned a report by bureau scientists on battery additives, chemical mixtures that promised to enhance the performance of automobile batteries. The report reviewed one such additive, called AD-X2, and judged that it had no value.[50] To the scientists, the report was a simple scientific conclusion. To the manufacturer of AD-X2, the report was unnecessary government interference in the marketplace. The manufacturer found enough sympathetic ears in Congress to provoke a debate on the report and put the National Bureau of Standards on the defensive. Allen Astin supported his scientists, much to the dismay of his boss, Secretary Weeks. Concluding that Astin was just one more example of "deadwood and poison oak,"[51] Weeks asked for his resignation.

Weeks did not anticipate the extent to which the scientific community would fight for its autonomy, its ability to make decisions without external interference. He also did not foresee the scale of political connections that the scientists could rally. The key players were not the scientists in the National Bureau of Standards, who might be expected to resign in protest, but outside researchers with an interest in the agency. The fight was led by the board of visitors, the scientists and industrialists who offered their advice on bureau operations. "The most influential member of the committee, who seemed to have Mr. Weeks' respect, was Mervyn Kelly, then President of Bell Telephone Laboratories," recalled Astin.[52] They also drew support from university presidents, academic scientists, and members of the press, such as *Washington Post* columnist Drew Pearson.[53]

Faced with a storm of bad publicity, Weeks asked Astin to return "on a temporary basis."[54] Astin used the opportunity to strengthen the hand of bureau scientists. He asked for a full review of the bureau by the National Academy of Sciences. The report, released in the summer of 1953, validated the position of the National Bureau of Standards in the AD-X2 matter, but it concluded that the bureau faced a fundamental conflict between its military research and its civilian duties. It recommended that the bureau return all military-sponsored research to the military agencies. Astin accepted the report and implemented its recommendations. He purged the bureau of all military research, one-third of its budget. The Institute for Numerical Analysis was one of the first units touched by the order. Though the institute appeared to be a civilian laboratory located on a university campus, it was actually financed by money from the air force and the navy. Astin announced that the institute would be closed in June 1954, that its equipment would be given to UCLA, and that its computing staff would be dispersed.[55]

Anticipating the demise of the Institute for Numerical Analysis, Gertrude Blanch resigned her position at the end of 1953. The computing office at the institute had started to shrink even before Astin announced the closure of the UCLA office. The engineers and researchers who needed computing services had begun acquiring calculating equipment of their own. The most common computing machines were the IBM 604 multiplying punch and its near relative, the Card-Programmed Calculator. The larger customers had ordered IBM's first electronic computer, the Model 701, which had arrived on the market a year before. Fully six of the first twelve customers for the 701 were companies that had once requested computations from the Institute for Numerical Analysis.[56]

Blanch could have sought a position with IBM, a job that probably would have taken her back to New York City, but instead she accepted an offer with the mathematics department of the ElectroData Company. ElectroData was located in nearby Pasadena and proclaimed that "mathematicians are the heroes of the new industrial revolution."[57] The firm had been founded by Herbert Hoover and had built electronic instruments during the Second World War. In 1954, it was trying to enter the computer business by building a small machine for engineering applications, the kind of work that was done by the firms that had once patronized the Institute for Numerical Analysis. The company had assembled a talented group of scientists, though none of them belonged to the inner circle of computer designers, the group that could trace their knowledge of computing machines back to the Moore School Lectures of 1946. In addition to Blanch, the office included Clifford Berry, an engineer who had helped John Atanasoff build his computing machine at Iowa State College, and Ted Glaser, a young physicist who was proving to be an exceptional computer scientist even though he had lost his sight as a child.[58]

The winter of 1954 was a poor time to be a government scientist and an equally inauspicious moment to join a small and inexperienced computer company. The managers of ElectroData were struggling to control the design of the machine, to keep the project on schedule, and to develop a customer base. Shortly after Blanch arrived at the company, the senior managers realized that their company needed computer programmers more than it needed mathematicians, and they replaced the "Mathematics Department" with the "Technical Services Group."[59] After a couple of weeks on the job, Blanch concluded that "the place in Pasadena wouldn't last too long," but she did not know where she might find another job.[60] Most of the old computing groups had been closed, and their leaders had retired. The dean of human computers, L. J. Comrie, had died. R. C. Archibald had resigned from *Mathematical Tables and Other Aids to Computation.*[61] H. T. Davis was preparing a final volume of mathematics tables, but his numbers had been calculated by his National Youth Administration computers and did not represent contemporary work. Clara Froelich had left Bell Telephone Laboratories, said good-bye to calculation, and departed for an extended vacation in Mexico.[62]

For Blanch, one of the few benefits of working for ElectroData was the contact it gave her with air force scientists. The company was a contractor for the air force and worked with many scientists who had known Blanch at the Institute for Numerical Analysis or the Mathematical Tables Project. One of these scientists was Knox Milsaps, the chief mathematician of the air force and a founder of the new Aerospace Research Laboratories at Wright Field in Ohio. During the war, Milsaps had worked with Blanch on several mathematical problems, notably one with an expression called the Mathieu function. In the winter of 1954, he was a regular visitor to the ElectroData offices, as he was planning on purchasing one of the new machines for the aerospace laboratories. As he passed through the building, he would often stop at Blanch's desk, chat about old times, ask about her new work, and inquire, "When are you coming to Wright Field?" At first, she deflected his questions. She was happy in California, and after her appearance before the Department of Commerce Loyalty Board, she was reluctant to expose herself to another security examination. Yet Milsaps was persistent, ElectroData was in financial turmoil, and Blanch recognized that her job was not interesting. "So one night I faced the floor," she recalled, "and decided to accept the job in Wright Field."[63]

While Blanch prepared to move to Ohio, the FBI began one more review of her background. The agents in charge of her case found the results of the earlier investigations and decided that they should reopen each of the charges. In addition to the five issues that had been dismissed by the Loyalty Board, they requested a study of one more concern: "Subject is not known to be married or have ever been married."[64] By this time, the political wars of the early 1950s were starting to wane. The Republicans were more comfortable with power, and the most violent of the anticommunist voices had lost some of their control over the public and their party. The Washington FBI office rejected the request, stating that a new investigation would have a "greater possibility of embarrassment to the Bureau . . . than [of] attaining any information of value from the subject."[65] They formally closed her file and granted Blanch the full security clearance needed for her work. She moved to Dayton, became head of mathematical research at the laboratory, and "learned to drive on icy streets the same as everybody else."[66]

Before Gertrude Blanch could settle in Dayton, before she could buy a house and learn the neighborhoods, she was contacted by Phil Morse about a project that he called the "handbook for the ordinary computer."[67] Morse had concluded that the new electronic computers would be of little help to his generation of scientists and engineers. He speculated that a decade would pass before the typical scientist would have easy access to them. From observing the machines at MIT, he had learned that "programming took weeks, not minutes," and that small and mid-sized problems could be handled much more quickly by hand.[68] Morse told Blanch that he wanted to hold a conference to discuss such computing problems and asked who should be invited to it. Blanch gave him a long list of "tablemakers," as she described the computers, who worked at the Aberdeen Proving Ground, the navy's proving ground at Dahlgren, Virginia, the Atomic Energy Commission, and the University of Illinois, though she was quick to characterize one Illinois faculty member as "anti-table."[69] She did not list Arnold Lowan, as she was not quite sure how he would feel about the invitation. Lowan had already received an invitation from Morse and had replied that "my chief interest in the conference in mathematical tables derives from the assumption that perhaps it is still possible to bring to life the 'Math Tables Project' in *New York*."[70] After Morse explained that he had no interest in reviving the Mathematical Tables Project, Lowan declined to come.

When the conference met in September 1954, most of those in attendance were senior computers of the Second World War: Gertrude Blanch, John Todd, Milton Abramowitz, and even Blanch's former critic Wallace J. Eckert, who had become an employee of the IBM Corporation. The list also included Nicholas Metropolis from the Manhattan Project; John Tukey, one of the Applied Mathematics Panel mathematicians from Princeton; and the recently fired John Curtiss, who had a temporary job at New York University. Adele Goldstine, who had worked with the computers at the University of Pennsylvania, came only as a spouse. Morse had invited her husband, Herman, because of his contributions to the ENIAC.[71]

As they discussed the problems of midsized calculations, the conference members quickly agreed that they needed to produce a book containing "tables of usually encountered functions" as well as graphs, mathematical analyses, and "other techniques useful to the occasional computer."[72] They gave the project a shorthand name, the "new Jahnke-Emde," a term which referred to the book *Funktionentafeln mit Formeln und Kurven* (*Tables of Functions with Formulae and Curves*), by Eugen Jahnke and Fritz Emde.[73] This book was nearly fifty years old but remained popular with those who worked with applied mathematics.[74] Twice during the 1940s, the National Bureau of Standards had suggested revising the book. The first time, in 1941, the bureau was preparing for war and could not find the money for a revision. The second time, in 1947, bureau scientists were organizing their new applied mathematics laboratory and were confidently predicting that the electronic computer would remove the need for such a book. Seven years later, when questioned by Phil Morse, the leaders of the bureau acknowledged that the electronic computers were not the solution to all calculating problems and that they could find some money for a new Jahnke and Emde.[75] The book would be prepared by the bureau's Computation Laboratory and would be edited by Milton Abramowitz and Irene Stegun, two of Blanch's closest lieutenants from the Mathematical Tables Project. It would contain twenty-two chapters, each written by a different volunteer. As the conference came to a close, Blanch agreed to write a chapter on Mathieu functions. After a little prodding, the sulking Arnold Lowan agreed to contribute a chapter as well.[76]

The project was formally named the *Handbook of Mathematical Functions with Formulas, Graphs, and Mathematical Tables*, but the mathematicians who worked on the book usually called it "the handbook" or "AMS 55," after its place in the National Bureau of Standards publication list, or "Abramowitz and Stegun," after its editors. Milton Abramowitz did the bulk of the preliminary work and recruited most of the contributors, but he died on a hot and sticky July day in 1957, when he imprudently attempted to mow his lawn in suburban Washington, D.C.[77] Irene Stegun retrieved the plans for the book from her partner's desk and finished the project. She corresponded with the contributors, corrected their chapters, and slowly pulled the material into a complete reference work.

The handbook required a decade to complete, a decade that marked a radical change in the electronic computer. In 1954, computers were handcrafted devices that could be found only in government laboratories and large businesses. By 1964, the year of the book's publication, computers were standard products that could be purchased from a dozen different vendors. The actual date of publication coincided with the announcement of the IBM System 360, a family of machines that IBM chairman Thomas J. Watson Jr. proudly called "the most significant product announcement in IBM history."[78] The System 360 would anchor IBM's product line for twenty-five years and would move computing into offices and laboratories that had never had access to the machines of 1954.[79] To those who were promoting the new electronic computers, the *Handbook of Mathematical Functions* seemed an anachronism, a tool for modern science that had been produced by the old human computers. Scientific calculation had become a small part of the subject known as computer science. Computer scientists were increasingly interested in databases, sorting methods, the manipulation of text, and the representation of human reasoning. The tables and formulas of the handbook appealed only to a small group of computer researchers.

Though it no longer represented the central issues in computation, the *Handbook of Mathematical Functions* ultimately validated the vision of Phil Morse. Neither the IBM 360 nor any of the other machines announced in 1964 was able to handle all of the small and midsized scientific computations that were found at universities and government research labs. Scientists, still having to do some calculations by hand, turned to the contents of the handbook for assistance. They gave good reviews to the book and purchased copy after copy for their laboratories. Within a few years, it became the most widely circulated scientific text ever published.[80] "I don't know what I'd do without it," wrote Blanch, who called it "one of the books I would keep, if I go anywhere, even if I don't look at it again."[81]

The *Handbook* sat on Blanch's desk until she left government service in 1967. It stayed with her in retirement as she wrote a textbook on calculation. In spite of her prediction, it was not taken on her tours of Europe, though one can easily imagine her discussing references to the book with her traveling companion, Ida Rhodes. It was found in her personal effects after her death. As she looked back on her career, the *Handbook of Mathematical Functions* was the symbol of all that she had accomplished. It appeared at the perihelion of her time in government service, her moment at the center of power. She had just finished a successful decade with the Aerospace Research Laboratories. She had published

44. Gertrude Blanch and the electronic computer

45. Gertrude Blanch and President Lyndon Johnson

twenty mathematical papers, been elected a fellow of the American Association for the Advancement of Science, been promoted to the highest rank possible for a government scientist, and been publicly recognized for her contributions to the air force.[82] Blanch had even learned a little about electronic computers, though she was never interested in programming. Programming could be left to assistants. As a newspaper article described her, Blanch was "the brain behind the mechanical brain,"[83] but a photo taken at the same time shows her glowering at a new electronic computer.

On March 3, 1964, just a few weeks before the *Handbook* was released to the public, Gertrude Blanch arrived at the White House in Washington, D.C., wearing the finest of her clothes and the tallest of shoes. Along with five other women, she was ushered through the east entrance, escorted down a long hallway, and asked to wait in a small public room with a presidential photographer and a few representatives of the press. President Lyndon Johnson soon walked into the room, greeted the women, and took his place at a podium. "I believe a woman's place is not only in the home," he began, "but also in the House and Senate."[84] Blanch and the other five women glanced at each other and smiled knowingly. None of them held an elective office, but all of them served in government. They had come to the White House as part of Lyndon Johnson's tribute to Eleanor Roosevelt, who had died the prior November. Johnson had wanted to emphasize the impact of the former First Lady by honoring six distinguished women in government service who had begun their careers when Roosevelt lived in the White House.[85] After the speech, Johnson moved around the room and shook the hands of his honored guests. He paused briefly with Gertrude Blanch, bending over to share a word or two. Like Blanch, he had started his professional career at a work relief agency. Between 1935 and 1937, he had served as director for the Texas office of the National Youth Administration and had distributed research funds to the universities of his state. Had there been more time for president and mathematician to speak, they might have found that they shared much in common, but Johnson quickly strode out of the room and vanished down the hall that led to the Oval Office.

From the White House, a car carried Blanch to a Washington air force base, where a dinner had been prepared in her honor. There, amidst friends and colleagues, she relaxed and smiled and danced. These were rare expressions of emotion for one who kept herself under tight discipline. That night, she enjoyed herself as if the whole world moved in orbit around her, for she had heard her president commend her as the top mathematician in the air force, as a founder of the scientific discipline called numerical analysis, as a patriotic citizen who had served in time of war for the Applied Mathematics Panel, as Gertrude Blanch, who had once worked for the WPA and had once managed a staff of human computers.

Final Passage: Halley's Comet 1986

> We cross our bridges when we come to them and burn them
> behind us, with nothing to show for our progress but the smell
> of smoke, and the presumption that our eyes once watered.
> Tom Stoppard, *Rosencrantz and Guildenstern*
> *Are Dead* (1967)

THE COMPUTATIONS for the 1986 return of Halley's comet began shortly
after Gertrude Blanch retired from scientific life in 1967. Though she was
not the last professional human computer, her departure coincided with
the final days of many computing offices. The National Bureau of Standards, now identified as the National Institute of Standards and Technology, closed its Computational Laboratory and reassigned the few remaining veterans of the Mathematical Tables Project to other divisions.
The American *Nautical Almanac* moved from punched card equipment
to electronic computers. Observatories were either acquiring their own
small computers or purchasing the services of larger machines. A few
businesses, such as insurance firms and petroleum refiners, retained small
staffs of calculating assistants, but these, too, were being replaced with
IBM 370s, DEC PDP-8s, Burroughs B-6500s, and other computers that
were powered by electricity and sported numerical names.

With the 1986 return, astronomers returned to the problem of testing
Newton's theory of gravitation in order that they might reduce Andrew
Crommelin's 1910 discrepancy of two days, sixteen hours, and forty-
eight minutes. The basic principles of Isaac Newton's gravitational theory
were never questioned, but researchers hoped that some slight modification might produce a more accurate prediction of the comet's perihelion.
One organization prepared ephemerides with a gravitational force that
was slightly weaker than the one identified by Newton. Another team
postulated the existence of one more giant outer planet, which they identified as "Planet X," and tried to find an orbit for the planet that would
account for Crommelin's missing two and a half days. A final group hy-
pothesized that the discrepancy was created by the comet itself. They suggested that the nucleus of the comet acted like a weak rocket engine because of a phenomenon known as outgassing. As the comet approached
the sun, its surface was warmed by the light until it boiled away as vapor.
This vapor created the tail and produced a gentle thrust that slowed the

comet's approach to the sun and sped its retreat into the distant spheres of the solar system.[1]

The ephemerides for the 1986 return, the fourth return since the death of Edmund Halley, were the first prepared in the United States. They were done at the Jet Propulsion Laboratory, an American research center that had flourished during the space race of the 1960s. The laboratory designed unmanned spacecraft and managed scientific space missions. It occupied a patch of flat land next to a Los Angeles freeway and resembled a bland contemporary office building. Like an observatory or an almanac office, the laboratory had a large central room, but this room was filled with displays and control panels rather than with telescopes or human computers. From this room, laboratory scientists controlled spacecraft in orbit around the earth, sitting on the face of the moon, and speeding across the solar system toward the planet Mars.

The calculations for the fourth return were prepared in a back office of the Jet Propulsion Laboratory by a young researcher named Donald Yeomans. Yeomans was a new addition to the laboratory and had recently finished a doctorate in astronomy. He prepared a mathematical model of the comet's orbit that combined the methods of Andrew Crommelin with an analysis of outgassing. Instead of preparing a computing plan, Yeomans created a computer program, a list of instructions for an electronic computer. The program was written in a language called FORTRAN IV, which was then popular among scientists and engineers. Programming was more exacting than planning, as electronic computers were less forgiving than their human counterparts. Unlike a good human computer, who could correct errors on computing sheets, an electronic computer would follow the instructions blindly, executing each operation even if the action made no sense.

Yeomans submitted his program to a central computer at the laboratory, a UNIVAC 1108. In the years that followed the Second World War, the name "UNIVAC" had been nearly synonymous with the term "electronic digital computer." The moniker had been invented by J. Presper Eckert and John Mauchly, the designers of the original ENIAC machine, as the name for their first commercial computer.[2] Eckert and Mauchly had stopped designing computers in 1958, but the name UNIVAC survived as the brand name for machines sold by the Sperry Rand Corporation.[3] The UNIVAC 1108 was considerably faster than a room of human computers, but it was shared by the entire staff of the Jet Propulsion Laboratory. "We were lucky to get one run per day," Yeomans recalled. "I had to come in nights and weekends to get the necessary turn around time."[4]

Yeomans needed repeated calculations in order to remove the 1910 discrepancy. He computed old orbits of the comet, including those that

46. Comet Halley, 1986, from Giotto Spacecraft

ended in 1682, 1758, and 1835. He compared his predictions of those early returns with the actual observations, adjusted the values in his mathematical model, and repeated the process. He finished the computation in 1977, reported that the next perihelion of Halley's comet should occur on February 9, 1986, at 15:50 (Universal time), and claimed that his result was accurate to within six hours.[5] He revised his prediction in 1982, after an earthbound telescope caught the first images of the comet. The actual perihelion occurred at 10:48, five hours and two minutes before Yeomans's original prediction.[6]

Even though the fourth anticipated return of Comet Halley had been well publicized, the event quickly lost its hold upon the public consciousness in the spring of 1986. The comet had skimmed low in the skies of the Northern Hemisphere and had not presented a dramatic picture to most observers. The professional astronomers clung to Halley's comet for only a little longer. They had observations to compare, theories to analyze, papers to write. During the next few years, they demonstrated how well modern astronomers could compute the paths of these irregular visitors.

In 1994, they followed the frightening Levy-Shoemaker comet, which slammed into Jupiter and thereby demonstrated the role that the outer planets have played in protecting the Earth. Three years later, astronomers computed the orbit of the stunning Hale-Bopp comet, which had a nucleus several orders of magnitude bigger than the core of Halley's comet and left a tail that stretched across the night sky.

As the astronomers put Halley's comet behind them, the scientists at the Jet Propulsion Laboratory discarded their old tools of computation. They removed the central UNIVAC 1108 and placed a small computer on the desk of each researcher. FORTRAN IV was replaced by more sophisticated languages and was eventually succeeded by programs for symbolic mathematics. These programs could handle many of the tasks that had once been undertaken by the mathematicians on planning committees, such as the manipulation of mathematical expressions, the operations of algebra, and even the solution of many problems in calculus. Not only could they calculate numerical answers, they could help a scientist derive a concise mathematical theory of a physical phenomenon. Had the comet reached its perihelion in 1996 instead of 1986, it would have been predicted by a much more refined computing process. "If I had access to the more sophisticated algorithms and code that we now employ on [electronic computers]," wrote Yeomans, "[then my] Halley prediction would have been even more accurate than the realized [value]."[7]

The scientists who convene every seventy-five years to compute the perihelion of Halley's comet have had trouble looking both forward and backward. They might project the next perihelion with some improvement in accuracy, but having computed the signs of the skies, they generally fail to foresee the signs of the times. Edmund Halley, who struggled to produce even a crude estimate of the first return of his comet, did not anticipate either the work of Clairaut, Lepaute, and Lalande or de Prony's radical division of labor or the factory system of computation. By contrast, Charles Babbage overestimated the demand for computation. Alone among those who witnessed the 1835 return, Babbage understood the nature and potential of computing machinery. He would have been surprised to learn that one century later, the largest computing organization was a work relief project that resembled an office of the 1790s. Andrew Crommelin, L. J. Comrie, and the others who saw the 1910 return had grand visions for computation, but they would not have appreciated that the computing machine would replace the mathematical table and that all their labor to correct table entries would eventually be forgotten.

In a like manner, the scientists have freely taken of the historical data and old computations, but they overlooked the people who handled the calculations: Nicole-Reine Étable de la Brière Lepaute of the Palais Luxembourg, George Airy's boy computers at the Greenwich Observatory, Oswald Veblen's staff at Aberdeen, the WPA workers of Gertrude Blanch, and even the table makers of the *Handbook of Mathematical Functions*. It seems likely that the scientists of the 2061 return of Halley's comet will shrink Yeomans's discrepancy to a few seconds, as they will have detailed observations of the comet through its entire orbit and computing tools beyond what we can now imagine.[8] It seems equally likely that those same scientists will know little of the daily lives of the computational workers who are so common in our age, the computer programmers, the network managers, and the Web designers. The generation of 2061 may be surprised when an elder of our time explains that she once worked with electronic computers, that she took pride in her skill, that she had been pleased to be part of a scientific endeavor, and that she had once opened a book and studied a subject called calculus.

Acknowledgments

ANYONE WHO has cherished the notion of personal authorship is always surprised to discover how many people are needed to write a book. I have a long list of people who graciously contributed their time, their expertise, and their interest to this project. I owe them much.

One cannot be too grateful for librarians, for it is upon their sturdy shoulders that any research is built. I particularly wish to express my thanks to Greg Skelton of the Naval Observatory Melvin Gillis Library, who has been ready at a moment's notice to find information for this book, but he shares this honor with many a good librarian and archivist, especially Marjorie Charliante, Ed Creech, Ann Commings, Mitchel Yokelson, and Shawn Aubitz of the National Archives and Records Administration, all of whom were incredibly patient with me as I struggled through federal records. I also need to thank Becky Jordan, Iowa State University Libraries; Alison Osborn, Smithsonian Institution Archives Center; Kevin Corbitt, Charles Babbage Institute, University of Minnesota; Kate Perry, Girton College Archives; Jane Knowles, Schlesinger Library, Harvard University; Janice Goldblum, National Academy of Sciences–National Research Council; Sheldon Hochheiser, ATT Corporate Archives; Andrea P. Mark Telli and Teresa Yoder, Chicago Public Library, Harold Washington Library Center; Elaine Engst, Cornell University; Jan Hermann, Naval Surgery Headquarters; Martha Mitchell, John Hay Library of Brown University; Brenda Corbin and Steven Dick, Naval Observatory Melvin Gillis Library; Elizabeth Harter and Ericka Gorder, New York University; George Washington University's Gelman Library; Donald Glassman, Barnard College Archives; Annie Accary, Paris Observatory; Bonnie Ludt, California Institute of Technology; Amey Hutchins, University of Pennsylvania Archives; Hollee Haswell, Low Library, Columbia University; Tony Lawless and Julie Archer-Gosnay, University College London; Shulamith Berger, Yeshiva University; Peter Hingley, Royal Astronomical Society; Karma Beal, National Institute of Standards and Technology Library; Ellen Alers, Smithsonian Archives; Dennis Bitterlich, University of California at Los Angeles Archives; Jessy Randall, Colorado College Library; Mark Renovitch, Franklin Roosevelt Presidential Library; Kevin Leonard, Northwestern University; G. David Anderson, George Washington University Archives; Paul Lascewicz, IBM Archives; Anne Frantilla and Nancy Barlett, Bentley Library, University of Michigan; Thomas Malefatto, Indiana University; Keir Sterling, U.S. Army; Carole Prietto, Washington University Archives; as well as librarians from the Library of Congress Manuscript Division, National Agricultural Library, National Oceanic and Atmospheric Administration Library, Naval Historical Center, U.S. Army Historical Center, Findley University Library, University of Chicago Archives, University of Iowa Archives, Dudley Observatory Archives, Federal Bureau of Investigation Freedom of Information Act Office, and the British Library.

I have been aided by an extraordinary group of research assistants, friends, and colleagues who have helped me with library work or specific details in the book, including Mark Brown, Elizabeth Gonser, Sarah Hillinsky, Johanna Osborne, Karie Orin, Erin Voelger, and Jonathan Wye. Emily Danyluk helped prepare the final manuscript and index. I am especially grateful to Jim Burks and David and Donald Sterrett, who bolstered my humble French skills. Three university registrars have given me key information that has shaped this book: Ed Loyer and Toni Kessler of the University of Michigan and Brian Selinsky of George Washington University.

Parts of the manuscript have been read by a number of friends who have added their comments and advice, including Jean Grier, Peter Grier (*il miglior fabbro*), Mary Croarken, Michael Olinick, Jim Burks, Peggy Kidwell, John Todd, Marcus Raskin, Tim Bergin, and Liz Harter. I am especially grateful for the careful reading and suggestions of Ted Porter. For their kindness, sharp eyes, and insightful suggestions, I absolve them of any responsibility for such errors as remain in the manuscript. I have also had guidance from a number of colleagues who have shared their thoughts on this project, including Horace Judson, Freeman Dyson, Susan Strasser, Anne Fitzpatrick, Bill Aspray, Martin Campbell-Kelly, Paul Ceruzzi, Jennifer Light, James Cortada, Tara Wallace, Ed McCord, Michael Moses, Steve Dick, Ian Bartky, Ruth Wallace, Chris Sterling, Ed Berkowitz, Howard Gillette, and Barney Mergan.

Appendix: Recurring Characters, Institutions, and Concepts

THIS BOOK contains a large number of characters, many of whom disappear after their initial introduction and return several chapters later in the narrative. This section is meant to be a guide to the most important recurring people, institutions, and concepts. It gives a brief description of key items; a more complete list of characters is found in the index.

ABERDEEN PROVING GROUND
> U.S. Army testing site for artillery, opened in 1918 with Princeton mathematician Oswald Veblen on staff

ABRAMOWITZ, MILTON (1913–1958)
> Planning committee member of Mathematical Tables Project, editor of *Handbook of Mathematical Functions*

ADMIRALTY COMPUTING SERVICE
> English computing organization of the Second World War led by John Todd

AIKEN, HOWARD (1900–1973)
> Harvard graduate student and designer of MARK I mechanical computing device

AIRY, GEORGE BIDDLE (1801–1892)
> English Astronomer Royal, reorganized Greenwich Observatory computers with principles of factory production

ALEMBERT, JEAN LE ROND D' (1717–1783)
> French mathematician and critic of calculations for first return of Halley's comet

APPLIED MATHEMATICS PANEL
> Division of Office of Scientific Research and Development concerned with mathematical research, led by Warren Weaver and headquartered in New York City

ARCHIBALD, RAYMOND CLAIRE (1875–1955)
> Editor of the journal *Mathematical Tables and Other Aids to Computation*

ATANASOFF, JOHN VINCENT (1903–1995)
> Iowa State College physics professor, designed small computer to solve simultaneous equations problems

BABBAGE, CHARLES (1791–1871)
> British mathematician and designer of computing machine called difference engine

BELL TELEPHONE LABORATORIES
American industrial research laboratory founded in 1925, supported computing groups and developed early electromechanical computer

BENNETT, A. A. (1888–1971)
Brown University mathematician, member of Aberdeen Proving Ground staff in First World War, chair of the Subcommittee on the Bibliography of Mathematical Tables and Other Aids to Computation

BIOMETRICS LABORATORY; *see* GALTON LABORATORY

BLANCH, GERTRUDE (1896–1996)
Mathematical leader of Mathematical Tables Project

BRITISH ASSOCIATION FOR THE ADVANCEMENT OF SCIENCE, MATHEMATICAL TABLES COMMITTEE
Committee of scientists who computed tables of higher mathematical functions

BRUNSVIGA
Mechanical calculator that was a favorite of Karl Pearson and the British statistical community

BUREAU DU CADASTRE
French survey organization, led by Gaspard de Prony, that computed decimal trigonometry tables during the French Revolution

BURROUGHS, WILLIAM SEWARD (1855–1898)
Early successful American adding machine manufacturer; machines later adapted as difference engines

BUSH, VANNEVAR (1890–1974)
MIT engineer, directed American scientific effort during Second World War

CAVE-BROWNE-CAVE, FRANCES (1876–1965)
Mathematics professor at Girton College, Cambridge, U.K., did calculations for Karl Pearson

CLAIRAUT, ALEXIS-CLAUDE (1713–1765)
French mathematician, computed perihelion of Halley's comet at first return in 1758

COAST (AND GEODETIC) SURVEY
First U.S. government scientific organization, maintained staff of computers to do survey and longitude calculations

COLUMBIA ASTRONOMICAL COMPUTING BUREAU; *see* THOMAS J. WATSON ASTRONOMICAL COMPUTING BUREAU

COMRIE, LESLIE JOHN (1893–1950)
Director of British Nautical Almanac and founder of Scientific Computing Service, key expert on methods of scientific computation

COWLES, ALFRED H. (1891–1984)
Industrialist and financier, founded Cowles Commission for Economic Research

COWLES COMMISSION FOR ECONOMIC RESEARCH
Organization founded in Colorado Springs (now at Yale) that maintained a computing staff to compile and analyze economic data

CROMMELIN, ANDREW CLAUDE DE LA CHEROIS (1865–1939)
English astronomer, computed perihelion of 1910 return of Halley's comet, developed general means of solving differential equations

CROWELL, PHILLIP (1879–1949)
Astronomer who worked with Crommelin (*see above*)

DAVIS, CHARLES HENRY (1807–1877)
Officer, U.S. Navy, founder and first director of American Nautical Almanac

DAVIS, HAROLD THAYER (1892–1974)
American mathematician and creator of encyclopedia of mathematical functions, worked with Cowles Commission and Subcommittee on the Bibliography of Mathematical Tables and Other Aids to Computation; no known relation to Charles Henry Davis

DE COLMAR, CHARLES XAVIER THOMAS (1785–1870)
Inventor of first commercially successful adding machine

DE PRONY, GASPARD CLAIR FRANÇOIS MARIE RICHE (1755–1839)
Leader of Bureau du Cadastre computing effort, created decimal trigonometry tables for French Metric Commission

DIFFERENCE ENGINE
Mechanical calculator that could interpolate functions, invented as special device by Charles Babbage but later adapted from commercial machines

DIFFERENTIAL ANALYZER
Electromechanical machine that could solve differential equations (*see below*), found at MIT and Aberdeen Proving Ground

DIFFERENTIAL EQUATIONS
Equations that describe physical motion; they generally express relationships among the position of an object, the direction of its motion, and its speed; can be solved by a Differential Analyzer (*see above*)

DOOLITTLE, MYRRICK (1830–1913)
Computer for Coast Survey, developed method for solving least squares and simultaneous equation problems

ECKERT, J. PRESPER (1919–1995)
Designer of ENIAC at University of Pennsylvania; no known relation to W. J. Eckert, below.

ECKERT, WALLACE J. (1902–1971)
Leader of punched card computing group at Columbia University and director of U.S. Nautical Almanac.

ENIAC
Electronic computing machine developed at University of Pennsylvania, often identified as precursor of modern computer

FROELICH, CLARA (B. 1892)
Computer at Bell Telephone Laboratories
FRY, THORNTON (1892–1991)
Mathematician at Bell Telephone Laboratories
GALTON LABORATORY
Statistical laboratory of Karl Pearson at University College London, employed substantial computing staff
GLOVER, JAMES W. (1868–1941)
Actuary, mathematician, and educator of female computers at University of Michigan
GOLDSTINE, ADELE (1920–1964)
Senior Computer, University of Pennsylvania
GREENWICH OBSERVATORY
Royal Observatory of England, maintained a staff of human computers for almost 200 years
HALLEY, EDMUND (1656–1742)
English astronomer, friend of Isaac Newton, identified Halley's comet as a returning comet, recognized that comet orbit calculations were difficult
HALLEY'S COMET (1758, 1835, 1910, 1986)
First major project for human computers because of the difficulties of tracking three or more bodies in space
IOWA STATE STATISTICAL LAB
Statistical laboratory run by George Snedecor that employed large computing staff, associated with Henry Wallace
LALANDE, JOSEPH-JÉRÔME LE FRANÇAIS DE (1732–1807)
French Astronomer Royal who worked on first calculation of Halley's comet
LANCZOS, CORNELIUS (1893–1974)
Hungarian mathematician who worked with Mathematical Tables Project
LEAST SQUARES
Method of estimating orbits, statistical quantities, and other numbers by minimizing the squared distance between data (such as astronomical observations) and the final solution; important least squares technique developed by Myrrick Doolittle
LEPAUTE, NICOLE-REINE ÉTABLE DE LA BRIÈRE (1723–1788)
French scientist, worked on calculation of first return of Halley's comet
LE VERRIER, URBAN JEAN JOSEPH (1723–1788)
Discoverer of Neptune
LORAN
Long-range navigation, a form of radio navigation developed by the United States during the Second World War

LOWAN, ARNOLD (1898–1962)
Director of Mathematical Tables Project

LUCASIAN PROFESSORSHIP
Mathematical professorship at Cambridge University in England, held, at different times, by Newton, Babbage, and Airy

MANHATTAN PROJECT
American atomic bomb effort in the Second World War

MASKELYNE, NEVIL (1732–1811)
British Astronomer Royal, founded Nautical Almanac

MATHEMATICAL TABLES AND OTHER AIDS TO COMPUTATION COMMITTEE/JOURNAL
National Research Committee chaired by R. C. Archibald; journal published by the same committee

MATHEMATICAL TABLES COMMITTEE
Not Mathematical Tables Project; *see* BRITISH ASSOCIATION FOR THE ADVANCEMENT OF SCIENCE

MATHEMATICAL TABLES PROJECT
WPA computing organization in New York City

MITCHELL, MARIA (1818–1889)
Early American computer and astronomer

MITCHELL, WILLIAM (1791–1869)
Amateur scientist and father of Maria Mitchell (*see above*)

MORSE, PHILIP (1903–1985)
MIT engineer and supporter of Mathematical Tables Project

MOULTON, FOREST RAY (1872–1952)
A leader of First World War ballistics computer effort with Oswald Veblen

NATIONAL BUREAU OF STANDARDS
American government research institute, sponsor of Mathematical Tables Project

NATIONAL DEFENSE RESEARCH COMMITTEE
Second World War committee for organizing scientific research, part of Office of Scientific Research and Development

NATIONAL RESEARCH COUNCIL
American committee for coordinating research, founded in First World War

NATIONAL YOUTH ADMINISTRATION
New Deal agency for employing high school and college youth, sponsored many small computing organizations

NAUTICAL ALMANAC, AMERICAN
American equivalent of British Nautical Almanac, founded in Cambridge, Mass., and moved to Washington, D.C.

NAUTICAL ALMANAC, BRITISH
Officially called Royal Nautical Almanac, prepared annual volume of navigation and astronomical tables

NAVAL OBSERVATORY
American National Observatory in Washington, D.C.

NEUMANN, JOHN VON (1903–1957)
American mathematician and key influence in development of modern electronic computer

NEW DEAL
Popular name for President Franklin Roosevelt's economic relief programs

NEWCOMB, SIMON (1835–1909)
Director of American Nautical Almanac and, for his time, America's most famous scientist

NEWTON, ISAAC (1642–1727)
An inventor of calculus and a friend of Edmund Halley

NEWTON, ISAAC (1837–1884)
Not to be confused with the above, first director of U.S. Department of Agriculture

NEYMAN, JERZY (1896–1981)
American statistician, worked on bombing problems in Second World War

PEARSON, KARL (1857–1934)
English statistician, founded computing organization and worked on bombing problems in First World War

PEIRCE, BENJAMIN (1809–1880)
American mathematician, friend of Charles Henry Davis, staff member of Nautical Almanac, director of Coast Survey

PICKERING, EDWARD (1846–1919)
Director, Harvard Observatory, hired large numbers of female computers

PONTÉCOULANT, PHILIPPE GUSTAVE LE DOULCET, COMTE DE (1795–1874)
Computed 1835 and 1910 returns of Halley's comet

PRINCIPIA
Isaac Newton's book on planetary motion

RICHARDSON, LEWIS FRY (1881–1953)
English meteorologist, envisioned truly massive computing laobratory

ROCKEFELLER FOUNDATION
Philanthropic organization of Rockefeller family, supported mathematical research

ROYAL ASTRONOMICAL SOCIETY
English scientific society organized in 1821 as an alternative to Royal

Society (*see below*); Babbage a member; supported computational work

ROYAL SOCIETY
England's first scientific society

SAUNDERS, RHODA (DATES UNKNOWN)
Computer at Harvard Observatory

SCHEUTZ, EDVARD (1821–1888) AND GEORGE (1785–1873)
Inventors of a difference engine following Babbage's ideas

SMITH, ADAM (1723–1790)
Scottish philosopher and economist

SNEDECOR, GEORGE (1881–1974)
Iowa statistician, student of George Glover

STIBITZ, GEORGE (1904–1995)
Staff member of Bell Telephone Laboratories, inventor of machine to do complex arithmetic with telephone relays

TAUSSKY-TODD, OLGA (1906–1995)
English mathematician and member of National Bureau of Standards staff

THOMAS J. WATSON ASTRONOMICAL COMPUTING BUREAU
Early punched card computing bureau at Columbia University

TOLLEY, HOWARD (1889–1958)
Mathematician and computer at U.S. Department of Agriculture

TRACTS FOR COMPUTERS
Computing pamphlets published by Karl Pearson

TRIPOS
Mathematical exams at Cambridge University in England; top students are known as First Wrangler, Second Wrangler, and so on

VEBLEN, OSWALD (1880–1960)
American mathematician, nephew of economist Thorstein Veblen, leader of American computing effort in First World War, and member of Applied Mathematics Panel during Second World War

WALLACE, HENRY A. (1888–1965)
American secretary of agriculture, vice president, and amateur mathematician, associated with computing groups at Iowa State College and U.S. Department of Agriculture

WATSON, THOMAS J., SR. (1874–1956)
First president of IBM

WEAVER, WARREN (1898–1978)
University of Wisconsin mathematician, chair of Applied Mathematics Panel in Second World War, scientific program director for Rockefeller Foundation

WIENER, NORBERT (1894–1964)

MIT mathematician, member of First World War ballistics computing effort

WILKS, SAMUEL (1906–1964)

Statistician at Institute for Advanced Study, member of Applied Mathematics Panel

WILSON, ELIZABETH WEBB (1896–1980)

Ballistics computer, First World War

WORK PROJECTS ADMINISTRATION (WPA) (1935–1943)

American economic relief program during Great Depression, organized and financed Mathematical Tables Project

YOWELL, EVERETT

Name of two computers, one for the U.S. Naval Observatory and the second with the Thomas J. Watson Astronomical Computing Bureau

Notes

Introduction
A Grandmother's Secret Life

1. Record Books for Mathematics 49 (1918), Mathematics 53 (1918), Mathematics 4B (1920), BENTLEY.
2. Class Records, 1917–21, MICHIGAN; Annual Reports, BENTLEY.
3. Karpinsky, "James W. Glover."
4. Glover, "Courses in Actuarial Mathematics."
5. Letters for Baillo, deVries, Hall, and McDonald, Alumni Directories, 1937, 1953, ALUM.
6. Barlow, *Barlow's Tables*, preface.
7. Babbage, *Economy of Machinery and Manufactures*, p. 191.
8. Croarken and Campbell-Kelly, *Table Making from Sumer to Spreadsheets*, preface; McLeish, *Number*, pp. 26, 65–66.
9. Galison and Hevly, *Big Science*.
10. See, for example, Galison and Hevly, *Big Science*.
11. Cardwell, *Norton History of Technology*, pp. 105, 106.
12. Ibid., p. 107.

Chapter One
The First Anticipated Return

1. Newton, *Principia*, preface.
2. Cook, *Edmund Halley*, p. 209.
3. Quoted in ibid., p. 210.
4. Ibid., p. 211.
5. Ibid., p. 212.
6. Edmund Halley to Isaac Newton, September 28, 1695, in MacPike, *Correspondence of Edmund Halley*, p. 92.
7. Edmond Halley to Isaac Newton, October 7, 1695, ibid., pp. 92–93.
8. Isaac Newton to Edmund Halley, October 17, 1695, ibid., pp. 93–94.
9. Cook, *Edmund Halley*, p. 211.
10. Halley, *Astronomiae Cometicae Synopsis* (1705).
11. Rigaud, *Some Account of Halley's Astronomiae Cometicae Synopsis*, pp. 3–23; Broughton, "The First Predicted Return of Comet Halley."
12. Halley, *Astronomical Tables* (1752); Broughton, "The First Predicted Return of Comet Halley," has noted that if Halley used the old-style calendar, in which the year changes at the March equinox, then Halley's prediction was very close to the actual date of March 13.
13. Halley, *Astronomical Tables* (1752).

14. Smith, A., "The Principles Which Lead and Direct Philosophical Enquiries" (1757), p. 48.

15. Messier and Maty, "A Memoir, Containing the History of the Return of the Famous Comet of 1682. . . ."

16. Gillispie, *Dictionary of Scientific Biography.*

17. Barker to Bradley, in *Philosophical Transactions (1683–1775).*

18. Alder, *The Measure of All Things* (2002), p. 78.

19. She is sometimes identified in the literature as Hortense Lepaute.

20. Lalande, *Astronomie* (1792), pp. 676–81.

21. Alder, *The Measure of All Things* (2002), p. 78.

22. Lalande, *Astronomie* (1792), pp. 676–81.

23. Wilson, "Clairaut's Calculation of Halley's Comet" (1993).

24. Lalande, *Astronomie* (1792), pp. 676–81.

25. Ibid.

26. Swift, *Gulliver's Travels,* Section 3.

27. Ibid.

28. Wilson, "Appendix: Clairaut's Calculation of the Comet's Return" (1995).

29. Wilson, "Clairaut's Calculation of Halley's Comet" (1993).

30. Lalande, *Astronomie* (1792), pp. 676–81.

31. Wilson, "Clairaut's Calculation of Halley's Comet" (1993).

32. Yeomans, "Comet Halley—The Orbital Motion" (1977).

33. Quoted in Gillispie, *Dictionary of Scientific Biography,* p. 283; Wilson, "Appendix: Clairaut's Calculation of the Comet's Return" (1995).

34. Wilson, "Appendix: Clairaut's Calculation of the Comet's Return" (1995).

35. Hobart and Schiffman, *Information Ages,* p. 166.

36. Jean d'Alembert quoted in Wilson (1993); Wilson (1995) gives a fairly complete account and assessment of the controversy.

37. Wilson, "Clairaut's Calculation of Halley's Comet" (1993).

38. Ibid.

39. Alexis Clairaut quoted in Wilson, "Appendix: Clairaut's Calculation of the Comet's Return" (1995).

40. Messier and Maty, "A Memoir, Containing the History of the Return of the Famous Comet of 1682. . . ."

41. Lalande, *Astronomie* (1792), pp. 676–81.

42. Ibid.

43. Stigler, "Stigler's Law of Eponymy" (1999), p. 277.

CHAPTER TWO
THE CHILDREN OF ADAM SMITH

1. Smith, A., "The Principles Which Lead and Direct Philosophical Enquiries" (1757).

2. Foley, *Social Physics of Adam Smith*, p. 34.

3. Smith, A., *Wealth of Nations* (1776), book 1, chapter 1.

4. Ibid.

5. Ibid.

6. Calendar of State Papers, Domestic, Car. II, 1675–76, p. 173, June 22, 1675, British Library, MS Birch 4393 f 104 r, v; Public Record Office, Kew, State Papers Domestic Entry Book 44, p. 10.

7. See Betts, *Harrison*, and Andrews, *The Quest for Longitude.*

8. Mayer, *Tabulae Motuum Solis et Lunae Novae et Correctae* (1770); Mayer, *Theoria Lunae Juxta Systema Newtonianum* (1767).

9. Leonhard Euler to Tobias Mayer, February 26, 1754, in Forbes (1971), *Connaissance des Temps pour l'Année 1761*, Paris, De l'Imprimerie Royale, 1761.

10. Sobel, *Longitude.*

11. Maskelyne, Nevil, "Memorial Presented to the Commissioners of the Longitude," February 9, 1765, in Mayer, *Tabulae Motuum Solis et Lunae Novae et Correctae* (1770), pp. cxvii–cxx; see also Betts, *Harrison* (1993), and Andrews, *The Quest for Longitude* (1996).

12. Croarken, "Tabulating the Heavens."

13. Ibid.

14. Maskelyne, *The British Mariner's Guide* (1763), pp. iv–v.

15. Howse, *Nevil Maskelyne*, p. 60.

16. See Nevil Maskelyne to Joshua Moore, September 30, 1788, MOORE.

17. Maskelyne, "Preface," *Nautical Almanac for 1767.*

18. Croarken, "Tabulating the Heavens."

19. Maskelyne, "Preface," *Nautical Almanac for 1767.*

20. Croarken, "Tabulating the Heavens."

21. Howse, *Nevil Maskelyne*, p. 86; Croarken, "Tabulating the Heavens."

22. Charles Talleyrand (1754–1838) quoted in Bradley, *Gaspard Clair François Marie Riche de Prony*, p. 16.

23. Porter, *Trust in Numbers* (1995), p. 24; Alder, "A Revolution to Measure" (2002), pp. 85–88.

24. Guillaume, *Procès-Verbaux du Comité d'Instruction Publique* (1897); Archibald, "Tables of Trigonometric Functions in Non-Sexagesimal Arguments" (1943).

25. Smith, C., "The Longest Run"; Bradley, *Gaspard Clair François Marie Riche de Prony*, pp. 5, 10.

26. Bradley, *Gaspard Clair François Marie Riche de Prony*, p. 10.

27. Smith, C., "The Longest Run."

28. Bradley, *Gaspard Clair François Marie Riche de Prony*, p. 11.

29. Quoted ibid., p. 17.

30. Quoted ibid.

31. Quoted in Archibald, "Tables of Trigonometric Functions in Non-Sexagesimal Arguments," (1943).

32. Quoted in Babbage, *Economy of Machinery and Manufactures* (1835), p. 193.

33. Smith, C., "The Longest Run"; see also Daston, "Enlightenment Calculations."

34. Quoted in Babbage, *Economy of Machinery and Manufactures*, p. 193.

35. Grattan-Guinness, "Work for the Hairdressers."

36. Quoted in Babbage, *Economy of Machinery and Manufactures*, p. 194.

37. Braverman, *Labor and Monopoly Capital*, p. 220.

38. Grattan-Guinness, "Work for the Hairdressers." The planners included Marc-Antoine Parseval (d. 1836), whose name is associated with "Parseval's inequality." Kline, *Mathematical Thought from Ancient to Modern Times*, p. 716.

39. Babbage, *Economy of Machinery and Manufactures*, pp. 193, 194.

40. Ibid., p. 194.

41. Grattan-Guinness, "Work for the Hairdressers."

42. Lalande, *Bibliographie Astronomique avec l'Histoire de l'Astronomie* (1802), pp. 743–44.

43. Bradley, *Gaspard Clair François Marie Riche de Prony*, pp. 19, 10.

44. See Guillaume, *Procès-Verbaux du Comité d'Instruction Publique* (1904), pp. 556–61.

45. Alder, "A Revolution to Measure" (1995).

46. Quoted in Bradley, *Gaspard Clair François Marie Riche de Prony*, p. 68.

47. Grattan-Guinness, "Work for the Hairdressers"; Charles Babbage to Sir Humphrey Davy, July 3, 1822, in Morrison and Morrison, *Charles Babbage and His Calculating Engines*, pp. 298–305.

48. Hyman, *Charles Babbage*, pp. 29, 39–40.

49. See Schaffer, "Babbage's Intelligence" (1994), p. 204.

50. Hyman, *Charles Babbage*, p. 34.

51. Ashworth, "The Calculating Eye."

52. "Prospectus for the Astronomical Society," 1820, RAS.

53. Babbage, *History of the Invention of the Calculating Engines*.

54. Babbage, *Passages from the Life of a Philosopher* (1864), p. 31.

55. See Swade, *The Cogwheel Brain*, pp. 26–27.

56. Hyman, *Charles Babbage*, pp. 40–41; Grattan-Guinness, "Work for the Hairdressers."

57. Babbage, *Economy of Machinery and Manufactures* (1835), p. 191.

58. Ibid., p. 187.

59. Williams, *A History of Computing Technology*, pp. 124, 129.

60. Cardwell, *Norton History of Technology*, p. 230.

61. Williams, *A History of Computing Technology*, p. 145.

62. Charles Babbage to Sir Humphrey Davy, July 3, 1822, in Morrison and Morrison, *Charles Babbage and His Calculating Engines*, p. 305.

63. Swift, *Gulliver's Travels*, section 3, chapter 5.

64. Charles Babbage to Sir Humphrey Davy, July 3, 1822, in Morrison and Morrison, *Charles Babbage and His Calculating Engines*, p. 305.

65. Dreyer and Turner, *History of the Royal Astronomical Society*, appendix: medal winners.

66. Baily, "On Mr. Babbage's New Machine" (1823).

67. Hyman, *Charles Babbage*, p. 65.

68. Babbage, *Economy of Machinery and Manufactures* (1835), p. 267.

69. Goldstine, *The Computer from Pascal to von Neumann*, p. 19.

70. Lovelace, "Notes by the Translator," in Morrison and Morrison, *Charles Babbage and His Calculating Engines*, p. 251.

71. Hyman, *Charles Babbage*, p. 165.

72. Babbage, *Economy of Machinery and Manufactures*, p. 169.

CHAPTER THREE
THE CELESTIAL FACTORY

1. Dickens, Charles, *Hard Times,* Project Gutenberg Edition, *"The Keynote."*
2. Yeomans, *Comets.*
3. Maunder, *The Royal Observatory Greenwich,* chapter 3.
4. Wilkins, "A History of H.M. Nautical Almanac Office," p. 56.
5. *Memoirs of the Royal Astronomical Society,* vol. 2, no. 2 (February 11, 1831), pp. 11–12.
6. Ibid., no. 25 (February 12,1830), p. 166.
7. Ashworth, "The Calculating Eye."
8. South, "Report on Nautical Almanac" (1831), pp. 459–71, p. 461.
9. Dunkin, *A Far Off Vision,* p. 45.
10. *Nautical Almanac for 1835,* London, John Murray, 1833, p. 493.
11. Yeomans, *Comets* (1991), pp. 256–57.
12. *Memoirs of the Royal Astronomical Society,* vol. 3, no. 20 (February 12, 1836), pp. 161–62.
13. Ibid.
14. Report by the Astronomer Royal to the Board of Visitors for 1837, London, 1838.
15. Smith, R., "A National Observatory Transformed"; Cannon, *Science and Culture,* p. 38.
16. Smith, R., "A National Observatory Transformed."
17. Quoted in Ashworth, "The Calculating Eye."
18. The Danish astronomer Tycho Brahe is generally credited with giving the task of reduction to human computers.
19. Smith, R., "A National Observatory Transformed"; Schaffer, "Astronomers Mark Time" (1988), pp. 115–45; Maunder, *The Royal Observatory Greenwich,* p. 117; Dunkin, *A Far Off Vision,* p. 72.
20. W. H. M. Christie quoted in Meadows, *Greenwich Observatory,* vol. 2, p. 9.
21. Dunkin, *A Far Off Vision,* p. 45.
22. Ibid., p. 72.
23. Ibid., p. 96.
24. Dickens, Charles, *Hard Times,* Project Gutenberg Edition, *"The Keynote."*
25. Maunder, *The Royal Observatory Greenwich,* p. 117.
26. Meadows, *Greenwich Observatory,* p. 11.
27. Dunkin, *A Far Off Vision,* p. 71.
28. Meadows, *Greenwich Observatory,* p. 7.
29. Chapman, "Sir George Airy," p. 332.
30. Smith, *Wealth of Nations,* book I, chapter 1.
31. See *Annual Reports of the Astronomer Royal to the Board of Visitors,* beginning in 1836.
32. George Airy to Henry Goulburn, September 16, 1842, 6-427, folder 68, GREENWICH.
33. Swade, *The Cogwheel Brain,* p. 153.

CHAPTER FOUR
THE AMERICAN PRIME MERIDIAN

1. Latrobe, *The History of Mason and Dixon's Line.*

2. Croarken, "Tabulating the Heavens" (2003); fragment labeled "Paper given at Columbian Institute," undated correspondence, MOORE; Joshua I. Moore to Thomas Jefferson, September 7, 1805; Thomas Jefferson to Joshua I. Moore, September 19, 1805, JEFFERSON.

3. Tocqueville, *Democracy in America,* book 2, chapter 9.

4. Dupree, *Science in the Federal Government,* p. 42.

5. Dick and Doggett, *Sky with Ocean Joined,* p. 169; see also Dick, *Sky with Ocean Joined.*

6. Theberge, *History of the Commissioned Corps of the National Oceanic and Atmospheric Administration,* "personnel policies," pp. 84ff. ("Investigation of 1842").

7. Will of James Smithson, quoted in Dupree, *Science in the Federal Government,* p. 66.

8. See *Congressional Globe,* 30th Cong., 2d sess., HR 699, "An Act Making appropriations for the naval service, for the year ending the thirtieth of June, 1850."

9. Turner, "The Significance of the Frontier in American History" (1893).

10. Matthew Fontaine Maury quoted in Waff, "Navigation vs. Astronomy."

11. Anonymous (1849); Waff concludes that Peirce or Davis was the most likely author (see Waff, "Navigation vs. Astronomy," p. 97).

12. Waff, "Navigation vs. Astronomy," p. 97.

13. Davis, C. H., II, *Life of Charles Henry Davis* (1899), pp. 4, 64, 89.

14. Ibid., p. 91.

15. Ibid., p. 58.

16. Ibid., pp. 4, 64, 89.

17. Emerson, Ralph Waldo, "Nature," section 4.

18. Peterson, "Benjamin Peirce," pp. 89–112.

19. Charles Henry Davis to William Ballard Preston, July 30, 1849, OBSERVATORY-LOC.

20. Davis, C. H., "Report of the Secretary of the Navy" (1852), p. 6.

21. Davis, C. H., "Report on the Nautical Almanac" (1852).

22. Charles Henry Davis to William Ballard Preston, July 30, 1849, box 15, OBSERVATORY-LOC.

23. Morando, "The Golden Age of Celestial Mechanics."

24. Gottfried Galle to Jean Joseph Le Verrier, September 25, 1846, quoted ibid.

25. Thoreau, *Walden,* section 1, "Where I Lived, and What I Lived For."

26. Charles Henry Davis to Charles Peirce, August 5, 1849, OBSERVATORY-LOC.

27. Tyler, *John David Runkle*; "Annual Report for the United States Navy for 1852," p. 8.

28. Gould, "Commemoration of Sears Cook Walker."

29. Ibid.

30. Benjamin Peirce to Matthew Fontaine Maury, January 21, 1846, OBSERVATORY-LOC.

31. Maury, Matthew, "Report on Leverrier," Nautical Almanac Correspondence, January 1847, OBSERVATORY-LOC.

32. Maury, Matthew, Report to Secretary of the Navy, February 7, 1847, OBSERVATORY-LOC.

33. Charles Henry Davis to William Ballard Preston, "Reply on employing subordinate computers," OBSERVATORY-LOC.

34. Jones and Boyd, *The Harvard College Observatory*, p. 384.

35. Fuller, Margaret, *Women in the Nineteenth Century* (1844), in *The Essential Margaret Fuller*, Jeffrey Steele, ed., New Brunswick, NJ, Rutgers University Press, 1992, p. 260. See also her story of Miranda, pp. 261ff.

36. Wright, *Sweeper in the Sky* (1949), p. 23.

37. Jones and Boyd, *The Harvard College Observatory*, pp. 384–85.

38. Charles Henry Davis to Maria Mitchell, January 5, 1851, OBSERVATORY-LOC; U.S. Nautical Almanac for 1852.

39. Charles Henry Davis to Maria Mitchell, December 24, 1849, OBSERVATORY-LOC.

40. Charles Henry Davis to Maria Mitchell, December 24, 1849, OBSERVATORY-LOC.

41. Newcomb, *The Reminiscences of an Astronomer*, pp. 1, 75.

42. Charles Henry Davis to Otis E. Kendall, April 24, 1851, OBSERVATORY-LOC.

43. Charles Henry Davis to William A. Graham, June 19, 1851, OBSERVATORY-LOC.

44. Charles Henry Davis to the Secretary of the Navy, October 14, 1850, OBSERVATORY-LOC.

45. Waff, "Navigation vs. Astronomy," p. 95.

46. Davis, C. H., *Remarks on an American Prime Meridian* (1849), p. 12.

47. Charles Henry Davis to William Preston, July 31, 1849, OBSERVATORY-LOC.

48. Charles Henry Davis to William Preston, July 31, 1849, OBSERVATORY-LOC.

49. Davis, C. H., *Remarks on an American Prime Meridian* (1849), pp. 32, 39.

50. Budgets for the Nautical Almanac Office, 1850, 1851, OBSERVATORY-LOC.

51. Comments of John P. Hale, *Congressional Globe*, n.s., no. 94, 32nd Cong., 1st sess., May 28, 1852, p. 1495.

52. Ibid.

53. Comments of George Badger, *Congressional Globe*, n.s., no. 94, 32nd Cong., 1st sess., May 28, 1852, p. 1495.

54. Davis, C. H., *Report of Lieutenant Charles H. Davis* (1852), pp. 7, 8; see also Davis, C. H., "Report on the Nautical Almanac" (1852).

55. Charles Henry Davis to August W. Smith, November 5, 1850, OBSERVATORY-LOC.

56. William Mitchell to Joseph Winlock, February 9, 1858, OBSERVATORY-LOC.

57. *American Ephemeris and Nautical Almanac* (1855–60), Washington, D.C., Government Printing Office.

58. Davis, C. H., *Life of Charles Henry Davis*, (1899), pp. 4, 102.

59. Newcomb, *The Reminiscences of an Astronomer*, p. 65.

60. Archibald, "P. G. Scheutz and Edvard Scheutz" (1947).

61. Gould, *Reply to the Statement of the Trustees of the Dudley Observatory* (1859), p. 142; James, *Elites in Conflict*, p. 61.

62. Gould, *Reply to the Statement of the Trustees of the Dudley Observatory* (1859), p. 141; see also *U.S. Naval Observatory Annual Report for 1858*.

63. Gould, *Reply to the Statement of the Trustees of the Dudley Observatory* (1859), p. 141; Dudley Observatory Annual Report for 1864, p. 42.

64. Gould, *Reply to the Statement of the Trustees of the Dudley Observatory* (1859), p. 141.

65. *U.S. Naval Observatory Annual Report for 1858*.

66. Ibid.

67. Gould, *Reply to the Statement of the Trustees of the Dudley Observatory* (1859), p. 221.

68. U.S. Nautical Almanac for 1859, p. 1.

69. U.S. Nautical Almanac for 1860.

70. Ibid.

71. Herman, *A Hilltop in Foggy Bottom*, p. 17.

72. C. H. Davis to his family, June 14, 1861, quoted in Davis, C. H., *Life of Charles Henry Davis* (1899), p. 121.

73. C. H. Davis to his family, July 21, 1861, quoted ibid., p. 151.

74. C. H. Davis to his family, July 21, 1861, quoted ibid.

75. C. H. Davis to his family, September 18, 1861, quoted ibid., p. 134; Theberge, *History of the Commissioned Corps of the National Oceanic and Atmospheric Administration*, p. 420.

CHAPTER FIVE
A CARPET FOR THE COMPUTING ROOM

1. Newcomb, *The Reminiscences of an Astronomer,* p. 342.

2. Davis, C. H., *Life of Charles Henry Davis* (1899), p. 102.

3. Lowell, James Russell, "The Present Crisis" (1856).

4. Davis, *The Coast Survey of the United States* (1849), p. 21.

5. "Annual Report of the U.S. Coast Survey for 1844," p. 29.

6. Theberge, *History of the National Oceanic and Atmospheric Administration*, pp. 424ff.

7. See boxes 24–28, Ordnance 1856–1866, DAHLGREN.

8. Dupree, *Science in the Federal Government*, p. 120.

9. Abbe, "Charles Schott."

10. Charles Saunders Peirce to Alexander Dallas Bache, August 11, 1862, Correspondence of the Director, COAST-SURVEY.

11. Annual Report of the U.S. Coast Survey for 1864, pp. 92–93, 222–23.

12. See, for example, John Dahlgren to Commander Morris, March 1852, Correspondence 1852, DAHLGREN.

13. An Act to Incorporate the National Academy of Sciences, March 3, 1863.

14. Dupree, *Science in the Federal Government*, pp. 141–47.

15. Quoted in Ebling, "Why Government Entered the Field of Crop Reporting and Forecasting."

16. Rasmussen and Baker, *The Department of Agriculture*, p. 6.

17. Report of the Smithsonian Institution, 1854–1855, pp. 30, 186.

18. Report of the Smithsonian Institution, 1851–1852, p. 168.

19. Report of the Smithsonian Institution, 1856–1857, p. 28; Nebeker, *Calculating the Weather*, p. 13.

20. "Statement of the Assistant to the Chief Signal Officer," in Testimony, pp. 113–30, 114.

21. Whithan, *A History of the United States Weather Bureau*, p. 19.

22. "Examination of Cleveland Abbe," in Testimony, pp. 247–63, 258.

23. Bartky, *Selling the True Time*, p. 33.

24. Ibid.

25. Sears Cook Walker; Theberge, *History of the Commissioned Corps of the National Oceanic and Atmospheric Administration*, "The American Method of Longitude Determination," note 12; Bartky, *Selling the True Time*, pp. 32ff.

26. Annual Report of the Harvard Observatory for 1859, p. 5.

27. Gauss, *Theory of the Motion of the Heavenly Bodies*.

28. Davis, C. H., *Life of Charles Henry Davis* (1899), p. 113.

29. Annual Report of U.S. Coast Survey for 1872, p. 50.

30. Annual Report of U.S. Coast Survey for 1868, p. 37.

31. Jarrold and Fromm, *Time—The Great Teacher*.

32. Doolittle obituary, *Evening Star*.

33. Ibid.

34. Ibid.

35. Annual Report of U.S. Naval Almanac for 1870.

36. Annual Report of U.S. Coast Survey for 1876, p. 81.

37. Doolittle obituary, *Evening Star*.

38. Charles Schott to Julius Hilgard, January 6, 1874, Report of the Computing Division 1869–1886, COAST-SURVEY.

39. Aron, "'To Barter Their Souls for Gold.'"

40. Rotella, *From Home to Office*, pp. 15ff., 29.

41. Arthur Searle to Charles W. Eliot, July 12, 1875, box 68, HARVARD ELIOT.

42. Jones and Boyd, *The Harvard College Observatory*, p. 386.

43. Ibid., "Anna Winlock," in Ogilvie and Harvey, *Biographical Dictionary of Women in Science*, pp. 1388–89.

44. Arthur Searle to Charles W. Eliot, July 12, 1875, box 68, HARVARD ELIOT.

45. Mack, "Strategies and Compromises: Women in Astronomy at Harvard College Observatory, 1870–1920."

46. Jones and Boyd, *The Harvard College Observatory*, pp. 386, 387.

47. Margaret Harwood quoted ibid., p. 390.

48. Annual Report for Radcliffe, 1879, pp. 6, 14.

49. Annual Report of the Harvard Observatory for 1898, p. 6.

50. "Reply to visitors," U.S. Naval Observatory, 1900, p. 10.

51. Welther, "Pickering's Harem."

52. "Reply to visitors," U.S. Naval Observatory, 1900, p. 11.

53. "Staff listing of the Naval Observatory," U.S. Naval Observatory, 1901.

54. Annual Report of the U.S. Coast Survey for 1893, p. 119; Carter, Cook, and Luzum, "The Contributions of Women to the Nautical Almanac Office, the First 150 Years."

55. Jones and Boyd, *The Harvard College Observatory*, p. 189.

56. Upton, "Observatory Pinafore," p. 1.

57. There are two versions of the manuscript. In one, Josephine is treated as a female, though she is clearly one of the male astronomers. In the other, male pronouns have been substituted.

58. Upton, "Observatory Pinafore," p. 7.

59. Ibid., p. 5.

60. Ibid., p. 3.

61. Ibid.

62. Ibid., p. 5.

63. Ibid., p. 9. The observatory history notes that there were six female computers in 1881 (Jones and Boyd, *The Harvard College Observatory*, p. 388).

64. Annual Report of the Harvard Observatory for 1880, p. 16.

65. Upton, "Observatory Pinafore, p. 16.

66. Ibid., p. 29.

CHAPTER SIX
LOOKING FORWARD, LOOKING BACKWARD

1. Hopp, *Slide Rules*.

2. Logarithm base 10.

3. Hopp, *Slide Rules*, Appendix 2, Key Dates in the History of Slide Rules.

4. Quoted in ibid., Appendix 2.

5. Ibid.; Riddell, *The Slide Rule Simplified*.

6. Williams, *A History of Computing Technology*, p. 128.

7. Cortada, *Before the Computer*, p. 35. See also Kidwell, "The Adding Machine Fraternity at St. Louis: Creating a Center of Invention," and "'Yours for improvement'—The Adding Machines of Chicago, 1884–1930."

8. Gray, "On the Arithmometer of M. Thomas (de Colmar)"; Johnston, "Making the Arithmometer Count"; Kidwell, "From Novelty to Necessity."

9. Jevons, "Remarks on the Statistical Use of the Arithmometer."

10. Dreieser, *Sister Carrie*.

11. Cortada, *Before the Computer*, pp. 31ff., 39ff.

12. U.S. Coast Survey Annual Report for 1890, p. 119.

13. Austrian, *Herman Hollerith*, p. 6.

14. Williams, *A History of Computing Technology*, pp. 248ff.

15. *Report of a Commission Appointed by the Honorable Superintendent of Census on Different Methods of Tabulating Census Data.*

16. Quoted in Porter, "The Eleventh Census."

17. *Chicago Tribune*, August 8, 1890, quoted in Austrian, *Herman Hollerith*, p. 62; ibid., pp. 61–62.

18. Handy, *Official Directory of the World's Columbian Exposition*, p. 157.

19. T. Talcott to H. Talcott, May 22, 1893, quoted in Austrian, *Herman Hollerith*, ibid., pp. 100–101.

20. U.S. Coast Survey Annual Report for 1892, p. 145.

21. Adams, *Education of Henry Adams*, chapter 12, "Chicago."

22. Ibid.

23. Handy, *Official Directory of the World's Columbian Exposition*, p. 199.

24. *The World's Congress Auxiliary of the World's Columbian Exposition of 1893*, WCE.

25. Veysey, *The Emergence of the American University*, p. 128.

26. "Program of the Congress on Mathematics and Astronomy," 1893, WCE.

27. Account Books of Artemas Martin, MARTIN.

28. Finkel, "Biography: Artemas Martin."

29. "Program of the Congress on Mathematics and Astronomy," WCE.

30. Kline, R., *Steinmetz: Engineer and Socialist*, p. 71.

31. Turner, "The Significance of the Frontier in American History" (1893).

32. Porter, T., *The Rise of Statistical Thinking*, p. 23.

33. Fitzpatrick, "Leading American Statisticians in the Nineteenth Century"; "Membership List, 1840" (American Statistical Association Membership).

34. "The International Statistical Institute at Chicago."

35. Ralph, "Chicago's Gentle Side"; *The World's Congress Auxiliary of the World's Columbian Exposition of 1893*, WCE.

36. Adams, *Education of Henry Adams*, chapter 4.

37. Ibid., chapter 12, "Chicago."

CHAPTER SEVEN
DARWIN'S COUSINS

1. Hamilton, *Newnham*, p. 136.

2. Pearl, "Karl Pearson."

3. Shaw, *Mrs. Warren's Profession*, act 1.

4. Ibid.

5. Ibid.

6. Rossiter, *Women Scientists in America*, pp. 52, 72.

7. "Maxims for Revolutionaries," in Shaw, *Man and Superman*.

8. Stigler, *History of Statistics*, p. 266; see also Porter, *The Rise of Statistical Thinking*, p. 271.

9. Quoted in Kelves, *In the Name of Eugenics*, p. 5.

10. Ibid.

11. Ibid., p. 6.

12. Francis Galton to Darwin Galton, February 23, 1851, in Pearson, *The Life, Letters and Labours of Francis Galton*, pp. 231–32.

13. Francis Galton to Darwin Galton, February 23, 1851, ibid.

14. Kelves, *In the Name of Eugenics*, p. 7.

15. Gillham, *A Life of Sir Francis Galton*, p. 148.

16. Stigler, *History of Statistics*, p. 268.

17. Galton, "Kinship and Correlation," pp. 419–31.

18. Stigler, *History of Statistics*, pp. 283–90. For an elementary modern treatment that shows the relationship between correlation coefficient and regression slope, see Freedman et al., *Statistics*.

19. Galton, "Regression towards Mediocrity in Hereditary Stature," p. 255.

20. Galton, "Kinship and Correlation."

21. Pearson, "Walter Frank Raphael Weldon," pp. 14, 24, 25.

22. Ibid., p. 18.

23. Ibid.

24. Porter, *Karl Pearson*, p. 3.

25. Haldane, "Karl Pearson"; Pearl, "Karl Pearson."

26. Porter, *Karl Pearson*, p. 12.

27. Walkowitz, "Science, Feminism, and Romance."

28. Pearson, "Walter Frank Raphael Weldon," p. 18.

29. Pearson, "Mathematical Contributions to the Theory of Evolution."

30. Pearson to Foster, "Draper's Company Grant," November 26, 1904, in Pearson, E., "Karl Pearson."

31. Pearson, "Cooperative Investigations on Plants."

32. Magnello, "The Non-correlation of Biometrics and Eugenics."

33. Pearson, "Cooperative Investigations on Plants." (1902).

34. Love, "Alice in Eugenics-Land."

35. Alice Lee to Pearson, December 1, 1895, and June 14, 1897, PEARSON, 01135, quoted ibid.

36. Love, "Alice in Eugenics-Land."

37. Pearson, *Life, Letters and Labour of Francis Galton*, vol. 3, p. 359.

38. Love, "Alice in Eugenics-Land."

39. Karl Pearson to Simon Newcomb, June 26, 1903, NEWCOMB.

40. Pearson, E., "An Appreciation of Some Aspects" (1938).

41. Pearson, K., "The Scope of *Biometrika*" (1901).

42. Karl Pearson to Beatrice Cave, November 25, 1907, 01137/1, PEARSON, University College London, quoted in Love, "Alice in Eugenics-Land."

43. Frances Cave-Browne-Cave, unsigned obituary, CBC.

44. Soper et al., "On the Distribution of the Correlation Coefficient"; Cave and Pearson, "Numerical Illustrations of the Variate Difference Correlation Method."

45. Frances Cave-Browne-Cave, unsigned obituary, CBC; Biographical Information forms for Frances and Beatrice Cave-Browne-Cave, GIRTON.

46. Cole, *Growing Up into Revolution*.

47. Cave-Brown-Cave and Pearson, "On the Correlation between the Barometric Height."

48. Cave-Brown-Cave, F., "On the Influence of the Time Factor on the Correlation."

49. Pearson, K., "On the Laws of Inheritance in Man," p. 136.

50. Pearson, E., "Karl Pearson," p. 199.

51. *Journal of Wilhamina Paton Fleming*, entry of March 4, 1900.

52. The bombing occurred on February 15, 1894; Taylor, "Propaganda by Deed—The Greenwich Observatory Bomb of 1894."

53. Quoted in Meadows, *Greenwich Observatory*, p. 14.

54. Taylor, "Propaganda by Deed—The Greenwich Observatory Bomb of 1894."

55. Conrad, *The Secret Agent*, chapter 2.

56. "Computers" (an announcement of the computing exam for 1906), December 19, 1905, OBSERVATORY-NARA; for vacancies, see Almanac and Observatory rosters for 1890–1905, OBSERVATORY-NARA.

57. "From the Unpopular Side."

58. Newcomb, *The Reminiscences of an Astronomer*, p. 223.

59. Henry Meier to Simon Newcomb, August 9, 1884, ALMANAC.

60. Newcomb, *The Reminiscences of an Astronomer*, pp. 223–24.

61. "Nautical Almanac Investigation."

62. See "Scientists at Sword's Points."

63. "Nautical Almanac Office Moved."

64. Karl Pearson to Simon Newcomb, May 26 , 1899, NEWCOMB.

65. Karl Pearson to Simon Newcomb, June 26, 1903, NEWCOMB.

66. Newcomb, "Abstract Science in America" (1876), p. 88; Simon Newcomb to Secretary of Carnegie Institution of Washington, May 12, 1906, 653/2, PEARSON.

67. Simon Newcomb to Secretary of Carnegie Institution of Washington, May 12, 1906, 653/2, PEARSON.

68. Newcomb,"The work of the Carnegie Institution," NEWCOMB.

69. Memo of Simon Newcomb, 1904, NEWCOMB.

70. H. H. Turner to Simon Newcomb, November 25, 1903, 653/2, PEARSON.

71. Simon Newcomb to Karl Pearson, November 21, 1904, 773/7, PEARSON.

72. Memo to Simon Newcomb, January 2, 1905, NEWCOMB.

73. Pearson, E., "An Appreciation of Some Aspects."

74. Ibid.

75. Kevles, *In the Name of Eugenics*, p. 34.

76. Pearson, E., "An Appreciation of Some Aspects."

CHAPTER EIGHT
BREAKING FROM THE ELLIPSE

1. Crommelin, "Note on the Approaching Return of Halley's Comet" (1906).

2. Ibid.

3. Ibid.

4. Report of the Astronomer Royal to the Board of Visitors of the Royal Observatory for 1906.

5. Employee Roster for 1910, box 10, ALMANAC.

6. Cowell and Crommelin, *The Return of Halley's Comet in 1910*, p. 11.

7. Ibid.

8. Ibid.

9. Ibid.

10. Twain, *Works.*

11. "Comet's Poisonous Tail," *New York Times*, February 8, 1910, p. 1.

12. Cowell and Crommelin, *The Return of Halley's Comet in 1910*, p. 11.

13. Whittaker and Robinson, *The Calculus of Observations*, p. v.

14. Erdélyi, "Edmund T. Whittaker."

15. Gibbs, *A Course in Interpolation* , pp. 1–2.

16. Whittaker and Robinson, *The Calculus of Observations*, p. v.

17. Kline, *Mathematical Thought*, p. 710.

18. Croarken, Early Scientific Computing in Britain, p. 25; Davis, *Tables of Higher Mathematical Functions*, vol. 1, p. 3.

CHAPTER NINE
CAPTAINS OF ACADEME

1. Karl Pearson to L. Gregory, February 23, 1916, 600, PEARSON.

2. Ibid.

3. Ibid.

4. Ibid.

5. Adelaide Davin to Karl Pearson, September 9, 1915, 674/9, PEARSON.

6. Karl Pearson, manuscript dated July 6, 1920, PEARSON.

7. Karl Pearson to L. Gregory, February 23, 1916, 600, PEARSON.

8. See correspondence between A. H. Webb and Karl Pearson, June–July 1916, 602, PEARSON.

9. McShane et al., *Exterior Ballistics*, p. 758.

10. Ibid., p. 778.

11. Di Scala, *Italy: From Revolution to Republic*, p. 121.

12. Siacci, "Rational and Practical Ballistics."

13. Bliss, *Mathematics for Exterior Ballistics*, p. 28; U.S. Army, *Ballisticians in War and Peace*, p. 3; Zabecki, *Steel Wind*, p. 13.

14. McShane et al., *Exterior Ballistics*, p. 783.

15. Bliss, *Mathematics for Exterior Ballistics*, p. 28.

16. Littlewood, *Collected Papers of J. E. Littlewood*, p. xxx.

17. Beatrice Cave-Browne-Cave to Karl Pearson, August 20, 1916, 606, PEARSON.

18. Beatrice Cave-Browne-Cave to Karl Pearson, July 21, 1916, 606, PEARSON.

19. Adelaide Davin to Karl Pearson, August 20, 1915, 674, PEARSON.

20. See correspondence between Kristina Smith and Karl Pearson, 1917, 857/6, PEARSON; Smith did research computation for Pearson during this period.

21. Karl Pearson to A. V. Hill, February 15, 1917, 606, PEARSON.

22. Karl Pearson to B. M. Cave, September 29, 1916, 909/8, PEARSON.

23. A. V. Hill to Karl Pearson, December 12, 1916, 606, PEARSON.

24. Mrs. Cain to Karl Pearson, March 30, 1917, 606, PEARSON.

25. Wimperis to Karl Pearson, August 23, 1916, 606; Huie to Karl Pearson,

March 7, 1917, 603; Douglas to Karl Pearson, July 7, 1917, 606; Hill to Karl Pearson, March 15, 1918, 606, PEARSON.

26. A. E. Moore to Pearson, September 25, 1917, 606, PEARSON.

27. Herbert W. Richmond to Karl Pearson, October 4, 1917, 606, PEARSON.

28. Herbert W. Richmond to Karl Pearson, October 7, 1917, 606, PEARSON.

29. A. V. Hill to Karl Pearson, October 1917, 606, PEARSON.

30. Herbert W. Richmond to Karl Pearson, October 7, 1917, 606, PEARSON.

31. Fowler to Karl Pearson, November 17, 1917, 606, PEARSON.

32. Pearson as summarized by A. V. Hill to Karl Pearson, December 28, 1917, 606, PEARSON.

33. A. E. Moore to Karl Pearson, April 10, 1918, 606b, PEARSON.

34. Kristina Smith to Karl Pearson, March 26, 1918, 209, PEARSON.

35. Mills, W., *Road to War*, p. 210.

36. Ernest Hemingway to his parents, October 18, 1918, in Villard and Nagel, *Hemingway in Love and War*, pp. 186–87.

37. Collins, *Princeton in the World War*, pp. xii–xiv.

38. Christman, *Sailors, Scientists and Rockets*, p. 18.

39. Moulton, *History of the Ballistics Branch*, p. 2; Oswald Veblen Diaries, March 23, 1917, VEBLEN.

40. E. H. Moore to Oswald Veblen, January 4, 1918, VEBLEN.

41. Service Record of Oswald Veblen, Records of Ordnance Officers, 1915–1919, vol. 6, ORDNANCE.

42. Crowell, *America's Munitions: 1917–1918*, pp. 548, 550.

43. History of Proving Grounds, Reports, Histories and Guides, E 522, ORDNANCE.

44. E. H. Moore to Oswald Veblen, January 4, 1918, VEBLEN.

45. Aberdeen Proving Ground, Technical Report 84, p. 1.

46. McShane et al., *Exterior Ballistics*, p. 234.

47. Aberdeen Proving Ground, Technical Report 12, p. 2 (Veblen presumed author).

48. Oswald Veblen 1918 Diary, VEBLEN; Aberdeen Proving Ground, Technical Report 12, p. 1.

49. *American Men of Science*, 5th ed., New York, Science Press, 1933, p. 937.

50. Farebrother, *Memoir on the Life of Myrrick Doolittle*; "Myrrick Haskell/ Doolittle," *Washington Evening Star*; Coast Survey Annual Report for 1911.

51. Aberdeen Proving Ground, Technical Report 12, p. 2.

52. Aberdeen Proving Ground, Technical Report 84, p. 1.

53. Aberdeen Proving Ground, Annual Report for 1919, Appendix 37, p. 13, ORDNANCE.

54. Aberdeen Proving Ground, Technical Report 84, pp. 3, 1.

55. Moulton, *History of the Ballistics Branch*, p. 71.

56. Order of March 1, 1918, Circular Orders of the Aberdeen Proving Ground, ORDNANCE.

57. John W. Langley, Member of Congress from Kentucky, quoted in Stevens, *Jailed for Freedom*, p. 135.

58. "Boston Woman Is Rated Insurance Expert Deluxe."

59. Dos Passos, *42nd Parallel*, p. 137.

60. Moulton, *History of the Ballistics Branch*, pp. 2–8.
61. Ibid., pp. 35–38.
62. Dunham Jackson, undated note (probably 1940s), WILSON PAPERS.
63. Moulton, *History of the Ballistics Branch*, pp. 6–7.
64. Ibid., p. 6.
65. Oswald Veblen Diaries, August 23, 1918, VEBLEN.
66. Wiener, *Ex-Prodigy*, p. 254.
67. Aberdeen Proving Ground, Technical Report 12, p. 2.
68. Moulton, *History of the Ballistics Branch*, p. 50.
69. Ibid., p. ii.
70. "Boston Woman Is Rated Insurance Expert Deluxe."
71. Wiener, *Ex-Prodigy*, p. 257.
72. Masani, *Norbert Wiener, 1894–1964*, p. 68.
73. Wiener, *Ex-Prodigy*, p. 258.
74. Fussell, *The Great War*, p. 161.
75. Richardson, *Weather Prediction by Numerical Process*, p. 219.
76. Ibid., p. vii.
77. Ibid., p. ix.
78. Ibid., p. 219.
79. Ibid.
80. Ibid., pp. 219–20.

CHAPTER TEN
WAR PRODUCTION

1. Crowell, *America's Munitions*, p. 16.
2. Cortada, *Before the Computer*, p. 81.
3. Pugh, *Building IBM*, p. 14.
4. Benedict, "Development of Agricultural Statistics in the Bureau of the Census."
5. Pugh, *Building IBM*, pp. 26–27.
6. Cortada, *Before the Computer*, pp. 80–81.
7. Pearl and Burger, "Retail Prices."
8. Ibid.
9. Ezekiel, "Reminiscences of Mordecai Ezekiel," p. 13; "Hog Astronomy." "The great progress which has been made in agricultural economics is doubtless due to the liberality of Congress toward the Department of Agriculture. Twenty years ago the total allowances for that department, including permanent appropriations were but little more than $5,000,000. For the current fiscal year, the total, including the road fund and the 'permanent' items, is $139,000,000, which accounts in part for the great strides made in what might be termed hog astronomy" ("Hog Astronomy," p. 51).
10. Hoover, "Testimony before Senate Committee on Agriculture."
11. See Food Administration Graphical Records, LOC.
12. Wallace, *Agricultural Prices* (1920), p. 30.
13. Friedberger, *Shake-Out: Iowa Farm Families in the 1980s*, p. 20.

14. Sinclair, *The Jungle*.

15. Culver and Hyde, *American Dreamer*, p. 16.

16. Winters, "The Hoover-Wallace Controversy," pp. 586–97.

17. Gen. 41:29-30.

18. Gen. 41:49, 57; Wallace, "Who Plays the Part of Joseph?" (1912).

19. Culver and Hyde, *American Dreamer*, p. 20.

20. Yule, *An Introduction to the Theory of Statistics*; Ezekiel, "Henry A. Wallace," p. 791.

21. Wallace, *Agricultural Prices* (1920), pp. 30, 34.

22. Ibid. He later went back through the data from the nineteenth century to verify his ideas.

23. Gen. 41:39-40.

24. Gen. 41:40.

25. Wallace, *Agricultural Prices* (1920), p. 34.

26. Culver and Hyde, *American Dreamer*, p. 49.

27. Moulton, *History of the Ballistics Branch*, p. 88.

28. Wiener, *Ex-Prodigy*, pp. 255–56.

29. De Weerd, "American Adoption of French Artillery 1917–1918," pp. 104–16.

30. Christman, *Sailors, Scientists and Rockets*, p. 27.

31. Kennedy, *Over Here*, p. 251.

32. Aberdeen Proving Ground, Technical Report 84, p. 2.

33. Wiener, *Ex-Prodigy*, pp. 255, 257.

34. Price, "American Mathematicians in World War I."

35. Dunham Jackson, undated note (probably 1940s), WILSON PAPERS.

36. Evans, "Elizabeth W. Wilson Has Won Distinction"; Radcliffe College Alumnae Information Form, 1937, WILSON PAPERS.

37. Orders for Major Oswald Veblen, USA, October 10 &16, 1918, Records of Ordnance Officers, 1915–19, vol. 6, ORDNANCE.

38. Oswald Veblen Diary, November 9, 1918, VEBLEN.

39. See Oswald Veblen Diaries, VEBLEN; Grier, "Dr. Veblen Gets a Uniform."

40. "Orders," April 23, 1919, VEBLEN.

41. Phil Schwartz to Oswald Veblen, August 10, 1919, VEBLEN.

42. Karl Pearson, manuscript dated July 6, 1920, PEARSON.

43. Croarken, *Early Scientific Computing in Britain* (1990), p. 24.

44. Porter, *Karl Pearson*, p. 3.

45. Pairman, *Tables of the Digamma*, p. 1.

46. Ibid.

47. Pearson, "On the Construction of Tables and on Interpolation." parts 1 and 2; Rhodes, E., "On Smoothing"; Irwin, "On Quadrature and Cubature."

48. Henderson, *Bibliotheca Tabularum Mathematicarum*, p. 2.

49. Pairman, *Tables of the Digamma*, pp. 1–2.

50. Thompson, *Logarithmetica Britannica*, p. 1.

51. Ibid., pp. 1–2.

52. Martin, *Die Rechenmaschinen*.

53. Thompson, *Logarithmetica Britannica*, p. 1.

54. See Archibald, "Reviews" (1921); Archibald, "Reviews" (1924).

CHAPTER ELEVEN
FRUITS OF THE CONFLICT

1. Cortada, *Before the Computer*, p. 81.
2. Tolley, "Interview."
3. Ibid.; Tolley and Ezekiel, "The Doolittle Method" (1927).
4. Tolley, "Interview."
5. Ibid.
6. Ibid.
7. Ibid.
8. Ibid.
9. Tolley, Memo on a course on least squares, November 16, 1922, TOLLEY.
10. Tolley and Ezekiel, "A Method of Handling Multiple Correlation Problems."
11. Brandt, "Uses of the Progressive Digit Method."
12. Tolley and Ezekiel, "A Method of Handling Multiple Correlation Problems."
13. "BAE News," vol. 8, no. 18, BAE.
14. Special Report of Howard Tolley, February 23, 1923, TOLLEY.
15. Ibid.
16. Cortada, *Before the Computer*, pp. 56–59.
17. "BAE News," vol. 8, no. 18, BAE.
18. Employee List of Tabulating Bureau, December 3, 1926, BAE.
19. "BAE News," vol. 8, no. 18, BAE.
20. Ezekiel, "Henry A. Wallace," p. 791.
21. Wallace, "What Is an Iowa Farm Worth?" (1924), pp. 1ff.
22. Ezekiel, "Henry A. Wallace," p. 791.
23. See Iowa State University, *Annual Report*, 1950–51, "John Evvard [*sic*]."
24. Henry Wallace to C. Cuthbert Hurd, February 21, 1965, NMAH.
25. Cox and Homeyer, "Professional and Personal Glimpses of George W. Snedecor."
26. Wallace and Snedecor, *Correlation and Machine Calculation*, p. 1.
27. Lush, "Early Statistics at Iowa State University," p. 220.
28. Culver and Hyde, *American Dreamer*, p. 147.
29. "Interview with Mary Clem by Uta Merzbach," June 27, 1969, pp. 1, 7–8, SMITHSONIAN.
30. Report of the Mathematics Department for 1928, Vice President for Research File, 6/1/1, ISU; see also Snedecor, "Uses of Punched Card Equipment."
31. Annual Report of Mathematics Department for 1928, Vice President for Research File, 6/1/1, ISU.
32. See Baehne, *Practical Applications of the Punched Card Method*, "Introduction and Table of Contents."
33. "Interview with Mary Clem by Uta Merzbach," June 27, 1969, pp. 1, 7–8, SMITHSONAIN.
34. "Mary Clem."
35. "Interview with Mary Clem by Uta Merzbach," June 27, 1969, pp. 22, 32, SMITHSONIAN.
36. MacDonald, *Henry Wallace*, p. 118.
37. Henry Wallace to George Snedecor, May 23, 1931, HAW.

38. Reich, *The Making of American Industrial Research*, pp. 163, 176.

39. "A Quarter Century of Transcontinental Telephone Service"; Mills, "The Line and the Laboratory."

40. Froelich, Clara, "Biographical Information Form," Barnard College Archives.

41. Price, "Award for Distinguished Service to Dr. Thornton Carl Fry."

42. Reich, *The Making of American Industrial Research*, p. 2.

43. "Mathematical Research."

44. Millman, *A History of Engineering and Science in the Bell System*, p. 352.

45. "Mathematical Research."

46. Comrie, L. J., "Inverse Interpolation and Scientific Applications of the National Accounting Machine" (1936).

47. Ibid.

48. Croarken, *Early Scientific Computing in Britain*, p. 24.

49. Ibid.

50. L. J. Comrie to T. D. Scott, November 28, 1924, Walter Dill Scott Papers, box 13, folder 10 (College of Liberal Arts: Department of Astronomy) Series 3/51/1, Northwestern University Archives.

51. Croarken, *Early Scientific Computing in Britain*, p. 25. For an interim period of six months, he served as an assistant in the almanac office.

52. Comrie, L. J., "Inverse Interpolation and Scientific Applications of the National Accounting Machine" (1936).

53. Ibid.

54. L. J. Comrie to Wallace Eckert, January 25, 1940, ECKERT.

55. Croarken and Campbell-Kelly, "Beautiful Numbers," pp. 44–61.

56. Ibid.

57. Ibid.; Croarken, "Case 5,656."

58. Croarken and Campbell-Kelly, "Beautiful Numbers"; Croarken, "Case 5,656."

59. Croarken and Campbell-Kelly, "Beautiful Numbers."

60. Ibid.; Croarken, "Case 5,656"; L. J. Comrie to Karl Pearson, June 1, 1933, 665/9, PEARSON.

61. Archibald, "BAASMTC Vol. 1."

CHAPTER TWELVE
THE BEST OF BAD TIMES

1. Executive Order of 1918, "National Research Council," in Cochrane, *The National Academy of Sciences* (1978), appendix.

2. Douglas Miller to Frank Schlesinger, June 27, 1930, Correspondence 1930–33, NRC-MTAC.

3. F. K. Richtmyer to D. C. Miller, October 9, 1930, Correspondence 1930–33, NRC-MTAC.

4. Frank Schlesinger to F. K. Richtmyer, June 19, 1930, Correspondence 1930–33, NRC-MTAC.

5. F. K. Richtmyer to Frank Schlesinger, June 24, 1930, Correspondence 1930–33, NRC-MTAC.

6. F. K. Richtmyer to D. C. Miller, October 9, 1933, Correspondence 1930–33, NRC-MTAC.

7. Bush, "The Differential Analyzer" (1931), n. 6.

8. Ibid.

9. D. C. Miller to F. K. Richtmyer, October 28, 1933, Correspondence 1930–33, NRC-MTAC.

10. Thornton Fry to F. K. Richtmyer, December 22, 1930, Correspondence 1930–33, NRC-MTAC.

11. Thornton Fry to F. K. Richtmyer, January 19, 1931, Correspondence 1930–33, NRC-MTAC.

12. Henry Reitz to F. K. Richtmyer, March 23, 1934, Correspondence 1934–36, NRC-MTAC.

13. Thornton Fry to F. K. Richtmyer, January 19, 1931, Correspondence 1930–33, NRC-MTAC.

14. National Research Council, *International Critical Tables* (1926), p. ii.

15. F. K. Richtmyer to Thornton Fry, January 21, 1931, Correspondence 1930–33, NRC-MTAC.

16. F. K. Richtmyer to Thornton Fry, May 5, 1931, Correspondence 1930–33, NRC-MTAC.

17. F. K. Richtmyer to Thornton Fry, June 4, 1931, Correspondence 1930–33, NRC-MTAC.

18. This image comes from book 35 of Pliny's *Natural History*. He would quote the proverb in Latin: "Ne sutor ultra crepidam."

19. Davis, H. T., *Adventures of an Ultra-Crepidarian* (1962), pp. 61–62.

20. Ibid., p. 98.

21. Ibid., pp. 174, 198.

22. Ibid., pp. 203, 294.

23. Ibid., p. 242.

24. Bryan, "The Life of the Professor."

25. William Rawles to W. L. Bryan, January 18, 1927, IU BRYAN; Davis, H. T., *Adventures of an Ultra-Crepidarian*, pp. 240–41, 271–72.

26. Thornton Fry to F. K. Richtmyer, January 19, 1931, Correspondence 1930–33, NRC-MTAC.

27. Davis, H. T., *Adventures of an Ultra-Crepidarian*, pp. 271–72.

28. Ibid.

29. Henry Reitz to F. K. Richtmyer, March 23, 1934, Correspondence 1934–36, NRC-MTAC.

30. Davis, H. T., *Adventures of an Ultra-Crepidarian*, p. 273.

31. Davis, H. T., *Tables of the Higher Mathematical Functions*, pp. xi–xiii.

32. Davis, H. T., *Adventures of an Ultra-Crepidarian*, p. 297.

33. Reiman, *The New Deal and American Youth*, p. 130.

34. Undated memo from Ardis Monk, head of the computing office, UC PHYSICS.

35. Davis, H. T., *Adventures of an Ultra-Crepidarian*, pp. 319, 337.

36. Fletcher et al., *Index of Tables*, pp. 804–6.

37. Comrie, L. J., "Tables of Higher Functions," *Mathematical Gazette*, vol. 20, 1936, pp. 225–27.

38. Comrie, "Inverse Interpolation" (1936); Fletcher et al., *Index of Tables*, pp. 804–6.

39. Fletcher et al., *Index of Tables*, pp. 804–6.

40. "History of the Arkansas, Louisiana & Mississippi Railroad Company," http://www.almrailroad.com/history.htm.

41. Cowles Commission, *Report of Research Activities*, July 1, 1964–June 30, 1967.

42. Christ, *History of the Cowles Commission* (1952), p. 7.

43. Carl Christ suggests that Davis may have known of some method to compute regression equations with tabulators (ibid., p. 8).

44. Davis, H. T., *Adventures of an Ultra-Crepidarian*, pp. 299–300.

45. Ibid., p. 302.

46. Dates unknown; she took her BA in 1927; Colorado College Library Special Collections; Davis, H. T., *Tables of the Higher Mathematical Functions*, p. 193.

47. Cowles, A., "Can Stock Market Forecasters Forecast?" (1933).

48. Davis, H. T., *Adventures of an Ultra-Crepidarian*, p. 302; Christ, *History of the Cowles Commission*, p. 4; Elizabeth Webb Wilson Scrapbooks, vol. 3, p. 40, WILSON PAPERS.

49. Christ, *History of the Cowles Commission*, p. 10.

50. Elizabeth Webb Wilson Scrapbooks, vol. 3, p. 40, WILSON PAPERS.

51. "Boston Woman Is Rated Insurance Expert Deluxe." See also Wilson, *Compulsory Health Insurance*.

52. Veblen, T., *The Higher Learning in America* (1918), p. 124.

53. Feffer, "Oswald Veblen."

54. Rodgers, *Think*, p. 134.

55. Pugh, *Building IBM* (1995), pp. 34–45.

56. Rodgers, *Think*, p. 135.

57. Benjamin Wood, interview by Henry Tropp, p. 1, SMITHSONIAN.

58. In his Smithsonian interview, Wood quotes Watson as saying that IBM had no installations in government offices before 1928 (ibid., p. 32), yet Wood clearly had Watson's attention.

59. Benjamin Wood, interview by Henry Tropp, p. 6, SMITHSONIAN.

60. Watson and Petre, *Father and Son and Company*, p. 37.

61. Benjamin Wood, interview by Henry Tropp, p. 11, SMITHSONIAN.

62. Watson and Petre, *Father and Son and Company*, p. 37.

63. Benjamin Wood, interview by Henry Tropp, p. 12, SMITHSONIAN.

64. Brennan, *The IBM Watson Laboratory*, p. 141.

65. Benjamin Wood, interview by Henry Tropp, p. 12, SMITHSONIAN.

66. Brown, E. W., *The Motion of the Moon*.

67. Comrie, L. J., "The Application of the Hollerith Tabulating Machine" (1932).

68. L. J. Comrie to Wallace J. Eckert, May 1, 1935, box 1.2, ECKERT. Strictly speaking, Comrie was using cards produced by the British Tabulating Machine Company, a firm that had IBM investment and had licensed IBM technology.

69. Eckert, "Astronomy" (1935); Comrie, L. J., "The Application of the Hollerith Tabulating Machine."

70. Eckert, "Astronomy."
71. Wallace J. Eckert to G. W. Baehne, January 9, 1934, box 1.2, ECKERT.
72. Baehne, *Practical Applications of the Punched Card Method*, preface.
73. Eckert, "Astronomy."
74. Brennan, *The IBM Watson Laboratory*, p. 9.
75. Grier, "The First Breach of Computer Security?" (2001). Numerov's death is described in P. G. Kulikousky, "Boris Numerov," *Dictionary of Scientific Biography*, ed. Charles Gillespie, New York, Scribners, 1974, pp. 158–60.
76. Eckert, *Punched Card Methods* (1940), p. 1.
77. Report on Annual Meeting of Board of Managers of Thomas J. Watson Astronomical Computing Bureau, April 27, 1940, ECKERT.
78. Henry Reitz to F. K. Richtmyer, February 19, 1934, Correspondence 1934–36, NRC-MTAC.
79. H. L. Reitz to F. K. Richtmyer, March 23, 1934, Correspondence 1934–36, NRC-MTAC.
80. U.S. Army, *Ballisticians in War and Peace*, p. 11; Aberdeen Proving Ground, Annual Reports for 1933–37.
81. U.S. Army, *Ballisticians in War and Peace*, p. 11.
82. R. A. Millikan to H. T. Davis, June 24, 1936, Correspondence 1934–36, NRC-MTAC.
83. L. J. Comrie to Robert Millikan, July 28, 1936, Correspondence 1937–39, NRC-MTAC.
84. A. A. Bennett to Henry Barton, October 10, 1936, Correspondence 1937–39, NRC-MTAC.
85. Croarken, "Case 5,656" (1999); see also Wilkins, "The History of H.M. Nautical Almanac Office," p. 59.
86. Croarken, "Case 5,656" (1999).
87. Ibid.
88. Comrie, "Scientific Computing Service Limited," pp. 1–3.

CHAPTER THIRTEEN
SCIENTIFIC RELIEF

1. *Polk's Washington (District of Columbia) City Directory*, Richmond, Va., R. L. Polk, 1936; "Malcolm Morrow."
2. When organized in 1935, the agency was originally called the Works Progress Administration. The Emergency Relief Act of 1939 changed its name to Work Projects Administration. To avoid confusion, this book will use the latter name throughout.
3. Bancroft, "Statistical Laboratory of the Iowa State University" (1966); Dedicatory Plaque, George Snedecor Hall, Iowa State University.
4. U.S. Federal Works Agency, *Final Report*, p. 65.
5. Works Progress Administration, *Index*, projects 3895, 3896, p. 13.
6. Ibid., p. iv.
7. Ibid., project 4273, p. 60.

8. "WPA Will Add 350,000 to Rolls."

9. Office Memorandum No. 433, October 21, 1937, SAB.

10. Ibid.

11. Lee, *To Kill a Mockingbird*, p. 250.

12. Paul Brockett to Frank Lillie, October 21, 1937, SAB.

13. Frank Lillie to Emerson Ross, October 29, 1937, SAB.

14. Office Memorandum No. 433, October 21, 1937, SAB.

15. Ibid.

16. Ickes, *Secret Diary of Harold Ickes*, p. 233; Kessner, *Fiorello H. La Guardia*, p. 418.

17. Office Memorandum No. 433, October 21, 1937, SAB.

18. Paul Brockett to Frank Lillie, November 6, 1937, SAB.

19. Paul Brockett to Frank Lillie, November 1, 1937, SAB.

20. Summary of Meeting, January 28, 1938, BRIGGS.

21. Lyman Briggs to the Secretary of Commerce, November 7, 1934, BRIGGS.

22. Civil Works Administration Projects folder #680, box 46, NBS.

23. Cochrane, *Measures for Progress* (1966), p. 332.

24. Paul Brockett to Frank Lillie, November 16, 1937, SAB.

25. "Lowan Information Sheet," LOWAN.

26. Snyder-Grenier, *Brooklyn*, p. 258.

27. Lowan WPA Employment Form, MTP WPA.

28. George Pegram to Oswald Veblen, June 13, 1933, LOWAN.

29. Arnold Lowan Biographical Record Form, LOWAN.

30. Ida Rhodes to Uta Merzbach, November 4, 1969, NMAH.

31. Gertrude Blanch, interview by Henry Thatcher in San Diego, March 17, 1989, STERN.

32. Howe, *World of Our Fathers*, p. 131.

33. Department of Justice, Immigration and Naturalization Service Certificate, March 16, 1966, STERN.

34. Gertrude Blanch, interview with Michael Stern, approximately 1989, STERN.

35. Blanch's library included Bennett, *Corporation Accounting*, New York, Ronald Press, 1919; Rosenthal, *Technical Procedure in Exporting and Importing*, New York, McGraw-Hill, 1922; and Mills, *Statistical Methods*, New York, Holt, 1924 (card file of book collection, STERN).

36. Transcript of Gittel Kaimowitz, BLANCH NYU.

37. Rossiter, *Women Scientists in America*, p. 173.

38. Hardy, *A Mathematician's Apology*, p. 70.

39. Director FBI to Assistant Attorney General, May 17, 1956, BLANCH FBI.

40. Order of King's County Court for Gittel Kaimowitz, February 9, 1932, STERN.

41. Gertrude Blanch, interview with Michael Stern, approximately 1989, STERN.

42. Gertrude Blanch, interview by Henry Thatcher in San Diego, March 17, 1989, STERN.

43. Ibid.

44. Blanch, *Properties of the Veneroni Transformation* (1934), p. i.

45. Records of the Alpha Chapter of Sigma Delta Epsilon, 1930–40, CORNELL.

46. Blanch, *Properties of the Veneroni Transformation* (1934), p. i.

47. "Field Mathematician Gets Top Job Rating."

48. Hazel Ellenwood to Gertrude Blanch, February 26, 1937, CORNELL.

49. Gertrude Blanch, interview with Michael Stern, approximately 1989, STERN.

50. Gertrude Blanch, interview by Henry Thatcher in San Diego, March 17, 1989, STERN.

51. Gertrude Blanch, interview with Michael Stern, approximately 1989, STERN.

52. Gertrude Blanch, interview by Henry Thatcher in San Diego, March 17, 1989, STERN.

53. Gertrude Blanch, interview by Henry Tropp, May 16, 1973, SMITHSONIAN.

54. Barlow, *Barlow's Tables*.

55. Mathematical Tables Project, *Tables of the Exponential Function* (1939).

56. Ibid., pp. 50–51.

57. Report of Meeting, January 28, 1938, BRIGGS.

58. Minutes of the Executive Committee of the Division of Physical Sciences, April 1, 1938, NAS.

59. U.S. Federal Works Agency, *Final Report*, p. 41.

60. Gertrude Blanch, interview with Henry Tropp, May 16, 1973, SMITHSONIAN.

61. Gertrude Blanch, interview by Henry Thatcher in San Diego, March 17, 1989, STERN.

62. Ida Rhodes to Uta Merzbach, November 4, 1969, NMAH.

63. Grier, "The Math Tables Project" (1998).

64. Slutz, "Memories of the Bureau of Standards SEAC."

65. Abraham Hillman, interview with the author, February 1996.

66. David Stern, interview with the author, January 2002.

67. Abraham Hillman, interview with the author, February 1996.

68. Gertrude Blanch, interview with Michael Stern, approximately 1989, STERN.

69. Gertrude Blanch, interview with Henry Tropp, May 16, 1973, SMITHSONIAN.

70. Weekly Report of the Mathematical Tables Project for September 15, 1941, MTP WPA.

71. Fletcher et al., *An Index of Mathematical Tables*, p. 881.

72. Blanch and Rhodes, "Table-Making at the National Bureau of Standards."

73. The procedure for reauthorizing is described in Office Memorandum No. 433, October 21, 1937, SAB.

74. The most complete series of this correspondence is found in Lowan Correspondence, 1940, box 13, Records Relating to Computing, NBS.

75. Lyman Briggs to Arnold Lowan, June 28, 1939, BRIGGS.

76. Arnold Lowan to Lyman Briggs, June 28, 1939, and Lyman Briggs to Arnold Lowan, July 27, 1939, BRIGGS.

77. John von Neumann to Arnold Lowan, September 19, 1940, NEUMANN.

78. Phil Morse to Arnold Lowan, September 21, 1938, W.P.A. Math Table Files, MORSE.

79. Arnold Lowan to Phil Morse, September 28, 1938, W.P.A. Math Table Files, MORSE.

80. Phil Morse to Arnold Lowan, October 18, 1938, W.P.A. Math Table Files, MORSE.

81. Howard, *The WPA and Federal Relief Policy*, p. 278.

82. Bethe and Critchfield, "The Formation of Deuterons by Proton Combination"; Bethe, "Energy Production in Stars."

83. Blanch et al., "The Internal Temperature Density Distribution of the Sun" (1942).

84. Hans Bethe to Arnold Lowan, February 14, 1939, MTP AMP.

85. Hans Bethe quoted in Bernstein, "Profiles (Hans Bethe Part I)."

86. Hans Bethe to Arnold Lowan, January 17, 1940, Bethe File, Records Relating to Computing, NBS.

87. Curtiss, "*Tables of the First Ten Powers of the Integers from 1 to 1000; Tables of the Exponential Function e^x*" (1941).

CHAPTER FOURTEEN
TOOLS OF THE TRADE

1. Lowan's letters regularly mention his search for calculators. See, for example, Lyman Briggs to Arnold Lowan, November 25, 1940, BRIGGS.

2. Ida Rhodes to Uta Merzbach, NMAH.

3. Interview of Gertrude Blanch with Henry Tropp, May 16, 1973, SMITHSONIAN.

4. Bulletin of Iowa State College for 1934–35, p. 326.

5. *American Men of Science*, 11th ed., New York, Bowker, 1965.

6. Stibitz, "Computer" (1940).

7. Kruger, "A Slide Rule for Vector Calculations."

8. Millman, *A History of Engineering and Science in the Bell System*, pp. 27–28; Fry, "Industrial Mathematics" (1941).

9. Stibitz, "Early Computers" (1980).

10. Notes of January 19, 1938, Research Case 20878, Reel FC-4618, ATT.

11. Stibitz, "Computer."

12. Andrews, E. C., "Telephone Switching and the Early Bell Laboratories Computers" (1982).

13. Stibitz, "Early Computers."

14. Irvine, "Early Digital Computers at Bell Telephone Laboratories."

15. See Iowa State College Budget for 1937–38, ISU-ADMIN; "Statement on Statistics," Statistics 1945 File, Records of the Vice p.resident for Research, ISU-ADMIN.

16. "Interview with Mary Clem by Uta Merzbach," June 27, 1969, p. 14, SMITHSONIAN.

17. Atanasoff, "Computing Machine for the Solution of Large Systems of Linear Algebraic Equations" (1984), p. 233.

18. The complexity increases with the cube of the size of the problem. Doubling the size of a problem increases the effort by a factor of $8 = 2^3$. Increasing a problem by a factor of 12 (from 2 to 24) demands $1738 = 12^3$ times the effort.

19. Atanasoff and Brandt, "Application of p.unched Card Equipment to the Analysis of Complex Spectra."

20. Atanasoff, "Advent of Electronic Digital Computing" (1984).

21. Grier, "Agricultural Computing and the Context for John Atanasoff" (2000); Iowa State College Budget for 1938–39, ISU-ADMIN.

22. Testimony of John V. Atanasoff, ENIAC Trial, vol. 11, CBI, quoted in Atanasoff, "Advent of Electronic Digital Computing," p. 239.

23. Atanasoff, "Advent of Electronic Digital Computing."

24. John Atanasoff to John Mauchly, May 31, 1941, Trial Record, *Honeywell v. Illinois Scientific*, CBI.

25. Atanasoff, "Computing Machine for the Solution of Large Systems of Linear Algebraic Equations."

26. J. V. Atanasoff to Warren Weaver, July 10, 1940, Feb. 26 folder 1, 13/20/51, ATANASOFF.

27. Deposition of Warren Weaver, *Honeywell v. Illinois Scientific*, Henry Hansen Papers, box 5, file 2, 3/6/72, ISU-ADMIN.

28. Diary of Warren Weaver, April 29, 1940, WEAVER.

29. J. V. Atanasoff to Warren Weaver, July 10, 1940, Feb. 26 folder 1, 13/20/51, ATANASOFF.

30. Mollenhoff, *Atanasoff, Forgotten Father of the Computer,* p. 52.

31. Atanasoff, "Computing Machine for the Solution of Linear Algebraic Equations," p. 233. Experience with a modern reconstruction has confirmed the challenges with the card punch and has also suggested that the machine could be a delicate device; see Grier, "Henry Wallace and the Start of Statistical Computing" (1999).

32. Stewart, R., "End of the ABC."

33. Mollenhoff, *Atanasoff, Forgotten Father of the Computer*, p. 10.

34. Herman Goldstine quoted in Stern, *From ENIAC to UNIVAC*, p. 34. The controversy has an extensive literature that includes Stern, *From ENIAC to UNIVAC*; Goldstine, *The Computer from Pascal to von Neumann*; the *IEEE Annals of the History of Computing*, vol. 6, no. 3; Burks and Burks, *The First Electronic Computer*; Burks, *Who Invented the Computer?*; McCartney, *ENIAC*.

35. Aiken, "Proposed Automatic Calculating Machine."

36. Aiken and Hopper, "The Automatic Sequence Controlled Calculator—I" (1946).

37. Aiken and Hopper, "The Automatic Sequence Controlled Calculator—III" (1946).

38. Cohen, *Howard Aiken* (1999), p. 63.

39. Ibid., pp. 12–13, 22.

40. Ibid., p. 42.

41. Kidwell, "From Novelty to Necessity" (1990); H. Shapley to L. J. Comrie, February 23, 1923, UA630.22 box 4, HARVARD OBS.

42. Aiken, "Proposed Automatic Calculating Machine."

43. Cohen, *Howard Aiken*, pp. 24, 85.

44. Quoted in Ceruzzi, *Reckoners* (1983), p. xi.

45. Bloch, "Programming Mark I," p. 82.

46. See Harvard University, *Annals of the Computation Laboratory*; Mathematical Tables Project, *Tables*.

CHAPTER FIFTEEN
PROFESSIONAL AMBITION

1. Wells, H. G., "The World of Tomorrow," *New York Times*, May 5, 1939, p. AS4.

2. "I.B.M. Convention Set," *New York Times*, April 29, 1939, p. 34.

3. "Debris Still Fills Pavilion of WPA," *New York Times*, May 4, 1939, p. 20.

4. "WPA Finds Friends at Its Fair Exhibit," *New York Times*, July 6, 1939, p. 18.

5. "WPA Exhibit Opens without Fanfare," *New York Times*, May 23, 1939, p. 18.

6. Lowan makes no mention of the fair in his correspondence, perhaps because he had little to show. See correspondence for 1939, MTP WPA.

7. A. A. Bennett quoted in Executive Committee Minutes, April 7, 1939, NRC-PS.

8. Aberdeen Proving Ground, Annual Report for 1936, ORDNANCE.

9. Bliss, *Mathematics for Exterior Ballistics*, p. 55.

10. R. C. Archibald to Simon Newcomb, July 2, 1902, NEWCOMB.

11. Resume, Raymond Claire Archibald, Archibald Correspondence, Electronic Computers, 1939–54, NBS.

12. Oswald Veblen to R. C. Archibald, January 12, 1928, VEBLEN.

13. R. C. Archibald to MTAC Committee, June 28, 1939, Correspondence, 1940, NBS-MTAC.

14. Albert Barrows to Luther p. Eisenhart, September 16, 1939, Correspondence, 1940, NRC-MTAC.

15. Albert Barrows to Luther p. Eisenhart, July 11, 1940, NRC-MTAC.

16. Archibald to Barrows, October 2, 1939, MTAC Correspondence, 1937–39, NRC-MTAC.

17. Archibald to Y. Tsuyi, National Research Council of Japan, Correspondence, 1940, NRC-MTAC.

18. W. J. Eckert to R. C. Archibald, August 16, 1940, ECKERT.

19. MTAC Membership List, 1940, MTAC Folder, ECKERT.

20. R. C. Archibald to Warren Weaver, Rockefeller Foundation, October 23, 1939, Correspondence, 1940, NRC-MTAC.

21. Lehmer, *Guide to Tables in the Theory of Numbers*.

22. L. P. Eisenhart to R. C. Archibald, July 20, 1940, Correspondence for 1940, NRC-MTAC.

23. Archibald to Eisenhart, July 16, 1940, Correspondence for 1940, NRC-MTAC.

24. Churchill Eisenhart, interview by William Aspray; Office Memorandum No. 433, 21 October, 1937, SAB.

25. L. P. Eisenhart to A. Barrows, July 20, 1940, NRC-MTAC.

26. R. C. Archibald to L. P. Eisenhart, July 29, 1940, NRC-MTAC.

27. R. C. Archibald to L. P. Eisenhart, August 1, 1940, NRC-MTAC.

28. See Correspondence for 1940, NRC-MTAC. The letter from Eisenhart to Albert Barrows, October 16, 1940, best describes the situation and the issues from Eisenhart's point of view.

29. Tocqueville, *Democracy in America*, vol. 2, chap. 5.

30. Gallup, "American Institute of Public Opinion Surveys" (1939), p. 599.

31. Quoted in Burns, *The Lion and the Fox*, p. 393.

32. Aberdeen Proving Ground, Annual Report for 1939, ORDNANCE.

33. Dick, "A History of the American Nautical Almanac Office," pp. 11–54.

34. L. J. Comrie to W. J. Eckert, January 25, 1940, ECKERT.

35. Weber, *The Naval Observatory*, pp. 42–43.

36. U.S. Naval Observatory, *Annual Report for 1944*, p. 2.

37. Capt. J. F. Hellweg, Superintendent, Naval Observatory, to Wallace Eckert, December 6, 1939, ECKERT; see Gutzwiller, "Wallace Eckert," pp. 150–51.

38. U.S. Naval Observatory, *Annual Report for 1941*, pp. 10, 11, 13.

39. "Final Report of the Math Tables Project," March 15, 1943, MTP WPA.

40. Lowan, "The Computational Laboratory at the National Bureau of Standards" (1949).

41. Baxter, *Scientists against Time*, p. 7.

42. Report of Arnold Lowan, June 13, 1941, MTP WPA.

43. Allegra Rogers to Arnold Lowan, June 9, 1941, MTP WPA.

44. He was discovering new groups as late as 1942; see Weekly Report of May 25, 1942, MTP WPA.

45. Lyman Briggs to Col. Loper, June 14, 1941, MTP WPA.

46. U.S. Federal Works Agency, *Final Report*, p. 79.

47. Arnold Lowan to Philip Morse, March 18, 1941, MORSE.

48. U.S. Federal Works Agency, *Final Report*, pp. 84, 85.

49. Arnold Lowan to Philip Morse, December 17, 1940, MORSE.

50. Ibid.

51. Arnold Lowan to John von Neumann, May 17, 1941, NEUMANN.

52. Plans were reviewed by an outside reviewer, the WPA office of New York, sponsor Lyman Briggs, the Washington WPA office, and the Central Statistical Office. This last group coordinated government statistical work and once had counted Howard Tolley of the Department of Agriculture among its members; see A. Lowan to p. Morse, October 19, 1938, MORSE.

53. Night Letter, August 29, 1940, BRIGGS.

54. R. C. Archibald to the Gentlemen of the Committee, December 31, 1941, NRC-MTAC.

55. Wallace Eckert to R. C. Archibald, January 7, 1942, ECKERT.

56. Lindbergh, *The War Within and Without*, pp. 241, 243.

57. Philip Morse to Arnold Lowan, January 19, 1942, MORSE.

58. Philip Morse to Arnold Lowan, January 12, 1942, and May 13, 1942, MORSE.

59. Memo of Karl Compton, January 29, 1942, MORSE.

60. Julius Stratton to Lyman Briggs, March 9, 1942, MORSE.

61. Florence Kerr to Lyman Briggs, April 8, 1942, MTP WPA.

62. Florence Kerr to R. C. Branion, May 19, 1942, MTP WPA; see also Lyman Briggs to General Huie, May 3, 1942, MTP WPA.

63. Florence Kerr to Lyman Briggs, April 8, 1942, MTP WPA.

64. Frank Culley to General Huie, July 1, 1942, MTP WPA.

65. Report from Cincinnati office of FBI, September 20, 1955, BLANCH FBI; Philip Morse to Lyman Briggs, May 30, 1943, MORSE.

66. "Summary of Supplemental Investigation," April 23, 1956, BLANCH FBI.

67. Grier, "The First Breach of Computer Security" (2001); see Stepanoff to Eckert, August 8, 1940, Eckert to Stepanoff, August 8, 1940, and Schilt to Eckert, August 9, 1940, ECKERT. For background on Amtorg, see Rhodes, R., *Dark Sun* (1995), pp. 57ff.

68. Phil Morse to Fewell of National Bureau of Standards, June 25, 1942, MORSE.

69. Pierce, *Long Range Navigation* (1948), p. 403.

70. Melville Eastman to Julius Furer, July 2, 1942, MTP ONR.

71. Ibid.

72. Lyman Briggs to General Huie, September 7, 1942, MTP WPA.

73. General Huie to Lyman Briggs, September 15, 1942, MTP WPA.

74. Philip Morse to Arnold Lowan, October 2, 1942, MORSE.

75. Internal Memo, "Simplified Outline of LORAN System Components," October 8, 1942, MTP ONR.

76. Fletcher Watson (MIT) to Warren Weaver (AMP), November 23, 1942, MORSE.

77. Arnold Lowan to Philip Morse, November 2, 1942, MORSE.

78. Ida Rhodes to Uta Merzbach, November 4, 1969, NMAH; a similar quote is given in Ida Rhodes, interview with Henry Tropp, March 21, 1973, SMITHSONIAN.

79. L. J. Comrie to Arnold Lowan, December 7, 1942 (copy), MORSE.

80. L. J. Comrie to Arnold Lowan, May 16, 1942, MTP WPA.

81. Arnold Lowan to Philip Morse, December 31, 1941, MORSE.

82. G. S. Bryan to New York City WPA Office, January 6, 1943, MTP WPA.

83. To the Commandant, 3rd Naval District, February 16, 1943, MTP ONR.

84. Regina Schlachter to Commandant, 3rd Naval District, February 15, 1944, MTP ONR.

85. NDRC Organizational Chart, August 1943, AMP.

86. Weaver, *Scene of Change*, p. 87.

87. Warren Weaver to Jerzy Neyman, October 21, 1941, NEYMAN.

88. Ibid.

89. See Minutes of Executive Committee, Applied Mathematics Panel, AMP.

90. Warren Weaver, "Report on the Proposed Applied Mathematics Panel," November 12, 1942, AMP.

91. Owens, "Mathematicians at War" (1996).

92. Diary of Warren Weaver, December 10, 1942, AMP.

93. Warren Weaver to James Conant, February 8, 1942, AMP.

94. Warren Weaver to Grace Hopper, February 16, 1942, AMP.

95. Hopper, "Commander Aiken and My Favorite Computer," pp. 185–94.

96. Warren Weaver to Lyman Briggs, February 23, 1942, AMP.

97. Arnold Lowan to Philip Morse, March 3, 1943, MORSE.

98. Lowan, "The Computational Laboratory at the National Bureau of Standards," p. 37.

99. Ida Rhodes to Uta Merzbach, November 4, 1969, NMAH.

CHAPTER SIXTEEN
THE MIDTOWN NEW YORK GLIDE BOMB CLUB

1. Beauclair, "Alwin Walther, IPM, and the Development of Calkulator/Computer Technology in Germany, 1930–1945."

2. Herbert Salzer, interview with the author.

3. Croarken, *Early Scientific Computing in Britain*, pp. 61–74.

4. Ceruzzi, "When Computers Were Human" (1991).

5. G. S. Bryan to New York City WPA Office, January 6, 1943, MTP WPA; E. C. Crittenden, Acting Director, ONR, to Lowan, October 23, 1943, MTP ONR; Warren Weaver to Applied Mathematics Panel, November 12, 1943, MTP AMP.

6. R. C. Archibald to MTAC Committee, August 6, 1942, ECKERT.

7. L. Eisenhart to A. Barrows, August 7, 1942; A. Barrows to L. Eisenhart, August 10, 1942, NRC-MTAC.

8. After his initial reaction, Weaver said that he would do nothing to stop the plan. W. Weaver to L. Eisenhart, August 14, 1942, NRC-MTAC.

9. W. Eckert to R. C. Archibald, October 28, 1942, ECKERT.

10. R. C. Archibald to MTAC Committee, August 6, 1942, ECKERT.

11. Archibald, "Introduction" (1943).

12. J. Brainerd to L. Briggs, February 9, 1943, Directors Correspondence, Records of the Director, UD E-6, NBS.

13. Memo from Arnold Lowan to Mathematical Tables Project Staff, October 16, 1946, Monte Carlo Computations File (verso used as scratch paper), Research on Electronic Computers, 1939–54, NBS.

14. Brainerd to J. A. Shohat, February 23, 1943, Course on Mathematical Ballistics, PENNSYLVANIA.

15. U.S. Army, *Ballisticians in War and Peace*, p. 11; Irven Travis, Oral History, pp. 2–3.

16. R. S. Zug to Major Gillon, "Report on work at Moore School, University of Pennsylvania, July 1–10, 1942," July 14, 1942, Course on Mathematical Ballistics, PENNSYLVANIA.

17. Ibid.

18. Ibid.

19. S. Reid Warren to J. Brainerd, January 9, 1945, Office of the Vice Dean, Correspondence for 1945, PENNSYLVANIA.

20. Herman Goldstine, interview with the author, July 2002; Goldstine, *The Computer from Pascal to von Neumann*, p. 133.

21. Herman Goldstine, interview with the author, July 2002.

22. Minutes of Moore School Meeting of December 18, 1944, Course on Mathematical Ballistics, PENNSYLVANIA; Report (unsigned), July 27, 1942, Office of the Vice Dean, Correspondence for 1943, PENNSYLVANIA; Program for Selection and Processing of Applicants, July 17, 1942, Office of the Vice Dean, Correspondence for 1942, PENNSYLVANIA.

23. Dean Pender to George Turner, September 21, 1942, Office of the Vice Dean, Correspondence for 1943, PENNSYLVANIA.

24. W. Weaver to Applied Mathematics Panel, November 12, 1943, MTP AMP.

25. Report (unsigned), July 27, 1942, Office of the Vice Dean, Correspondence for 1943, PENNSYLVANIA.

26. See, for example, letters of July 10, 15; November 9, 14, 18, 1942; January 22, 23, 24, 28; March 27, 1943, Moore School of Electrical Engineering, Office of the Vice Dean, Records 1931–1948, UPD 8.1, PENNSYLVANIA.

27. Minutes of Moore School Meeting of December 18, 1944, Course on Mathematical Ballistics, PENNSYLVANIA.

28. Adele Goldstine to J. G. Brainerd, May 27, 1943, Office of the Vice Dean, Correspondence for 1943, PENNSYLVANIA.

29. Comrie, L. J., "Computing Machines," *Mathematical Tables and Other Aids to Computation,* vol. 1, no. 2 (April 1943), pp. 63–64; Shannon, C. E., "Mathematical Theory of the Differential Analyzer," *Mathematical Tables and Other Aids to Computation,* vol. 1, no. 2 (April 1943), p. 64.

30. Goldstine, *The Computer from Pascal to von Neumann,* p. 149.

31. Croarken, *Early Scientific Computing in Britain,* pp. 62, 64.

32. "The Missions of a Convert," undated manuscript, TODD.

33. John Todd, Interview, p. 9, SMITHSONIAN.

34. Ibid.; John Todd, interview with the author.

35. Taussky, "How I Became a Torchbearer for Matrix Theory."

36. Croarken, *Early Scientific Computing in Britain* (1990), p. 68; a similar statement was made by Todd in an interview with the author, January 2002.

37. John Todd, interview with the author.

38. Sadler and Todd, "Mathematics in Government Service and Industry" (1946).

39. Frank Olver, interview with the author, March 2002.

40. Olga Taussky-Todd to Frances Cave-Browne-Cave, August 29, 1947, CBC.

41. John Todd, interview with the author.

42. Todd, "John von Neumann and the National Accounting Machine" (1974).

43. John von Neumann to John Todd, November 17, 1947, quoted ibid.; see also Aspray, *John von Neumann* (1990), pp. 27–28.

44. For much of the war, the office of the Applied Mathematics Panel was in the Empire State Building, but all committee meetings were held at Rockefeller Center.

45. Minutes of Executive Committee, December 13, 1943, AMP.

46. Minutes of Executive Committee, May 24, 1943, AMP.

47. Minutes of Executive Committee, April 26, 1943, AMP.

48. Diary of J. D. Williams, February 9, 1944, AMP.

49. Minutes of Executive Committee, February 7, 1944, AMP.

50. Minutes of Executive Committee, November 22, 1943, AMP.

51. See, for example, the discussion of the Coast Artillery for a specialized theory of ballistics, Minutes of Executive Committee, September 20, 1943, AMP.

52. Applied Mathematics Panel to Chairs of Mathematics Departments, December 15, 1943, Correspondence 1940–45, IOWA MATH.

53. Reports of Columbia SRG, AMG-C, and BRG, Minutes of Executive Committee, May 4, 1944, AMP; MacLane, "Appendix: Roster of People."

54. Isaacson, "The Origin of Mathematics of Computation and Some Personal Recollections."

55. Budget of the Applied Mathematics Panel for 1944, Diary of Warren Weaver, January 10, 1944, AMP.

56. See AMP Correspondence for September–October 1943, November 1943, especially Equitable Life Insurance to Warren Weaver, October 27, 1943, and Diary of Mina Rees for November 6, 1944, AMP.

57. Minutes of Executive Committee, June 28, 1943, AMP.

58. Warren Weaver to Harold V. Gaskill, Iowa State College, November 4, 1943, Diary of Mina Rees, AMP.

59. Stewart, "End of the ABC."

60. Rosser, "Mathematics and Mathematicians in World War II."

61. Abraham Hillman, interview with the author, February 1996.

62. Weekly Reports of Mathematical Tables Project, 1943–44, AMP.

63. Officer in Charge of New York Project to Hydrographer, October 9, 1944, MTP ONR.

64. Warren Weaver to Lyman Briggs, March 31, 1944, AMP.

65. Craven and Gate, *The Army Air Forces in World War II*, p. 169.

66. Reid, *Neyman from Life*, p. 190.

67. Owens, "Mathematicians at War: Warren Weaver and the Applied Mathematics Panel, 1942–1945" (1989).

68. Diary of Warren Weaver, November 16, 1943, AMP.

69. See "Excerpt from Diary of J. Neyman, Washington, DC and Eglin Field, Florida," December 3–19, 1942, Jerzy Neyman Correspondence, AMP.

70. Reid, *Neyman from Life*, p. 183.

71. Jerzy Neyman to Warren Weaver, August 6, 1942, Jerzy Neyman Files, AMP.

72. Ibid.

73. Linus Pauling to Warren Weaver, February 5, 1942, Jerzy Neyman Files, AMP.

74. Warren Weaver to Jerzy Neyman, July 2, 1942, Jerzy Neyman Files, AMP.

75. Executive Committee Minutes, November 29, 1943, AMP.

76. Jerzy Neyman to Warren Weaver, December 17, 1943, AMP. This letter was written from California after the Mathematical Tables Project had finished work in New York, but Neyman had yet to learn this.

77. Diary of Warren Weaver, December 17, 1943, AMP.

78. Craven and Gate, *The Army Air Forces in World War II*, p. 169.

79. Commander Bramble quoted in Executive Committee Minutes, December 20, 1943, AMP.

80. Andrews, E. C., "Telephone Switching and the Early Bell Laboratories

Computers" (1982); Williams, *A History of Computing Technology*, pp. 225–27; Cesareo, "The Relay Interpolator."

81. Executive Committee Minutes, December 20, 1943, AMP.

82. There are many discussions of the ENIAC. The canonical accounts are Stern, *From ENIAC to UNIVAC*, and Goldstine, *The Computer from Pascal to von Neumann*. Two versions that attempt to address the machine in the context of its human computers are Fritz, "The Women of ENIAC," and Bergin, *Fifty Years of Army Computing*.

83. Arthur Burks, interview conducted by William Aspray, June 20, 1987, OH 136, CBI.

84. Eckstein, "J. Presper Eckert."

85. Cohen, I. B., *Howard Aiken* (1999), pp. 115, 119; Campbell, "Mark II, an Improved Mark I."

86. Executive Committee Minutes, March 6, 1944, AMP.

87. Ibid.

88. Mina Rees to Oswald Veblen, June 9, 1945, Applied Mathematics Panel Correspondence, AMP.

89. See, for example, Executive Committee Minutes, March 1, 1943, AMP.

90. Quoted in Stachel, "Lanczos's Early Contributions to Relativity and His Relationship with Einstein."

91. Blanch and Rhodes, "Table-Making at the National Bureau of Standards."

92. Executive Committee Minutes, November 6, 1944, AMP.

93. Undated memo from Ardis Monk, head of the computing office, UC PHYSICS.

94. Metropolis and Nelson, "Early Computing at Los Alamos"; see also Gleick, *Genius*, pp. 175–84.

95. Everett Yowell, Interview, SMITHSONIAN.

96. *Register of the United States for 1901,* Washington, D.C., Government Printing Office, 1901.

97. Everett Yowell, interview with the author, December 1998.

98. Everett Yowell, Interview, SMITHSONIAN.

99. Metropolis and Nelson, "Early Computing at Los Alamos."

100. Feynman, "Los Alamos from Below."

101. Metropolis and Nelson, "Early Computing at Los Alamos"; Gleick, *Genius*, pp. 175–84.

102. Cohen, Portrait of a Computer Pioneer, p. 164.

CHAPTER SEVENTEEN
THE VICTOR'S SHARE

1. Stibitz, "Lecture" (1946), p. 15; Comrie, "Careers for Girls" (1944).

2. Diary of J. B. Williams, December 2, 1944, AMP. The term is used in a reference to John Tukey of Princeton University. Tukey, credited with inventing the word "bit" to refer to a binary digit, was known to be inventive with language and is probably the source of the term "kilogirl." Stibitz referred to "girl years"; Stibitz, "Lecture" (1946).

3. Comrie, "Careers for Girls" (1944).

4. Ida Rhodes to Uta Merzbach, November 4, 1969, NMAH; Abraham Hillman, interview with the author, February, 1996; Golemba, *Women in Aeronautical Research*, p. 41.

5. Minutes of Executive Committee, September 25, 1944, AMP; see Reingold, "Vannevar Bush's New Deal for Research" (1987).

6. Stewart, I., *Organizing Scientific Research for War*, p. 299.

7. Zachary, *Endless Frontier*, p. 218.

8. Warren Weaver to Lyman Briggs, October 2, 1944, General Correspondence 7/1/44 to 21/31/44, AMP.

9. Minutes of Executive Committee, September 25, 1944, AMP; see Stewart, I., *Organizing Scientific Research for War*, pp. 299–309.

10. Warren Weaver to J. G. Brainerd of the University of Pennsylvania, December 19, 1944, General Correspondence 7/1/44 to 21/31/44, AMP.

11. Minutes of Executive Committee, September 11, 1944, AMP.

12. Ibid.

13. Officer in Charge of New York Project to Hydrographer, October 9, 1944, MTP ONR.

14. Minutes of Executive Committee, August 28, 1944, AMP.

15. Notes of Thornton Fry, October 18, 19, 1944, AMP.

16. Minutes of Executive Committee, November 6, 1944, AMP.

17. Minutes of Executive Committee, March 15, 1945, AMP.

18. Ibid.

19. Minutes of Executive Committee, April 2, 1945, AMP.

20. Ibid.

21. Richard Courant, "Some Thoughts on the Research Board for National Security" (ca. March 1945), AMP.

22. Minutes of Executive Committee, April 2, 1945, AMP.

23. McCullough, *Truman*, p. 349.

24. Minutes of Executive Committee, April 16, 1945, AMP.

25. Minutes of Executive Committee, June 2, 1945, AMP.

26. Minutes of Executive Committee, April 2, 1945, AMP.

27. Lyman Briggs to General Huie, April 13, 1942, MTP WPA.

28. Todd, "Oberwolfach—1945" (1983).

29. Beauclair, "Alwin Walther, IPM, and the Development of Calkulator/Computer Technology in Germany, 1930–1945," p. 342.

30. Todd, "Oberwolfach—1945" (1983).

31. Ibid.

32. Süss, "The Mathematical Research Institute Oberwolfach through Critical Times."

33. Beauclair, "Alwin Walther, IPM, and the Development of Calkulator/Computer Technology in Germany, 1930–1945," p. 347.

34. Todd, "Applied Mathematical Research in Germany with Particular Reference to Naval Applications" (1945).

35. Beauclair, "Alwin Walther, IPM, and the Development of Calkulator/Computer Technology in Germany, 1930–1945," p. 340.

36. John Todd, interview with the author, January 2002.

37. Todd, "Oberwolfach—1945" (1983).

38. H. M. MacNeille to Dorothy Weeks, June 3, 1945, Lyman Briggs File, AMP.

39. Arnold Lowan to Mina Rees, June 26, 1945, AMP.

40. Thornton Fry to Mina Rees, July 9, 1945, AMP.

41. Lowan, Arnold, "Report on Math Tables Project work done during the War," December 4, 1945, AMP.

42. Abraham Hillman, interview with the author.

43. Mina Rees to Arnold Lowan, June 21, 1945, Correspondence with Mathematical Tables Project, AMP.

44. Minutes of Executive Committee, September 24, 1945, AMP.

45. Ibid.

46. Arnold Lowan to R. W. Smith, December 18, 1945, BRIGGS.

47. R. C. Archibald to Churchill Eisenhart, June 23, 1945, General Correspondence, 1940–46, NRC-MTAC; Archibald, "Conference on Advanced Computation Techniques" (1946).

48. Shannon, "Mathematical Theory of the Differential Analyzer."

49. Archibald, "Conference on Advanced Computation Techniques"(1946).

50. S. Charp to Adele Goldstine, May 14, 1945, Course on Mathematical Ballistics, PENNSYLVANIA.

51. Grier, "ENIAC, the Verb 'to Program' and the Emergence of Digital Computers."

52. Herman Goldstine, interview with the author, July 2002.

53. There is an extensive literature on the stored program concept and the role of the ENIAC; see Stern, *From ENIAC to UNIVAC*; Williams, *A History of Computing Technology*, pp. 266–83; Ceruzzi, *A History of Modern Computing*, pp. 20–27.

54. Campbell-Kelly and Williams, *The Moore School Lectures*, pp. xv–xvi.

55. Those who might have represented human computers were present as machine designers. Among these were George Stibitz and Herman Goldstine (ibid.).

56. Travis, "The History of Computing Devices."

57. Campbell-Kelly and Williams, *The Moore School Lectures*, p. 18.

58. See Owens, "Mathematicians at War," and Owens, "The Counterproductive Management of Science in the Second World War."

59. Minutes of Executive Committee, April 25, 1946, AMP.

60. Mina Rees to Oswald Veblen, June 9, 1945, Future of Science Folder, AMP.

61. See National Bureau of Standards, "Activities in Applied Mathematics, 1946–1947," pp. 12–13; compare National Bureau of Standards, *Projects and Publications of the National Applied Mathematics Laboratories*, 1947–1949, Particle Interaction (June 30, 1948), Reactor Turbine Design (December 31, 1947, p. 16), Radiation (March 31, 1949, p. 63).

62. Irvin Stewart to Warren Weaver, April 18, 1945, Correspondence File, MTP AMP.

63. Ida Rhodes to Lyman Briggs, November 21, 1945, BRIGGS.

64. John Curtiss to Arnold Lowan, April 12, 1946, Directors Correspondence, NBS.

65. Todd, "John Hamilton Curtiss, 1909–1977" (1980).

66. National Bureau of Standards, *The National Applied Mathematics Laboratories—A Prospectus,* p. 7; see also Aspray and Gunderloy, "Early Computing and Numerical Analysis at the National Bureau of Standards."

67. *The National Applied Mathematics Laboratories,* p. 7.

68. Quoted in Arnold Lowan to Philip Morse, August 15, 1947, MORSE.

69. Arnold Lowan to Philip Morse, August 15, 1947, MORSE.

70. Everett Yowell, interview with the author, December 1998.

71. Ibid.

72. Arnold Lowan to Philip Morse, August 15, 1947, MORSE.

73. Arnold Lowan to Philip Morse, July 26, 1947; Philip Morse to Arnold Lowan, August 12, 1947, MORSE.

74. John von Neumann to John Curtiss, May 29, 1948, MORSE.

75. Arnold Lowan to Philip Morse, December 15, 1947, December 28, 1947, January 31, 1948, MORSE.

76. Arnold Lowan to Philip Morse, March 18, 1948, MORSE.

77. Arnold Lowan to Edward Condon, May 5, 1948, MORSE.

78. Arnold Lowan to Edward Condon, May 5, 1948; Samuel Finkelstein to Edward Condon, July 19, 1948, MORSE.

79. Lyman Briggs to Arnold Lowan, July 17, 1939, BRIGGS.

80. Samuel Finkelstein to Philip Morse, June 22, 1948, MORSE.

81. Samuel Finkelstein to R. C. Archibald, April 9, 1948, and attached memos (n.d.), NRC-MTAC.

82. Ibid.

83. Miller and Gillette, *Washington Seen,* p. 52.

84. R. C. Archibald to R. C. Gibbs, April 27, 1948, NRC-MTAC.

85. John von Neumann to John Curtiss, May 13, 1948, MORSE.

86. Arnold Lowan to Philip Morse, May 20, 1948, MORSE.

87. Arnold Lowan to Philip Morse, July 1, 1948, MORSE.

88. Samuel Finkelstein to Philip Morse, June 22, 1948, MORSE.

89. Dantzig, "Reminiscences about the Origins of Linear Programming" (1982); Dantzig, "Origins of the Simplex Method" (1990).

90. Dorfman, "The Discovery of Linear Programming."

91. Dantzig quoted ibid.

92. Stigler, G., "The Cost of Subsistence" (1945).

93. George Dantzig to John von Neuman, April 28, 1948, Correspondence "D," NEUMANN.

94. Stigler, G., "The Cost of Subsistence" (1945).

95. George Dantzig to John von Neuman, April 28, 1948, NEUMANN. The timings for the ENIAC do not conform to the values given in Goldstine and Goldstine, "The Electronic Numerical Integrator and Computer (ENIAC)." That paper gives values which are about twice as large. Using these values, the problem would have taken slightly more than 18 hours.

96. National Bureau of Standards, *Projects and Publications of the National Applied Mathematics Laboratories,* April–June 1948, Project 48S2-15, p. 20.

97. Gurer, "Women's Contributions to Early Computing" (1996); Fritz, "The Women of ENIAC" (1996).

98. Arnold Lowan to Philip Morse, January 5, 1949, MORSE.

99. Philip Morse to Arnold Lowan, January 24, 1949, MORSE.

CHAPTER EIGHTEEN
I ALONE AM LEFT TO TELL THEE

1. Aspray and Williams, "Arming American Scientists."
2. See Report No. 7 of the Association for Computing Machinery, May 30, 1949, BERKELEY.
3. Farewell card to Gertrude Blanch (n.d.), STERN.
4. Federal Bureau of Investigation file on Gertrude Blanch, BLANCH FBI.
5. Pursell, "A Preface to Governmental Support of Research."
6. Curtiss, Interview, p. 20, SMITHSONIAN.
7. Curtiss, "The National Applied Mathematics Laboratory" (1947).
8. Curtiss, Interview, SMITHSONIAN.
9. John Todd, interview with the author, January 2002.
10. Olga Taussky-Todd to Frances Cave-Browne-Cave, August 29, 1947, CBC.
11. John Todd, interview with the author, January 2002.
12. Hestenes and Todd, *NBS-INA—The Institute for Numerical Analysis.*
13. Curtiss, *Problems for the Numerical Analysis of the Future* (1951), p. xi.
14. Hestenes and Todd, *NBS-INA—The Institute for Numerical Analysis*, appendix F.
15. The request came from Samuel Herrick of UCLA and was published in *Tables for Rocket and Comet Orbits* (AMS 20), Washington, DC, National Bureau of Standards, 1953.
16. Huskey, "SWAC"; see also Rutland, *Why Computers Are Computers*, pp. 24–25.
17. Randell, *The Origins of Digital Computers*, p. 193.
18. Albert Cahn to George F. Taylor, March 16, 1949, UCLA ADMIN.
19. National Bureau of Standards, *Projects and Publications of the National Applied Mathematics Laboratories*, September 1948.
20. Everett Yowell, interview with the author, December 30, 1998.
21. Sheldon and Tatum, "The IBM Card-Programmed Electronic Calculator"; Bashe et al., *IBM's Early Computers*, pp. 68–72.
22. Hestenes and Todd, *NBS-INA—The Institute for Numerical Analysis*, p. 19.
23. Ibid., pp. v, 7, 8, 11, 29.
24. Edward Condon to Robert Sproul, May 7, 1951, UCLA ADMIN.
25. Huskey, "SWAC."
26. Hestenes and Todd, *NBS-INA—The Institute for Numerical Analysis*, p. 5.
27. Halberstam, *The Fifties*, p. 4.
28. Rhodes, *Dark Sun* (1995), p. 365.
29. Quoted ibid., p. 363.
30. Wang, "Science, Security and the Cold War."
31. Quoted ibid.
32. Memo from R. L. Randell, Personnel Officer, December 5, 1949, NBS DIRECTOR.
33. Rhodes, *Making of the Atomic Bomb* (1986), p. 397.
34. July 17 Petition, Harrison-Bundy File, folder 76, MANHATTAN.
35. Ian Bartky, communication with the author concerning his father, Walter Bartky.

36. Memo to Raymond Allen, University Provost, January 1952, UCLA ADMIN.

37. SAC Memo Regarding Gertrude Blanch Security Matter—C, September 20, 1955, p. 1, BLANCH FBI.

38. Miriam Gaylin, interview with the author.

39. Whyte, *Organization Man*, pp. 4, 5.

40. Memo to Director, FBI, August 30, 1955, BLANCH FBI.

41. Lillian Hellman to John S. Wood, May 19, 1952, reprinted in Hellman, *Scoundrel Time*, pp. 89–91.

42. Security Case of Gertrude Kaimowitz, aka Gertrude Blanch, aka Gertrude Blanch Cassidy, last dated August 30, 1955, p. 14, BLANCH FBI.

43. Director, FBI, to SAC Cincinnati, April 22, 1956, BLANCH FBI.

44. "Weeks Sees Ousting of 'Holdovers' Here," *Washington Evening Star*.

45. Todd, "John Hamilton Curtiss" (1980).

46. Huskey, "SWAC."

47. See U.S. Senate, "Testimony of Robert J. Ryan."

48. John Curtiss to Allen V. Astin, March 8, 1953, ASTIN.

49. Wang, "Science, Security and the Cold War."

50. Cochrane, *Measures for Progress* (1966), p. 484; Perry, *The Story of Standards*, pp. 197–201.

51. "Weeks Ends Silence in Forecasting Swing of Ax on Deadwood," *Washington Evening Star*.

52. Astin, Oral History, p. 11.

53. Hestenes and Todd, *NBS-INA—The Institute for Numerical Analysis*, p. 37; see also "President Backs Weeks on Ouster," *New York Times*; "Weeks Sees Ousting of 'Holdovers' Here," *Washington Evening Star*.

54. Astin, Oral History, p. 11.

55. Cochrane, *Measures for Progress* (1966), pp. 484–86, 497.

56. Hestenes and Todd, *NBS-INA—The Institute for Numerical Analysis*, p. 24.

57. ElectroData News Release, March 24, 1954, ELECTRODATA.

58. See Personnel Lists, Mathematics Office, ELECTRODATA.

59. ElectroData Staff Minutes, March 3, 1954, box 24, ELECTRODATA.

60. Gertrude Blanch, interview with Michael Stern, approximately 1989, STERN.

61. See Grier, "The Rise and Fall of the Committee on Mathematical Tables and Other Aids to Computation" (2001).

62. "Clara Froelich," *The Reporter*.

63. Gertrude Blanch, interview with Michael Stern, approximately 1989, STERN.

64. SAC Cincinnati to Director, FBI, August 30, 1955, BLANCH FBI.

65. Director, FBI, to SAC Cincinnati, September 20, 1955, BLANCH FBI.

66. Gertrude Blanch, interview with Michael Stern, approximately 1989, STERN.

67. "Report on Mathematical Tables," 1952, Mathematical Tables Committee File, MORSE.

68. Morse, *In at the Beginning*, p. 282.

69. Blanch to Morse, January 4, 1954, MORSE.

70. Arnold Lowan to Philip Morse, September 1, 1954; Philip Morse to Arnold Lowan, September 7, 1954, MORSE.

71. Philip Morse, Manuscript Report on Conference on Mathematical Tables, September 15–16, 1954, MORSE.

72. Ibid.

73. Jahnke and Emde, *Tables of Functions*.

74. Fletcher et al., *An Index of Mathematical Tables*, pp. 863–64.

75. National Bureau of Standards, *Projects and Publications of the National Applied Mathematics Laboratories*, July–December 1947, Project 47D2-4, p. 9.

76. "Clara Froelich," *The Reporter*; Nancy Persily, interview with the author, June 4, 1998; Murray Pfefferman to John von Neumann, March 5, 1952, NEUMANN; Farebrother, *A Memoir on the Life of Harold Thayer Davis*; Croarken, *Early Scientific Computing in Britain* (1990), p. 23.

77. Abramowitz and Stegun, *Handbook of Mathematical Functions*, p. vi; "Dr. Abramowitz, Standards Unit Mathematician," *Washington Evening Star*; Wrench, "Handbook of Mathematical Functions with Formulas, Graphs and Mathematical Tables."

78. Quoted in Pugh, *Building IBM*, p. 275.

79. Ceruzzi, *A History of Modern Computing*, p. 145.

80. The book is uncopyrighted and has been reprinted in many forms and many languages. The two most popular editions are distributed by Dover Press and by the U.S. Government Printing Office. Between these two editions, the book has sold about 1.2 million copies.

81. Gertrude Blanch, interview by Henry Thatcher in San Diego, March 17, 1989, STERN.

82. Eisenman, *History of Mathematical Statistics Research*, p. 6; Sterling, "Blond Fashion Designer"; "Biography for Gertrude Blanch, Federal Woman's Award," LBJ.

83. McLendon, "She Corrects Computers."

84. Public Papers of the Presidents of the United States, NARA, Papers of Lyndon Johnson, vol. 1, 1963, p. 330.

85. Federal Woman's Award.

EPILOGUE:
FINAL PASSAGE

1. Yeomans, "Comet Halley—The Orbital Motion" (1977).

2. Donald Yeomans, personal communication with the author, July 2002.

3. Stern, *From ENIAC to UNIVAC,* p. 149.

4. Donald Yeomans, personal communication with the author.

5. Yeomans, "Comet Halley—The Orbital Motion"; Hughes, D. W., "The History of Halley's Comet."

6. Hughes, D. W., "The History of Halley's Comet."

7. Donald Yeomans, personal communication with the author.

8. MacRobert, "Halley in the Distance."

Research Notes and Bibliography

ABBREVIATIONS

CBI   Charles Babbage Institute
CORNELL   Rare and Manuscript Collections, Carl Kroch Library, Cornell
   University
GIRTON   Girton College Archives
HUA   Harvard University Archives
ISU   Iowa State University Special Collections and University Archives
LBJ   Lyndon Baines Johnson Presidential Library
LOC   Manuscript Collections, Library of Congress
MTAC   Mathematical Tables and Other Aids to Computation
NARA   National Archives and Records Administration
NAS   National Academy of Sciences Archives
RAS   Royal Astronomical Society Library

PRIMARY SOURCES AND MANUSCRIPT COLLECTIONS

ALMANAC   Records of the Nautical Almanac, Record Group 78.4, NARA
ALUM   Alumnae Letters, Alumni Directories, Bentley Historical Library, University of Michigan
AMP   Records of the Applied Mathematics Panel, 1942–46, Record Group 227,
   NARA
ASTIN   Papers of Allen Astin, LOC
ATANASOFF   Papers of John Vincent Atanasoff, ISU
ATT   Research Records of Bell Telephone Laboratories, ATT Archives
BAE   Records of the Bureau of Agricultural Economics, Record Group 83,
   NARA
BENTLEY   Records of the Mathematics Department, Bentley Historical Library, University of Michigan
BERKELEY   Papers of Edmund Berkeley, CBI
BLANCH FBI   Federal Bureau of Investigation File: Gertrude Blanch, Freedom
   of Information Act Request
BLANCH NYU   Records of Washington Square College, New York University
   Archives
BRIGGS   General Correspondence, Lyman J. Briggs, 1931–62, Office Files,
   Records of National Bureau of Standards, Record Group 167, NARA
CBC   Papers of Frances Cave-Browne-Cave, GIRTON
COAST-SURVEY   Records of the Coast Survey, Record Group 23, NARA
DAHLGREN   Papers of John Adolphus Dahlgren, LOC
ECKERT   Papers of Wallace J. Eckert, CBI 9, CBI

ELECTRODATA   Records of ElectroData Corporation, Burroughs Corporation Collection, CBI

FROELICH   Papers of Clara Froelich, Barnard College Archives

GREENWICH   Papers of the Royal Greenwich Observatory, Cambridge University

HARVARD ELIOT   Records of President Eliot, UA 1.5.150, HUA

HARVARD OBS   Records of the Harvard Observatory, UA 630, HUA

HAW   Papers of Henry Wallace, University of Iowa

IOWA MATH   Records of the University of Iowa Mathematics Department, University of Iowa Archives

ISU-ADMIN   Administrative Records of Iowa State University, ISU

IU BRYAN   Papers of W. L. Bryan, Indiana University Archives

JEFFERSON   Thomas Jefferson Papers, LOC

LOWAN   Papers of Arnold Lowan, Yeshiva University Archives

MANHATTAN   Records of the Chief of Engineers, Manhattan Engineer District, Record Group 77, NARA

MARTIN   Papers of Artemas Martin, Special Collections, American University Library

MAUCHLY   Papers of John Mauchly, University of Pennsylvania

MICHIGAN   Records of the Registrar, University of Michigan

MOORE   Papers of Joshua Moore, LOC

MORSE   Papers of Philip Morse, MIT Archives

MTP AMP   Files of Certain Contractors, Applied Mathematics Panel Records, 1942–46, E-153, Records of the Office for Scientific Research and Development, Record Group 227, NARA

MTP ONR   Records of the Office of Naval Research, box 77, Record Group 298, NARA

MTP WPA   Administrative Records of the Math Tables Project of New York City, 1940–42, Project 365-97-3-11 and 765-97-3-10, Records of the WPA (FERA), Record Group 69, NARA

NBS   Records of National Bureau of Standards, Record Group 167, NARA

NBS DIRECTOR   Directors Correspondence, Records of the Director, Records of the National Bureau of Standards, Record Group 167, NARA

NEUMANN   Papers of John von Neumann, LOC

NEWCOMB   Papers of Simon Newcomb, LOC

NEYMAN   Files relating to Jerzy Neyman, Applied Mathematics Panel, Records of the Office of Scientific Research and Development, Record Group 227, E 153, NARA

NMAH   Files of Computing Collection, National Museum of American History, Smithsonian Institution

NORTHWEST   Papers of Walter Dill Scott, Northwestern University Archives

NRC-MATH   Files on the Committee on Applied Mathematics, Physical Science Division Records, National Research Council, NAS

NRC-MTAC   Files on the Subcommittee on the Bibliography of Mathematical Tables and Other Aids to Computation, Physical Science Division Records, National Research Council, NAS

NRC-PS   Files of the Executive Council, Physical Science Division Records, National Research Council, NAS

NWU DAVIS   Papers of Harold T. Davis, Northwestern University Archives

OBSERVATORY-LOC   Papers of the U.S. Naval Observatory, LOC

OBSERVATORY-NARA   Records of the U.S. Naval Observatory, Record Group 78.4, NARA

ORDNANCE   Records of the Department of Ordnance, Record Group 156, NARA

PEARSON   Papers of Karl Pearson, University College Special Collections, University of London

PENNSYLVANIA   Records of the Moore School, UPD 8, University of Pennsylvania Archives

SAB   Files of the Scientific Advisory Board, National Research Council Records, NAS

SHAW   Papers of George Bernard Shaw, British Library

SMITHSONIAN   Interviews with Computer Pioneers, Record Group 196, Smithsonian Archives Center, National Museum of American History

STERN   Papers of Gertrude Blanch, Stern Family Collection, CBI

STIBITZ PAPERS   Papers of George Stibitz, Dartmouth College Library

TODD   Papers of John Todd, California Institute of Technology Archives

TOLLEY   Papers of Howard Tolley, 1923–28, Entry 125, BAE

UCLA ADMIN   Records of the University Chancellor, UCLA, Record Series 359, University Archives, UCLA

UC PHYSICS   Records of the University of Chicago Physics Department, University of Chicago Archives

UPTON PAPERS   Papers of Winslow Upton, John Hay Library, Brown University

VEBLEN   Papers of Oswald Veblen, LOC

WCE   Records of the World's Columbian Exposition, Special Collections, Harold Washington Library Center

WEAVER   Papers of Warren Weaver, Rockefeller Foundation Archives

WILSON PAPERS   Papers of Elizabeth Webb Wilson, Schlesinger Library, Harvard University

INTERVIEWS

Bartky, Ian
Dyson, Freeman
Gaylin, Miriam
Goldstine, Herman H.
Hillman, Abraham
Issacson, Eugene
Lozier, Daniel
Olver, Frank
Persily, Nancy

Salzer, Herbert
Stegun, Irene
Stern, David and Debra
Todd, John
Yowell, Everett

SECONDARY SOURCES AND BIBLIOGRAPHY

Abbe, Cleveland, "Charles Schott," *Biographical Memoirs, National Academy of Sciences*, vol. 8, 1915, pp. 87–133.

Abbott, David, *Astronomers*, New York, Peter Bedrick, 1984.

Aberdeen Proving Ground, Annual Reports, Records of the Department of Ordnance, Record Group 156, NARA.

Aberdeen Proving Ground, Technical Report 12, "The range firing section of the proof department, Aberdeen Proving Ground. Its objects, its development and its accomplishments," Aberdeen, Md., Army Ballistics Research Laboratory, Technical Reports of Aberdeen Proving Ground, 1918–20, ORDNANCE.

Aberdeen Proving Ground, Technical Report 84, *History of the Range Firing Section of the Proof Department of the Aberdeen Proving Ground*, Aberdeen, Md., Army Ballistics Research Laboratory, Technical Reports of Aberdeen Proving Ground, 1918–20, ORDNANCE. (Veblen presumed author.)

Abramowitz, Milton, and Irene Stegun, *Handbook of Mathematical Functions with Formulas, Graphs and Mathematical Tables*, Washington, D.C., National Bureau of Standards, 1964.

Adams, Henry, *Education of Henry Adams*, Project Gutenberg, 2000, ftp://ibiblio.org/pub/docs/books/gutenberg/etext00/eduha10.txt.

Addams, Jane, *Twenty Years at Hull House* (New York, MacMillan, 1912), Project Gutenberg, 1998, ftp://ibiblio.org/pub/docs/books/gutenberg/etext98/20yhh10.txt.

Aiken, Howard, "Proposed Automatic Calculating Machine" (unpublished memorandum, 1938), in Cohen and Welch, pp. 9–30.

Aiken, Howard, and Grace Hopper, "The Automatic Sequence Controlled Calculator—I" (1946), in Randell, pp. 203–9.

Aiken, Howard, and Grace Hopper, "The Automatic Sequence Controlled Calculator—III" (1946), in Randell, pp. 216–22.

Alder, Ken, *Engineering the Revolution*, Princeton, N.J., Princeton University Press, 1997.

Alder, Ken, *The Measure of All Things*, New York, Free Press, 2002.

Alder, Ken, "A Revolution to Measure: The Political Economy of the Metric System in France," in *The Values of Precision*, Wise, M. Norton, ed., Princeton, N.J., Princeton University Press, 1995, pp. 39–71.

American Statistical Association, "Membership List, 1840," *Journal of the American Statistical Association*, vol. 35, no. 209, pt. 2, March 1940, pp. 305–8.

Andrews, E. C., "Telephone Switching and the Early Bell Laboratories Computers," *IEEE Annals of the History of Computing*, vol. 4, no. 1, January 1982, pp. 13–19.

Andrews, William, ed., *The Quest for Longitude*, Cambridge, Mass., Harvard University, 1996.

Archibald, R. C., "BAASMTC vol. 1," *MTAC*, vol. 2, no. 16, 1946, p. 122.

Archibald, R. C., "Conference on Advanced Computation Techniques," *MTAC*, vol. 2, no. 14, April 1946, pp. 65–68.

Archibald, R. C., "Introduction," *MTAC*, vol. 1, no. 1, January 1943, pp. 1–2.

Archibald, Raymond Claire, *Mathematical Tables Makers*, New York, Scripta Mathematica Series Number 3, 1948.

Archibald, R. C., "The New York Mathematics [*sic*] Tables Project," *Science*, vol. 96, no. 2491, September 25, 1942, pp. 294–96.

Archibald, Raymond Claire, "P. G. Scheutz, Publicist, Author, Scientific Mechanician and Edvard Scheutz—Biography and Bibliography," *MTAC*, vol. 18, April 1947, pp. 238–45.

Archibald, Raymond Claire, "Reviews," *American Mathematical Monthly*, vol. 28, nos. 6/7, June–July 1921, pp. 266–67.

Archibald, Raymond Claire, "Reviews," *American Mathematical Monthly*, vol. 31, no. 7, September 1924, pp. 348–49.

Archibald, Raymond Claire, "Tables of Trigonometric Functions in Non-Sexagesimal Arguments," *MTAC*, vol. 1, no. 2, April 1943, pp. 33–44.

Aron, Cindy, "'to Barter Their Souls for Gold': Female Clerks in Federal Government Offices, 1862–1890," *Journal of American History*, vol. 67, no. 4, March 1981, pp. 835–53.

Ashworth, William, "The Calculating Eye: Baily, Herschel, Babbage and the Business of Astronomy," *British Journal of the History of Science*, vol. 27, pp. 409–41.

Aspray, William, ed., *Computing before Computers*, Ames, Iowa, Iowa State University Press, 1990.

Aspray, William, *John von Neumann and the Origins of Modern Computing*, Cambridge, Mass., MIT Press, 1990.

Aspray, William, and Michael Gunderloy, "Early Computing and Numerical Analysis at the National Bureau of Standards," *IEEE Annals of the History of Computing*, vol. 11, no. 1, 1989, pp. 3–12.

Aspray, William, and Bernard Williams, "Arming American Scientists: NSF and the Provision of Scientific Computing Facilities for Universities, 1950–1973," *IEEE Annals of the History of Computing*, vol. 16, no. 4, 1994, pp. 60–74.

Astin, Allen V., Oral History, Columbia University Oral History Project.

Atanasoff, John V., "Advent of Electronic Digital Computing," *Annals of the History of Computing*, vol. 6, no. 3, July 1984, pp. 229–82.

Atanasoff, John V., "Computing Machine for the Solution of Large Systems of Linear Algebraic Equations" (unpublished memorandum, 1940), in Randell, pp. 315–35.

Atanasoff, J. V., and A. E. Brandt, "Application of Punched Card Equipment to the Analysis of Complex Spectra," *Journal of the Optical Society of America*, vol. 26, no. 2, pp. 83–88.

Austrian, Geoffrey, *Herman Hollerith: Forgotten Giant of Information Processing*, New York, Columbia University Press, 1982.

Babbage, Charles, *Economy of Machinery and Manufactures*, London, Charles Knight, 1835.

Babbage, Charles, *History of the Invention of the Calculating Engines*, Buxton Papers, Oxford Science Museum, p. 10, quoted in Hyman, p. 49.

Babbage, Charles, *Passages from the Life of a Philosopher*, 1864, reprint New Brunswick, N.J., Rutgers University Press, 1994.

Bacon, Margaret Hope, *Mothers of Feminism: The Story of Quaker Women in America*, San Francisco, Calif., Harper & Row, 1986.

Baehne, George W., ed., *Practical Applications of the Punched Card Method in Colleges and Universities*, New York, Columbia University Press, 1935.

Baily, Francis, "On Mr. Babbage's New Machine for Calculating and Printing Mathematical and Astronomical Tables" (1823), in Bromley, pp. 225–35.

Baily, Francis, "Report of Committee," *Memoirs of the Royal Astronomical Society*, vol. 5, London, Priestley and Weale, 1833, pp. 341–48.

Bancroft, T. A., "Statistical Laboratory of the Iowa State University," *Bulletin of the Institute of Statistical Research and Training*, vol. 1, no. 1, December 1966, pp. 14–20.

Bancroft, T. A., ed., *Statistical Papers in Honors of George W. Snedecor*, Ames, Iowa, Iowa State University Press, 1972.

Barker, Thomas, "Extract of a Letter of Thomas Barker, esq., To the Reverend James Bradley DD, Astronomer Royal and FRS concerning the Return of the Comet Expected in 1757 or 1758," *Philosophical Transactions (1683–1775)*, vol. 49 (1755–56), pp. 347–50.

Barlow, Peter, *Barlow's Tables of Squares, Cubes, Square Roots, Cube Roots and Reciprocals of All Integer Numbers up to 12,500*, Brooklyn, Chemical Publishing, 1941.

Bartky, Ian, *Selling the True Time*, Stanford, Calif., Stanford University Press, 2000.

Bashe, Charles J., L. R. Johnson, J. H. Palmer, and Emerson Pugh, *IBM's Early Computers*, Cambridge, Mass., MIT Press, 1986, pp. 68–72.

Baxter, James Phinney, III, *Scientists against Time*, Boston, Little, Brown, 1946.

Beauclair, Wilfried de, "Alwin Walther, IPM, and the Development of Calkulator/Computer Technology in Germany, 1930–1945," *IEEE Annals of the History of Computing*, vol. 8, no. 4, October 1986, pp. 334–50.

Beckenbach, E. F., and W. Walter, eds., *General Inequalities 3, Proceedings of 3rd International Conference on General Inequalities, Oberwolfach, April 26–May 2, 1981*, Basel, Birkhäuser Verlag, 1983.

Benedict, Murray, "Development of Agricultural Statistics in the Bureau of the Census," *Journal of Farm Economics*, vol. 21, no. 4, November 1939, pp. 735–60.

Beninger, James, *The Control Revolution*, Cambridge, Mass., Harvard University Press, 1986.

Bergin, Thomas, ed., *Fifty Years of Army Computing*, Army Research Laboratory, 2000.

Bernstein, Jeremy, "Profiles (Hans Bethe Part I)," *New Yorker*, December 3, 1979, pp. 50–107.

Bernstein, Peter, *Against the Gods*, New York, John Wiley, 1996.

Bethe, Hans A., "Energy Production in Stars," *Physical Review*, vol. 55, 1939, pp. 434–56.

Bethe, Hans A., and C. L. Critchfield, "The Formation of Deuterons by Proton Combination," *Physical Review*, vol. 54, 1938, pp. 248–54.

Betts, Jonathan, *Harrison*, London, National Maritime Museum, 1993.

"Biography for Gertrude Blanch, Federal Woman's Award" (n.d.), LBJ.

Blanch, Gertrude, interview with Michael Stern, approximately 1989, STERN.

Blanch, Gertrude, interview by Henry Thatcher in San Diego, March 17, 1989, STERN.

Blanch, Gertrude, interview with Henry Tropp, May 16, 1973, SMITHSONIAN.

Blanch, Gertrude, *Properties of the Veneroni Transformation in S_4*, PhD diss., 1934, Cornell University Mathematics Department Library.

Blanch, G., A. N. Lowan, R. E. Marshak, and H. A. Bethe, "The Internal Temperature Density Distribution of the Sun," *Journal of Astrophysics*, 1942, pp. 37–45.

Blanch, Gertrude, and Ida Rhodes, "Table-Making at the National Bureau of Standards," in Scaife, pp. 1–6.

Bliss, Gilbert, *Mathematics for Exterior Ballistics*, New York, John Wiley, 1944.

Bloch, Richard, "Programming Mark I," in Cohen and Welch, pp. 77–110.

"Boston Woman Is Rated Insurance Expert Deluxe," *Boston Traveler*, December 10, 1935.

Bradley, Margaret, *A Career Biography of Gaspard Clair François Marie Riche de Prony, Bridget-Builder, Educator and Scientist*, Lewiston, N.Y., Edwin Mellen Press, 1998.

Brandt, A. E., "Uses of the Progressive Digit Method," in Baehne, pp. 423–36.

Braverman, Harry, *Labor and Monopoly Capital*, New York, Monthly Review Press, 1974.

Brennan, Jean Ford, *The IBM Watson Laboratory at Columbia University: A History*, New York, IBM, 1971.

Bromley, Allan, ed., *Babbage's Calculating Engines*, Los Angeles, Computer Society Press, 1982.

Broughton, Peter, "The First Predicted Return of Comet Halley," *Journal of the History of Astronomy*, vol. 16, 1985, pp. 123–33.

Brown, Ernest W., *The Motion of the Moon*, Edinburgh, Neill & Co., 1933.

Bryan, William, "The Life of the Professor," *National Association of State Universities in the United States Transactions and Proceedings*, 1912, pp. 32–33.

Burks, Alice R., *Who Invented the Computer?* Amherst, N.Y., Prometheus Books, 2003.

Burks, Alice R., and Arthur Burks, *The First Electronic Computer: The John Atanasoff Story*, Ann Arbor, University of Michigan Press, 1988.

Burks, Arthur, interview conducted by William Aspray, June 20, 1987, OH 136, CBI.

Burns, James MacGregor, *The Lion and the Fox*, New York, Harcourt & Brace, 1956.

Bush, Vannevar, "The Differential Analyzer: A New Machine for Solving Differential Equations," *Journal of the Franklin Institute*, vol. 212, 1931, pp. 447–88.

Byerly, W. E., "Reminiscences," in Cohen (1980), pp. 5–7.

Campbell, Robert, "Mark II, an Improved Mark I," in Cohen and Welch, pp. 111–28.

Campbell-Kelly, Martin, and WilliamAspray, *Computer*, New York, Basic Books, 1996.

Campbell-Kelly, Martin, Mary Croarken, Raymond Flood, and Eleanor Robson, eds., *The History of Mathematical Tables: From Sumer to Spreadsheets*, Oxford, U.K., Oxford University Press, 2003.

Campbell-Kelly, Martin, and Michael Williams, *The Moore School Lectures*, Cambridge, Mass., MIT Press, 1985.

Cannon, Susan F., *Science and Culture: The Early Victorian Period*, New York, Science History Publications, 1978.

Cardwell, Donald, *The Norton History of Technology*, New York, W. W. Norton, 1995.

Carter, Mary Sue, Phyllis Cook, and Brian Luzum, "The Contributions of Women to the Nautical Almanac Office: The First 150 Years," in Fiala and Dick, pp. 165–77.

Casson, Herbert N., *The History of the Telephone*, 1910, reprint Freeport, N.Y., Books for Libraries Press, 1971.

Cave, Beatrice, and Karl Pearson, "Numerical Illustrations of the Variate Difference Correlation Method," *Biometrika*, vol. 10, nos. 2/3, November 1914, pp. 340–55.

Cave-Browne-Cave, Frances, "On the Influence of the Time Factor on the Correlation between Barometric Heights at Stations More than 1000 Miles Apart," *Proceedings of the Royal Society of London*, vol. 74, 1904–5, pp. 403–13.

Cave-Browne-Cave, Frances, and Karl Pearson, "On the Correlation between the Barometric Height at Stations on the Eastern Side of the Atlantic," *Proceedings of the Royal Society of London*, vol. 70, 1902, pp. 465–70.

Ceruzzi, Paul, *A History of Modern Computing*, Cambridge, Mass., MIT Press, 1999.

Ceruzzi, Paul, *Reckoners: The Prehistory of the Digital Computer, from Relays to the Stored Program Concept, 1935–1945*, Westport, Conn., Greenwood Press, 1983.

Ceruzzi, Paul, "When Computers Were Human," *IEEE Annals of the History of Computing*, vol. 13, no. 3, 1991, pp. 237–44.

Cesareo, O., "The Relay Interpolator," in Randell, pp. 247–52.

Chandler, Alfred D., Jr., *Scale and Scope: The Dynamics of Industrial Capitalism*, Cambridge, Mass., Belknap Press of Harvard University Press, 1999.

Chandler, Alfred D., Jr., *The Visible Hand: The Managerial Revolution in American Business*, Cambridge, Mass., Belknap Press of Harvard University Press, 1977.

Chapman, Allan, "Sir George Airy (1801–1892) and the Concept of International Standards in Science, Timekeeping and Navigation," *Vistas in Astronomy*, vol. 28, 1985, pp. 321–28.

Christ, Carl, "History of the Cowles Commission, 1932–52," in *Economic Theory and Measurement: A Twenty Year Research Report*, Chicago, Cowles Commission, 1952, pp. 3–65.

Christman, Albert, *Sailors, Scientists and Rockets*, Washington, D.C., Naval History Center, 1971.

"Clara Froelich," *The Reporter*, August 1953, p. 22, ATT.

Clem, Mary, "Interview with Mary Clem by Uta Merzbach," June 27, 1969, SMITHSONIAN.

Cochrane, Rexmond, *Measures for Progress*, Washington, D.C., Government Printing Office, 1966.

Cochrane, Rexmond, *The National Academy of Sciences*, Washington, D.C., National Academy of Sciences, 1978.

Cohen, I. Bernard, ed., *Benjamin Peirce, "Father of Pure Mathematics in America,"* New York, Arno Press, 1980.

Cohen, I. Bernard, *Howard Aiken: Portrait of a Computer Pioneer*, Cambridge, Mass., MIT Press, 1999.

Cohen, I. Bernard, and Gregory Welch, eds., *Makin' Numbers: Howard Aiken and the Computer*, Cambridge, Mass., MIT Press, 1999.

Cole, Margaret, *Growing Up into Revolution*, London, Longmans, Green, 1949.

Collins, V. Lansing, *Princeton in the World War*, Princeton, N.J., Princeton University Press, 1932.

Comrie, L. J. "The Application of the Hollerith Tabulating Machine to Brown's Tables of the Moon," *Monthly Notices of the Royal Astronomical Society*, vol. 92, May 1932, pp. 694–706.

Comrie, L. J., "The Applications of Calculating Machines to Astronomical Computing," *Popular Astronomy*, vol. 33, 1925, pp. 243–46.

Comrie, L. J., "Careers for Girls," *Mathematical Gazette*, vol. 23, 1944, pp. 90–95.

Comrie, L. J., "Inverse Interpolation and Scientific Applications of the National Accounting Machine," *Supplement to the Journal of the Royal Statistical Society*, vol. 3, no. 2, 1936, pp. 87–114.

Comrie, L. J., *Scientific Computing Service Limited: A Description of Its Activities, Equipment and Staff*, 1938.

Comrie, L. J., "Tables of Higher Functions," Mathematical Gazette, vol. 20, 1936, pp. 225–27.

Conrad, Joseph, *The Secret Agent* (1907), Project Gutenberg, 1997, ftp://ibiblio.org/pub/docs/books/gutenberg/etext97/agent10.txt.

Cook, Alan, *Edmond Halley: Charting the Heavens and the Seas*, New York, Oxford University Press, 1998.

Cortada, James, *Before the Computer*, Princeton, N.J., Princeton University Press, 1993.

Cowell, P. H., and A. C. D. Crommelin, *The Return of Halley's Comet in 1910*, Von der Astronomischen gesellschaft mit dem von herrn A. F. Lindemann gestifteten preise gekrönte und veröffentlichte preisschrift, Leipzig, W. Engelmann (Publikation der Astronomischen gesellschaft XXIII), 1910.

Cowles, Alfred, "Can Stock Market Forecasters Forecast?" *Econometrica*, vol. 1, no. 3, July 1933, pp. 309–24.

Cowles Commission for Economic Research, *Report*.

Cox, Gertrude, and Paul Homeyer, "Professional and Personal Glimpses of George W. Snedecor," *Biometrics*, vol. 31, June 1975, pp. 265–301.

Craven, Wesley Frank, and James Lea Gate, *The Army Air Forces in World War II, Volume 3, Europe: Argument to V-E Day, January 1944 to May 1945*, Washington, D.C., Office of Air Force History, 1983.

Croarken, Mary, "Astronomical Labourers: Maskelyne's Assistants at the Royal Observatory, Greenwich, 1765–1811," *Notes and Records of the Royal Society*, vol. 57, no. 3, pp. 285–98.

Croarken, Mary, "Case 5,656: L. J. Comrie and the Origins of the Scientific Computing Service Ltd.," *IEEE Annals of the History of Computing*, vol. 21, no. 4, October–December 1999, pp. 69ff.

Croarken, Mary, *Early Scientific Computing in Britain*, Oxford, U.K., Oxford University Press, 1990.

Croarken, Mary, "Tabulating the Heavens: First Computers of the Nautical Almanac," *IEEE Annals of the History of Computing*, vol. 25, no. 3, July 2003, pp. 48–61.

Croarken, Mary, and Martin Campbell-Kelly, "Beautiful Numbers: The Rise and Decline of the British Association Mathematical Tables Committee," *IEEE Annals of the History of Computing*, vol. 22, no. 4, October–December 2000, pp. 44–61.

Crommelin, A. C. D., "Note on the Approaching Return of Halley's Comet," *Memoirs of the Royal Astronomical Society*, vol. 67, December 1906, pp. 137–38.

Crowell, Benedict, *America's Munitions: 1917–1918*, Washington, D.C., Government Printing Office, 1919.

Culver, John C., and John Hyde, *American Dreamer*, New York, W. W. Norton, 2000.

Curtiss, John, "Forward," in *Problems for the Numerical Analysis of the Future*, AMS 15, Washington, D.C, National Bureau of Standards, June 29, 1951.

Curtiss, John, Interview, SMITHSONIAN.

Curtiss, John, "The National Applied Mathematics Laboratory" (1947), reprinted in *IEEE Annals of the History of Computing*, vol. 11, no. 1, November 1989, pp. 13–30.

Curtiss, John, "Tables of the First Ten Powers of the Integers from 1 to 1000; Tables of the Exponential Function e^x," *American Mathematical Monthly*, vol. 48, no. 1, January 1941, pp. 56–57.

Daniels, George, "The Process of Professionalization in American Science: The Emergent Period, 1820–1860," *Isis*, vol. 58, 1967, pp. 151–66.

Dantzig, George, "Origins of the Simplex Method," in Nash, pp. 141–51.

Dantzig, George, "Reminiscences about the Origins of Linear Programming," *Operations Research Letters*, vol. 1, no. 2, April 1982, pp. 43–48.

Daston, Loraine, "Enlightenment Calculations," *Critical Inquiry*, vol. 24, no. 1, August 1994, pp. 182–202.

Davis, Charles Henry, *The Coast Survey of the United States*, Cambridge, Mass., Metcalf and Company, 1849.

Davis, Charles Henry, *Remarks on an American Prime Meridian*, Washington, D.C., 1849.

Davis, Charles Henry, *A Report of Lieutenant Charles H Davis, the Officer Charged with the Superintendence of the Preparation of the American Nautical Almanac*, Cambridge, Mass., Metcalf and Company, 1852.

Davis, Charles Henry, *Report of the Secretary of the Navy Communicating in Answer to a Resolution of the Senate a Report of Lieutenant Charles H Davis, the Officer Charged with the Superintendence of the Preparation of the American Nautical Almanac*, Cambridge, Mass., Metcalf and Company, 1852.

Davis, Charles Henry, "Report on the Nautical Almanac," *American Journal of Science and Arts*, 2nd ser., vol. 14, 1852, pp. 317–35.

Davis, Charles Henry, II, *Life of Charles Henry Davis*, Boston, Houghton, Mifflin and Co., 1899.

Davis, Harold T., *The Adventures of an Ultra-Crepidarian*, San Antonio, privately printed, 1962.

Davis, Harold T., *Tables of the Higher Mathematical Functions*, vol. 1, Bloomington, Ind., Principia Press, 1933.

"Debris Still Fills Pavilion of WPA," *New York Times*, May 4, 1939, p. 20.

De Weerd, H. A., "American Adoption of French Artillery 1917–1918," *Journal of the American Military Institute*, vol. 3, no. 2, 1939, pp. 104–16.

Dick, Steven, "A History of the American Nautical Almanac Office," in Faila and Dick, pp. 11–54.

Dick, Steven, *Sky with Ocean Joined*, Oxford, U.K., Oxford University Press, 2003.

Dick, Steven, and LeRoy Doggett, *Sky with Ocean Joined: Proceedings of the Sesquicentennial Symposia of the U.S. Naval Observatory*, Washington, D.C., U.S. Naval Observatory, 1983.

Di Scala, Spencer, *Italy: From Revolution to Republic, 1700 to the Present*, Boulder, Colo., Westview Press, 1998.

Dorfman, Robert, "The Discovery of Linear Programming," *Annals of the History of Computing*, vol. 6, no. 3, July 1984, pp. 283–95.

Dos Passos, John, *42nd Parallel*, New York, Library of America, 1996.

"Dr. Abramowitz, Standards Unit Mathematician," *Washington Evening Star*, July 6, 1958.

Dreiser, Theodore, *Sister Carrie*, Project Gutenberg, 1995, ftp://ibiblio.org/pub/docs/books/gutenberg/etext95/scarr10.txt.

Dreyer, J. L. E., and H. H. Turner, *History of the Royal Astronomical Society, 1820–1920*, London, Royal Astronomical Society, 1923.

Dudley Observatory, *Report of the Astronomer in Charge of the Dudley Observatory for the 1863*, Albany, N.Y., J. Munsell, 1864.

Dunkin, Edwin, *A Far Off Vision*, Cornwall, Royal Institution of Cornwall, 1999.

Dupree, A. Hunter, *Science in the Federal Government*, Cambridge, Mass., Belknap Press of Harvard University Press, 1957.

Duren, Peter, ed., *A Century of Mathematics in America*, Providence, R.I., American Mathematical Society, 1989.

Ebling, Walter H., "Why Government Entered the Field of Crop Reporting and Forecasting," *Journal of Farm Economics*, vol. 21, no. 4, November 1939, pp. 718–34.

Eckert, Wallace J., "Astronomy," in Baehne, pp. 389–96.

Eckert, Wallace John, *Punched Card Methods in Scientific Computation*, New York, Thomas J. Watson Astronomical Computing Bureau, 1940.

Eckstein, Peter, "J. Presper Eckert," *IEEE Annals of the History of Computing*, vol. 18 no. 1, Spring 1996, pp. 25–44.

Eisenhart, Churchill, interview by William Aspray, July 10, 1984, "The Princeton Mathematics Community in the 1930s," Princeton University Library.

Eisenman, Harry, *History of Mathematical Statistics Research at the Aeronautical Research Laboratories*, Washington, D.C., Office of Information, Office of Aerospace Research, 1962.

Erdélyi, A., "Edmund T. Whittaker," *MTAC*, vol. 11, no. 57, January 1957, p. 53.

Evans, Jessie Fant, "Elizabeth W. Wilson Has Won Distinction in Mathematics," *Washington Evening Star*, March 13, 1938.

Ezekiel, Mordecai, "Henry A. Wallace, Agricultural Economist," *Journal of Farm Economics*, vol. 48, no. 4, pt. 1, November 1966, pp. 789–802.

Ezekiel, Mordecai, "The Reminiscences of Mordecai Ezekiel," transcript of interview conducted by Dean Albertson for the Oral History Research Office of Columbia University, 1957.

Farebrother, R. W., "A Memoir on the Life of Harold Thayer Davis (1892–1974)," n.d., NWU DAVIS.

Farebrother, R. W., "Memoir on the Life of Myrrick Doolittle," *Bulletin of the Institute of Mathematics and Its Applications*, vol. 23, nos. 6/7, June/July 1987, p. 102.

Federal Women's Award, Description, March 3, 1964, LBJ.

Feffer, Loren Butler, "Oswald Veblen and the Capitalization of American Mathematics: Raising Money for Research, 1923–1928," *ISIS*, vol. 89, September 1998, pp. 474–97.

Feynman, Richard, "Los Alamos from Below," *IEEE Annals of the History of Computing*, vol. 10, no. 4, 1989, pp. 342–45.

Fiala, Alan, and Steven J. Dick, eds., *Proceedings: Nautical Almanac Office Sesquicentennial Symposium, U.S. Naval Observatory, March 3–4, 1999*, Washington, D.C., U.S. Naval Observatory, 1999.

"Field Mathematician Gets Top Job Rating," *Dayton Daily News*, March 2, 1962.

Finkel, B. F., "Biography: Artemas Martin," *American Mathematical Monthly*, vol. 1, no. 4, April 1894, pp. 108–11.

Fitzpatrick, Paul, "Leading American Statisticians in the Nineteenth Century," *Journal of the American Statistical Association*, vol. 52, no. 279, September 1957, pp. 301–21.

Fleming, Wilhamina, *Journal of Wilhamina Paton Fleming*, HUA 900.11, HUA.

Fletcher, A., J. C. P. Miller, L. Rosenhead, and L. J. Comrie, *An Index of Mathematical Tables*, 2nd ed., Reading, Mass., Addison-Wesley Publishing, 1962.

Foley, Vernard, *The Social Physics of Adam Smith*, West Lafayette, Ind., Purdue University Press, 1976.

Forbes, Eric, *The Euler-Mayer Correspondence (1751–1755)*, New York, Elsevier, 1971.

Foucault, Michel, *Birth of the Clinic*, New York, Vintage, 1995.

Foucault, Michel, *Discipline and Punish*, New York, Vintage, 1995.

"Frances Evelyn Cave-Browne-Cave," *Times* (London), April 2, 1965.

Freedman, David, Robert Pisani, and Roger Purves, *Statistics*, 2nd ed., New York, Norton, 1990.

Friedberger, Mark, *Shake-Out: Iowa Farm Families in the 1980s*, Lexington, University Press of Kentucky, 1989.

Fritz, Barkley, "The Women of ENIAC," *IEEE Annals of the History of Computing*, vol. 18, no. 3, Fall 1996, pp. 13–28.

"From the Unpopular Side," *Washington Post*, February 5, 1895, p. 6.

Fry, Thornton, "Industrial Mathematics," *American Mathematical Monthly*, vol. 48, no. 6, pt. 2, 1941, pp. 1–38.

Fry, Thornton, "Mathematical Research," *Bell Laboratories Record*, vol. 1, 1925–26, pp. 15–18.

Fry, Thornton, *Probability and Its Engineering Uses*, New York, D. Van Nostrand, 1928.

Fussell, Paul, *The Great War in Modern Memory*, Oxford, U.K., Oxford University Press, 1975.

Galison, Peter, and Bruce Hevly, eds., *Big Science*, Stanford, Calif., Stanford University Press, 1992.

Gallup, George, "American Institute of Public Opinion Surveys, 1938–1939," *Public Opinion Quarterly*, vol. 3, no. 4, October 1939, pp. 581–607.

Galton, Francis, "Kinship and Correlation," *North American Review*, vol. 150, no. 400, March 1890, pp. 419–31.

Galton, Francis, "Regression towards Mediocrity in Hereditary Stature," *Journal of the Anthropological Institute*, vol. 15, 1895, pp. 246–63.

Gauss, Carl Friedrich, *Theory of the Motion of the Heavenly Bodies Moving about the Sun in Conic Sections*, trans. Charles Henry Davis, Boston, Little, Brown and Co., 1857.

Gibbs, David, *A Course in Interpolation and Numerical Integration, for the Mathematical Laboratory*, London, G. Bell & Sons, 1915.

Gillham, Nicholas, *A Life of Sir Francis Galton*, Oxford, U.K., Oxford University Press, 2001.

Gillispie, Charles, ed., *Dictionary of Scientific Biography*, New York, Charles Scribner's Sons, 1971.

Gleick, James, *Genius: The Life and Science of Richard Feynman*, New York, Pantheon, 1992.

Glover, James, "Courses in Actuarial Mathematics," in *The University of Michigan: An Encyclopedic Survey*, Wilfred Byron Shaw, ed., Ann Arbor, University of Michigan Press, 1942, pp. 654–55.

Goldstine, Herman H., *The Computer from Pascal to von Neumann*, Princeton, N.J., Princeton University Press, 1972.

Goldstine, H. H., and A. Goldstine, "The Electronic Numerical Integrator and Computer (ENIAC)," *MTAC*, vol. 2, 1946, pp. 97–110, reprinted in Randell, pp. 359–74.

Golemba, Beverly, "The Women in Aeronautical Research" (manuscript, 1994), Iowa State University Archives.

Gould, Benjamin, "An Address in Commemoration of Sears Cook Walker," *Proceedings of the American Association for the Advancement of Science*, vol. 8, 1854, pp. 18–45.

Gould, Benjamin, *Reply to the Statement of the Trustees of the Dudley Observatory*, Albany, N.Y., Charles van Benthuysen, 1859.

Grattan-Guinness, I., "Work for the Hairdressers: The Production of de Prony's Logarithmic and Trigonometric Tables," *Annals of the History of Computing*, vol. 12, no. 3, Summer 1990, pp. 177–85.

Gray, Peter, "On the Arithmometer of M. Thomas (de Colmar) and Its Application in the Construction of Life Contingency Tables," London, Institute of Actuaries, 1874.

Grier, David Alan, "Agricultural Computing and the Context for John Atanasoff," *IEEE Annals of the History of Computing*, vol. 22, no. 1, January 2000, pp. 48–61.

Grier, David Alan, "Atanasoff-Berry Computer Replica," *IEEE Annals of the History of Computing*, vol. 20, no. 1, 1998, pp. 77–78.

Grier, David Alan, "Dr. Veblen Gets a Uniform: Mathematics in the First World War," *American Mathematical Monthly*, vol. 108, no. 10, December 2001, pp. 922–31.

Grier, David Alan, "ENIAC, the Verb 'to Program' and the Emergence of Digital Computers," *IEEE Annals of the History of Computing*, vol. 18, no. 1, Spring 1996, pp. 51–55.

Grier, David Alan, "The First Breach of Computer Security," *IEEE Annals of the History of Computing*, vol. 23, no. 2, 2001, pp. 78–79.

Grier, David Alan, "Henry Wallace and the Start of Statistical Computing," *Chance*, vol. 1, no. 2, May 1999, pp. 14–20.

Grier, David Alan, "The Math Tables Project: The Reluctant Start of the Computing Era," *IEEE Annals of the History of Computing*, vol. 20, no. 3, 1998, pp. 33–50.

Grier, David Alan, "Nineteenth Century Observatories and the Chorus of Computers," *IEEE Annals of the History of Computing*, vol. 21, no. 1, January 1999, pp. 45–46.

Grier, David Alan, "The Rise and Fall of the Committee on Mathematical Tables and Other Aids to Computation," *IEEE Annals of the History of Computing*, vol. 23, no. 2, 2001, pp. 38–49.

Guillaume, M. J., ed., *Procès-Verbaux du Comité d'Instruction Publique de la Convention Nationale*, vol. 3 (November 21, 1793–March 20, 1794), Paris, Imprimerie Nationale, 1897, pp. 605–6.

Guillaume, M. J., ed., *Procès-Verbaux du Comité d'Instruction Publique de la

Convention Nationale, vol. 5 (September 3, 1794–March 20, 1795), Paris, Imprimerie Nationale, 1904, pp. 551–62.

Gurer, Denise, "Women's Contributions to Early Computing at the National Bureau of Standards," *IEEE Annals of the History of Computing*, vol. 18, no. 3, Fall 1996, pp. 29–25.

Gutzwiller, Martin, "Wallace Eckert, Computers and the Nautical Almanac Office," in Fiala and Dick, pp. 147–63.

Hainaut, O., R. M. West, B. G. Marsden, A. Smette, and K. Meech, "Post-perihelion Observations of Comet P/Halley," *Astronomy and Astrophysics*, vol. 293, 1995, pp. 941–47.

Halberstam, David, *The Fifties*, New York, Fawcett, 1993.

Haldane, J. B. S., "Karl Pearson, 1857–1957," *Biometrika*, vol. 44, nos. 3/4, December 1957, pp. 303–13.

Halley, Edmund, *Astronomiae Cometicae Synopsis*, London, John Senex, 1705.

Halley, Edmund, *Astronomical Tables with Precepts Both in English and Latin*, London, 1752.

Hamilton, Mary Agnes, *Newnham*, London, Faber and Faber, 1936.

Handy, Moses P., ed., *Official Directory of the World's Columbian Expedition*, Chicago, W. D. Gonkey, 1893.

Hardy, G. H., *A Mathematician's Apology*, Cambridge, U.K., Cambridge University Press, 1967.

Harvard College Observatory, *Annual Report of the Director of the Astronomical Observatory of Harvard College*, Cambridge, Mass., Harvard University.

Harvard University, *Annals of the Computation Laboratory of Harvard University*, Cambridge, Mass.

"Health Bill Foe Debunked," *The Nation*, May 23, 1946.

Hellman, Lillian, *Scoundrel Time*, New York, Little, Brown, 1976.

Henderson, James, *Bibliotheca Tabularum Mathematicarum, Being a Descriptive Catalogue of Mathematical Tables*, Tracts for Computers, no. 13, Cambridge, U.K., Cambridge University Press, 1926.

Herman, Jan, *A Hilltop in Foggy Bottom*, Washington, D.C., Department of the Navy, 1996.

Hestenes, Magnus, and John Todd, *NBS-INA—The Institute for Numerical Analysis—UCLA 1947–1954*, Washington, D.C., U.S. Department of Commerce, NIST Special Publication 730.

Hobart, Michael, and Zachary Schiffman, *Information Ages*, Baltimore, Johns Hopkins University Press, 1998.

"Hog Astronomy," *Washington Post*, December 5, 1926, p. S1 (editorial).

Hoover, Herbert, *Testimony before Senate Committee on Agriculture, June 19, 1917*, Washington, D.C., Government Printing Office, 1917.

Hopp, Peter, *Slide Rules: Their History, Models and Makers*, Mendham, N.J., Astragal Press, 1999.

Hopper, Grace, "Commander Aiken and My Favorite Computer," in Cohen and Welch, pp. 185–94.

Howard, Donald S., *The WPA and Federal Relief Policy*, New York, Da Capo Press, 1943.

Howe, Irving, *World of Our Fathers*, New York, Simon and Schuster, 1976.

Howse, Derek, *Nevil Maskelyne: The Seaman's Astronomer*, Cambridge, U.K., Cambridge University Press, 1989.

Hughes, D. W., "The History of Halley's Comet," *Philosophical Transactions of the Royal Society of London, Series A, Mathematical and Physical Sciences*, vol. 323, no. 1572, September 30, 1987, pp. 349–66.

Huskey, Harry, "SWAC—Standards Western Automatic Computer," *IEEE Annals of the History of Computing*, vol. 19, no. 2, 1997, pp. 51–61.

Hyman, Anthony, *Charles Babbage*, Princeton, N.J., Princeton University Press, 1980.

"I.B.M. Convention Set," *New York Times*, April 29, 1939, p. 34.

Ickes, Harold, *Secret Diary of Harold Ickes*, vol. 2, New York, Simon and Schuster, 1954.

"The International Statistical Institute at Chicago," *Journal of the Royal Statistical Society*, vol. 57, no. 1, March 1894, pp. 168–71.

Iowa State University, *Annual Report*, ISU.

Iowa State University, *Bulletin of Iowa State University*, Ames, Iowa.

Irvine, M. M., "Early Digital Computers at Bell Telephone Laboratories," *IEEE Annals of the History of Computing*, vol. 23, no. 3, July–September, 2001, pp. 22–43.

Irwin, Joseph Oscar, *On Quadrature and Cubature; or, On Methods of Determining Approximately Single and Double Integrals*, Tracts for Computers, no. 10, Cambridge, U.K., Cambridge University Press, 1924.

Isaacson, Eugene, "The Origin of Mathematics of Computation and Some Personal Recollections," in Nash, pp. 211–16.

Jahnke, Eugen, and Fritz Emde, *Tables of Functions with Formulae and Curves*, Leipzig and Berlin, Teubner, 1909.

James, Mary Ann, *Elites in Conflict: The Antebellum Clash over the Dudley Observatory*, New Brunswick, N.J., Rutgers University Press, 1987.

Jarrold, Rachel, and Glenn Fromm, *Time—The Great Teacher (A History of One Hundred Years of the New Jersey State Teachers College at Trenton, 1855–1955)*, Princeton, N.J., Princeton University Press, 1955.

Jevons, W. Stanley, "Remarks on the Statistical Use of the Arithmometer," *Journal of the Statistical Society of London*, vol. 41, no. 4, December 1878, pp. 597–601.

Johnston, Stephen, "Making the Arithmometer Count," *Bulletin of the Scientific Instrument Society*, 1997.

Jones, Bessie Zaban, and Lyle Boyd, *The Harvard College Observatory: The First Four Directorships, 1839–1919*, Cambridge, Mass., Belknap Press of Harvard University Press, 1971.

Karpinsky, Louis, "James W. Glover," *Science*, vol. 94, no. 2433, August 15, 1941, pp. 156–57.

Kelves, Daniel, *In the Name of Eugenics*, Cambridge, Mass., Harvard University Press, 1985.

Kelves, Daniel, *The Physicists*, Cambridge, Mass., Harvard University Press, 1971.

Kennedy, David M., *Over Here*, New York, Oxford University Press, 1980.

Kessner, Thomas, *Fiorello H. La Guardia and the Making of Modern New York*, New York, McGraw-Hill, 1989.

Kidwell, Peggy, "The Adding Machine Fraternity at St. Louis: Creating a Center of Invention," *IEEE Annals of the History of Computing*, vol. 22, no. 2, April–May 2000, pp. 4–21.

Kidwell, Peggy, "From Novelty to Necessity," *IEEE Annals of the History of Computing*, vol. 12, no. 1, 1990, pp. 31–40.

Kidwell, Peggy, "Three Women of American Astronomy," *American Scientist*, vol. 78, 1990, pp. 244–50.

Kidwell, Peggy, "'Yours for Improvement'—The Adding Machines of Chicago, 1884–1930," *IEEE Annals of the History of Computing*, vol. 23, no. 3, July–September 2001, pp. 3–21.

Kline, Morris, *Mathematical Thought from Ancient to Modern Times*, Oxford, U.K., Oxford University Press, 1972.

Kline, Ronald, *Steinmetz: Engineer and Socialist*, Baltimore, Johns Hopkins University Press, 1992.

Kohlstedt, Sally Gregory, "Creating a Forum for Science: AAAS in the Nineteenth Century," in Kohlstedt et al., pp. 7–49.

Kohlstedt, Sally Gregory, Michael M. Sokal, and Bruce V. Lewenstein, eds., *The Establishment of Science in America: 150 Years of the American Association for the Advancement of Science*, New Brunswick, N.J., Rutgers University Press, 1999.

Kruger, M. K., "A Slide Rule for Vector Calculations," *Bell Laboratories Record*, vol. 7, 1929, pp. 405–10.

Lalande, Joseph-Jérôme Le Français de, *Astronomie*, 3rd ed., Paris, Desaint, 1792.

Lalande, Joseph-Jérôme Le Français de, *Bibliographie Astronomique avec l'Histoire de l'Astronomie depuis 1781 jusqu'à 1802*, Paris, Imprimerie de la Republique, 1802.

Latrobe, John, *The History of Mason and Dixon's Line; Contained in an Address, Delivered by John H. B. Latrobe, of Maryland, Before the Historical Society of Pennsylvania, November 8, 1854*, Philadelphia, Lippincott, Grambo, and Co., 1855.

Lee, Harper, *To Kill a Mockingbird*, New York, J. P. Lippingcott, 1960.

Leffingwell, William Henry, *Office Management Principles and Practice*, Chicago, A. W. Shaw, 1927.

Lehmer, D. H., *Guide to Tables in the Theory of Numbers*, Washington, D.C., National Research Council, 1941.

Light, Jennifer, "When Computers Were Women," *Technology and Culture*, vol. 40, no. 3, July 1999, pp. 455–83.

Lindbergh, Anne Morrow, *The War Within and Without: Diaries and Letters of Anne Morrow Lindbergh, 1939–1944*, New York, Harcourt Brace Jovanovich, 1980.

Littlewood, John E., *Collected Papers of J. E. Littlewood, Edited by a Committee Appointed by the London Mathematical Society*, Oxford, Clarendon Press, 1982.

Love, Rosaleen, "Alice in Eugenics-Land: Feminism and Eugenics in the Scientific

Careers of Alice Lee and Ethel Elderton," *Annals of Science*, vol. 36, no. 2, 1979, pp. 145–58.

Lovelace, Ada, "Notes by the Translator" (of "Sketch of the Analytical Engine Invented by Charles Babbage," by L. F. Menabrea), in Morrison and Morrison, pp. 245–97.

Lowan, Arnold N., "The Computational Laboratory at the National Bureau of Standards," *Scripta Mathematica*, vol. 15, 1949, pp. 33–63.

Lush, J., "Early Statistics at Iowa State University," in Bancroft (1972), pp. 211–26.

MacDonald, Dwight, *Henry Wallace: The Man and the Myth*, New York, Garland Publishing, 1947.

Mack, Pamela, "Strategies and Compromises: Women in Astronomy at Harvard College Observatory, 1870–1920," *Journal of the History of Astronomy*, vol. 21, 1990, pp. 65–75.

MacLane, Saunders, "The Applied Mathematics Group at Columbia in World War II," in Duren, pp. 485–515.

MacPike, Eugene Fairfield, ed., *The Correspondence of Edmond Halley*, Oxford, U.K., Clarendon Press, 1932.

MacRobert, Alan, "Halley in the Distance," *Sky and Telescope*, vol. 106, no. 6, December 2003, p. 24.

Magnello, M. Eileen, "The Non-correlation of Biometrics and Eugenics: Rival Forms of Laboratory Work in Karl Pearson's Career at University College London, Part 1," *History of Science*, vol. 37, no. 2, 1999, pp. 79–106.

"Malcolm Morrow," *Washington Post*, April 23, 1982, p. B18.

Marcus, L., "A Mathematical Match: Olga and John Todd," *Caltech News*, no. 25, February 1991, pp. 6–7.

Marsden, Brian, "Crommelin, A. C. D.," *Dictionary of Scientific Biography*, New York, Scribners, 1970, vol. 3, pp. 472–73.

Martin, Ernst, *Die Rechenmaschinen*, 1925, translated and edited by Peggy Kidwell and Michael Williams as *The Calculating Machines: Their History and Development*, Cambridge, Mass., MIT Press, 1992.

"Mary Clem," *Iowa State University Faculty Newsletter*, vol. 25, no. 20, February 2, 1979, p. 179.

Masani, Pesi Rustom, *Norbert Wiener, 1894–1964*, Basel, Birkhäuser, 1990.

Maskelyne, Nevil, *The British Mariner's Guide Containing Complete and Easy Instructions for the Discovery of Longitude at Sea*, London, 1763.

Maskelyne, Nevil, "Preface," *Nautical Almanac for 1767*, London, J. Nourse, J. Mount and T. Page, 1767.

"Mathematical Research," *Bell Laboratories Record*, vol. 1, 1925–26, pp. 15–18.

Mathematical Tables Project, *Tables of the First Ten Powers*, New York, WPA, 1939.

Maunder, E. Walter, *The Royal Observatory Greenwich*, London, Religious Tract Society, 1900.

Mayer, Tobias, *Tabulae Motuum Solis et Lunae Novae et Correctae*, London, J. Nourse, J. Mount and T. Page, 1770.

Mayer, Tobias, *Theoria Lunae Juxta Systema Newtonianum*, London, J. Nourse, J. Mount and T. Page, 1767.

McCartney, Scott, *ENIAC: The Triumphs and Tragedies of the World's First Computer*, New York, Walker, 1999.

McCullough, David, *Truman*, New York, Simon and Schuster, 1992.

McLeish, John, *Number*, New York, Fawcett, 1991.

McLendon, Winzola, "She Corrects Computers," *Washington Post*, March 19, 1962, p. B3.

McShane, Edward, John L. Kelly, and Franklin Reno, *Exterior Ballistics*, University of Denver Press, 1953.

Meadows, A. J., *Greenwich Observatory*, vol. 2, London, Taylor and Francis, 1975.

Messier, Charles, and Matthew Maty, "A Memoir, Containing the History of the Return of the Famous Comet of 1682 . . . ," *Philosophical Transactions (1683–1775)*, vol. 55, 1765, pp. 264–325.

Metropolis, N., and E. C. Nelson, "Early Computing at Los Alamos," *Annals of the History of Computing*, vol. 4, no. 4, October 1982, pp. 348–57.

Metropolis, Nicholas, J. Howlett, and Gian-Carlo Rota, eds., *A History of Computing in the Twentieth Century*, London, Academic Press, 1980.

Miller, Frederick, and Howard Gillette, *Washington Seen: A Photographic History, 1875–1965*, Baltimore, Johns Hopkins University Press, 1995.

Millman, S., ed., *A History of Engineering and Science in the Bell System*, vol. 1, New York, ATT, 1984.

Mills, John, "The Line and the Laboratory," *Bell Telephone Quarterly*, vol. 19, no. 1, January 1940, pp. 5–21.

Mills, John, "A Quarter Century of Transcontinental Telephone Service," *Bell Telephone Quarterly*, vol. 19, no. 1, January 1940, pp. 1–2.

Mills, Walter, *Road to War*, Boston, Houghton Mifflin, 1935.

Mollenhoff, Clark, *Atanasoff, Forgotten Father of the Computer*, Ames, Iowa State University Press, 1988.

Moorehead, E. J., *Our Yesterdays: The History of the Actuarial Profession in North American, 1809–1979*, Schaumburg, Ill., Society of Actuaries, 1989.

Morando, Bruno, "The Golden Age of Celestial Mechanics," in Taton and Wilson, pp. 221–39.

Morrison, Philip, and Emily Morrison, eds., *Charles Babbage and His Calculating Engines*, New York, Dover, 1966.

Morse, Philip, *In at the Beginnings: A Physicist's Life*, Cambridge, Mass., MIT Press, 1977.

Moulton, Forest Ray, "History of the Ballistics Branch of the Artillery Ammunition Section, Engineering Division of the Ordnance Department for the Period April 6, 1918, to April 2, 1919" (manuscript, 1919), U.S. Military History Institute, Carlisle, Pa.

Moulton, Forest Ray, *New Methods in Exterior Ballistics*, Chicago, Chicago University Press, 1926.

"Myrrick Haskell Doolittle," *Washington Evening Star*, Saturday, June 28, 1913.

Nash, Stephen, ed., *A History of Scientific Computing*, New York, ACM Press, 1990.

National Bureau of Standards, "Activities in Applied Mathematics, 1946–1947" (report, 1947), pp. 12–13.

National Bureau of Standards, *The National Applied Mathematics Laboratories—A Prospectus*, Washington, D.C., National Bureau of Standards, 1947.

National Bureau of Standards, *Projects and Publications of the National Applied Mathematics Laboratories*, Washington, D.C.

National Research Council, *International Critical Tables of Numerical Data, Physics, Chemistry and Technology*, New York, McGraw-Hill, 1926.

"Nautical Almanac Investigation," *Washington Post*, July 19, 1893, p. 4.

"Nautical Almanac Office Moved," *Washington Post*, October 19, 1893, p. 8.

Nebeker, Frederik, *Calculating the Weather: Meteorology in the 20th Century*, New York, Academic Press, 1995.

Newcomb, Simon, "Abstract Science in America, 1776–1876, " *North American Review*, vol. 122, no. 250, 1876, pp. 88–124.

Newcomb, Simon, *The Reminiscences of an Astronomer*, Boston and New York, Houghton, Mifflin and Company, 1903.

Newton, Isaac, *Philosophiae Naturalis Principia Mathematica*, London, Royal Society, 1687.

O'Connor, J. J., and E. F. Robertson, "Francesco Siacci," MacTutor History of Mathematics, http://turnbull.mcs.st-and.ac.uk/~history/Mathematicians/Siacci.html.

Ogilvie, Marilyn Bailey, and Joy Dorothy Harvey, *Biographical Dictionary of Women in Science*, vol. 2, New York, Routledge, 2000.

Olinick, Michael, *An Introduction to Mathematical Models in the Social and Life Sciences*, Reading, Mass., Addison-Wesley, 1978.

Owens, Larry, "The Counterproductive Management of Science in the Second World War: Vannevar Bush and the Office of Scientific Research and Development," *Business History Review*, vol. 68, no. 4, 1994, pp. 515–76.

Owens, Larry, "Mathematicians at War: Warren Weaver and the Applied Mathematics Panel, 1942–1945," in Rowe and McCleary, pp. 287–305.

Owens, Larry, "Where Are We Going, Phil Morse? Changing Agendas and the Rhetoric of Obviousness in the Transformation of Computing at MIT, 1939–1957," *IEEE Annals of the History of Computing*, vol. 18, no. 4, October–December 1996, pp. 42–48.

Pairman, Eleanor, *Tables of the Digamma and Trigamma Functions*, Tracts for Computers, no. 1, London, Cambridge University Press, 1919.

Pearl, Raymond, "Karl Pearson, 1857–1936," *Journal of the American Statistical Association*, vol. 31, no. 196, December 1936, pp. 653–64.

Pearl, Raymond, and Magdalen Burger, "Retail Prices of Food during 1917 and 1918," *Publications of the American Statistical Association*, vol. 16, no. 127, September 1919, pp. 411–39.

Pearson, Egon S., "An Appreciation of Some Aspects of His Life and Work," *Biometrika*, vol. 29, nos. 3/4, February 1938, pp. 161–248.

Pearson, Egon S., "Karl Pearson: An Appreciation of Some Aspects of His Life and Work," *Biometrika*, vol. 28, nos. 3/4, December 1936, pp. 193–257.

Pearson, Karl, "Cooperative Investigations on Plants: I. On Inheritance in the Shirley Poppy," *Biometrika*, vol. 2, no. 1, November 1902, pp. 56–100.

Pearson, Karl, *The Life, Letters and Labours of Francis Galton*, Cambridge, U.K., Cambridge University Press, 1914–30, vol. 3.

Pearson, Karl, "Mathematical Contributions to the Theory of Evolution, III: Regression, Heredity and Panmixia," *Philosophical Transactions of the Royal Society of London (A)*, vol. 187, 1896, pp. 253–318.

Pearson, Karl, *On the Construction of Tables and on Interpolation: Part I. Univariate Tables*, Tracts for Computers, no. 2, Cambridge, U.K., Cambridge University Press, 1920.

Pearson, Karl, *On the Construction of Tables and on Interpolation: Part II. Bivariate Tables*, Tracts for Computers, no. 3, Cambridge, U.K., Cambridge University Press, 1920.

Pearson, Karl, "On the Laws of Inheritance in Man: II. On the Inheritance of the Mental and Moral Characters in Man, and Its Comparison with the Inheritance of the Physical Characters," *Biometrika*, vol. 3, nos. 2/3, March–July 1904, pp. 131–90.

Pearson, Karl, "The Scope of *Biometrika*," *Biometrika*, vol. 1, no. 1, October 1901, pp. 1–2.

Pearson, Karl, "Walter Frank Raphael Weldon, 1860–1906," *Biometrika*, vol. 5, nos. 1/2, October 1906, pp. 1–52.

Perry, John, *The Story of Standards*, New York, Funk & Wagnalls, 1955.

Peterson, Sven R., "Benjamin Peirce: Mathematician and Philosopher," in Cohen (1980), pp. 89–112.

Pierce, John Alvin, ed., *Long Range Navigation*, New York, McGraw-Hill, 1948.

Polachek, Harry, "History of the Journal *Mathematical Tables and Other Aids to Computation*, 1959–1965," *IEEE Annals of the History of Computing*, vol. 17, no. 3, 1995, pp. 67–74.

Porter, Robert, "The Eleventh Census," *Publications of the American Statistical Association*, vol. 2, no. 15, September 1891, pp. 321–79.

Porter, Theodore, *Karl Pearson: The Scientific Life in a Statistical Age*, Princeton, N.J., Princeton University Press, 2004.

Porter, Theodore, *The Rise of Statistical Thinking*, Princeton, N.J., Princeton University Press, 1986.

Porter, Theodore, *Trust in Numbers*, Princeton, N.J., Princeton University Press, 1995.

"President Backs Weeks on Ouster," *New York Times*, April 3, 1953.

Price, G. Baley, "American Mathematicians in World War I," in Duren, vol. 1, pp. 267–73.

Price, G. Baley, "Award for Distinguished Service to Dr. Thornton Carl Fry," *American Mathematical Monthly*, vol. 89, no. 2, 1982, pp. 81–83.

Public Papers of the Presidents of the United States, NARA, Papers of Lyndon Baines Johnson, 1963.

Pugh, Emerson, *Building IBM*, Cambridge, Mass., MIT Press, 1994.

Pursell, Carrol, "A Preface to Governmental Support of Research," *Technology and Culture*, vol 9, no. 2, 1968, pp. 145–64.

"A Quarter Century of Transcontinental Telephone Service," *Bell Telephone Quarterly*, vol. 19, no. 1, January 1940, pp. 1–2.

Radcliffe College, *Private Collegiate Instruction for Women in Cambridge, Mass-*

achusetts. *Courses of Study . . . with Requisitions for Admission and Report of the . . . Year 1879–80*, Cambridge, Mass., Harvard University, 1880.

Ralph, Julian, "Chicago's Gentle Side," *Harper's New Monthly Magazine*, vol. 87, no. 518, July 1893, pp. 286–98.

Randell, Brian, ed., *The Origins of Digital Computers*, New York, Springer-Verlag, 1982.

Rasmussen, Wayne D., and Gladys L. Baker, *The Department of Agriculture*, New York, Praeger, 1972.

Reich, Leonard, *The Making of American Industrial Research*, Cambridge, U.K., Cambridge University Press, 1985.

Reid, Constance, *Neyman from Life*, Springer-Verlag, 1982.

Reiman, Richard, *The New Deal and American Youth*, Athens, University of Georgia Press, 1992.

Reingold, Nathan, ed., *Papers of Joseph Henry*, Washington, D.C., Smithsonian Institution Press, 1979.

Reingold, Nathan, *Science American Style*, New Brunswick, N.J., Rutgers University Press, 1991.

Reingold, Nathan, "Vannevar Bush's New Deal for Research; or, The Triumph of the Old Order," *Historical Studies in the Physical and Biological Sciences*, vol. 17, no. 2, 1987, pp. 299–344.

Rhodes, Edmund Cecil, *On Smoothing*, Tracts for Computers, no. 6, Cambridge, U.K., Cambridge University Press, 1921.

Rhodes, Ida, interview with Henry Tropp, March 21, 1973, SMITHSONIAN.

Rhodes, Richard, *Dark Sun*, New York, Simon and Schuster, 1995.

Rhodes, Richard, *The Making of the Atomic Bomb*, New York, Simon and Schuster, 1986.

Richardson, Lewis Fry, *Weather Prediction by Numerical Process*, Cambridge, U.K., Cambridge University Press, 1922.

Riddell, Robert, *The Slide Rule Simplified*, Philadelphia, 1881.

Rigaud, S. P., *Some Account of Halley's Astronomiae Cometicae Synopsis*, Oxford, 1835.

Rodgers, William, *Think: A Biography of the Watsons and IBM*, New York, Stein and Day, 1969.

Rosser, J. Barkley, "Mathematics and Mathematicians in World War II," *Notices of the American Mathematical Society*, vol. 29, 1982, pp. 509–15.

Rossiter, Margaret, *Women Scientists in America: Struggles and Strategies to 1940*, Baltimore, Johns Hopkins University Press, 1992.

Rotella, Elyce, *From Home to Office: U.S. Women at Work, 1870–1930*, Ann Arbor, Mich., UMI Research Press, 1981.

Rowe, David, and John McCleary, eds., *The History of Modern Mathematics: Proceedings of the Symposium on the History of Modern Mathematics, Vassar College, Poughkeepsie, New York, 20–24 June 1989*, vol. 2, *Institutions and Applications*, Boston, Academic Press, 1989.

Royal Observatory, *Report of the Astronomer Royal to the Board of Visitors of the Royal Observatory*, Greenwich.

Rutland, David, *Why Computers Are Computers: The SWAC and the PC*, Philomath, Ore., Wren Publishing, 1995.

Sadler, D. H., and John Todd, "Mathematics in Government Service and Industry," *Nature*, no. 3992, May 4, 1946.

Scaffer, Simon, "Babbage's Intelligence," *Critical Inquiry*, vol. 21, no. 1, August 1994, pp. 203–27.

Scaife, B. K. P., ed., *Studies in Numerical Analysis*, New York, Academic Press, 1974.

Schaffer, Simon, "Astronomers Mark Time: Discipline and the Personal Equation," *Science in Context*, vol. 2, no. 1, 1988, pp. 115–45.

"Scientists at Sword's Points," *Washington Post*, July 9, 1893, p. 7.

Shannon, Claude E., "Mathematical Theory of the Differential Analyzer," *MTAC*, vol. 1, no. 2, April 1943, pp. 63–64.

Shaw, George Bernard, *Man and Superman* (Cambridge, Mass., University Press, 1903), Bartleby.com, 1999, http://www.bartleby.com/157/.

Shaw, George Bernard, *Mrs. Warren's Profession* (1894), Project Gutenberg, 1997, ftp://ibiblio.org/pub/docs/books/gutenberg/etext97/wrpro10.txt.

Sheldon, John, and Liston Tatum, "The IBM Card-Programmed Electronic Calculator," *Review of Electronic Digital Computers, Joint AIEE-IRE Conference, December 10–12, 1951*, New York, American Institute of Electrical Engineers, 1952, pp. 30–36, reprinted in Randell, pp. 233–40.

Siacci, Francesco, "Rational and Practical Ballistics—New Methods for Solving Problems of Fire," trans. O. B. Mitcham, in U.S. Ordnance Department, *Report of the Chief of Ordnance*, Washington, D.C., Government Printing Office, 1881, pp. 218–42.

Sinclair, Upton, *The Jungle* (1906), Project Gutenberg, 1994, ftp://ibiblio.org/pub/docs/books/gutenberg/etext94/jungl10.txt.

Slutz, Ralph, "Memories of the Bureau of Standards SEAC," in Metropolis et al., pp. 471–77.

Smith, Adam, *An Inquiry into the Nature and Causes of the Wealth of Nations* (London, W. Strahan and T. Cadell, 1776), Project Gutenberg, 2002, ftp://ibiblio.org/pub/docs/books/gutenberg/etext02/wltnt10.txt.

Smith, Adam, "The Principles Which Lead and Direct Philosophical Enquiries: Illustrated by the History of Astronomy" (1757), in *The Early Writings of Adam Smith*, ed. J. Ralph Lindgren, New York, Augustus Kelley, 1967, pp. 32–108.

Smith, Betty, *A Tree Grows in Brooklyn*, New York, Harper and Row, 1947.

Smith, Cecil, "The Longest Run: Public Engineers and Planning in France," *American Historical Review*, vol. 95, no. 3, June 1990, pp. 657–92.

Smith, Robert, "A National Observatory Transformed: Greenwich in the Nineteenth Century," *Journal of the History of Astronomy*, vol. 17, 1991, pp. 5–20.

Smithsonian Institution, *Smithsonian Institution Research Reports*, Washington, D.C.

Snedecor, George W., "Uses of Punched Card Equipment in Mathematics," *American Math Monthly*, vol. 35, 1928, pp. 161–69.

Snyder-Grenier, Ellen, *Brooklyn*, Philadelphia, Temple University Press, 1996.

Sobel, Dava, *Longitude*, New York, Walker, 1995.

"Some Remarks upon an American Nautical Almanac," *American Journal of Science and Arts*, 2nd ser., vol. 7, January 1849, pp. 123–35.

Soper, H. E., A. W. Young, B. M. Cave, A. Lee, and K. Pearson, "On the Distribution of the Correlation Coefficient in Small Samples: Appendix II to the Papers of 'student' and R. A. Fisher," *Biometrika*, vol. 11, no. 4, May 1917, pp. 328–413.

South, James, "Report of the Committee of the Astronomical Society of London," November 19, 1830, in *The Nautical Almanac and Astronomical Ephemeris for the Year 1834*, London, John Murray, 1833, pp. xii–xxii.

South, James, "Report on Nautical Almanac," *Memoirs of the Royal Astronomical Society*, vol. 55, 1831, pp. 459–71.

Spinoza, Benedict de, *The Ethics* (1677; trans. R. H. M. Elwes, 1883), Project Gutenberg, 2003, ftp://ibiblio.org/pub/docs/books/gutenberg/etext03/ethic10.txt.

Stachel, J., "Lanczos's Early Contributions to Relativity and His Relationship with Einstein," in *Proceedings of the Cornelius Lanczos International Centenary Conference*, Philadelphia, Society for Industrial and Applied Mathematics, 1994, pp. 201–21.

Sterling, Pauline, "Blond Fashion Designer Has Eye on Space," *Detroit Free Press*, October 20, 1965.

Stern, Nancy, *From ENIAC to UNIVAC*, Maynard, Mass., Digital Press, 1981.

Stevens, Doris, *Jailed for Freedom*, Troutdale, Ore., NewSage Press, 1975.

Stewart, Irvin, *Organizing Scientific Research for War: The Administrative History of the Office of Scientific Research and Development*, Boston, Little, Brown, 1948.

Stewart, Robert, "End of the ABC," *Annals of the History of Computing*, vol. 6, no. 3, July 1984, p. 317.

Stibitz, George, "An Application of Number Theory to Gear Ratios," *American Mathematical Monthly*, vol. 45, no. 1, January 1938, pp. 22–31.

Stibitz, George, "Computer" (Bell Telephone Laboratories memo, 1940), printed in Randell, pp. 246–52.

Stibitz, George, "Early Computers," in Metropolis et al., pp. 479–83.

Stibitz, George, "Lecture," in Campbell-Kelly and Williams, pp. 5–16.

Stigler, George, "The Cost of Subsistence," *Journal of Farm Economics*, vol. 27, May 1945, pp. 303–14.

Stigler, Stephen, *The History of Statistics*, Cambridge, Mass., Belknap Press of Harvard University, 1986.

Stigler, Stephen, "Stigler's Law of Eponymy," in *Statistics on the Table*, Cambridge, Mass., Harvard University Press, 1999, pp. 277–90.

Strom, Sharon Hartman, *Beyond the Typewriter: Gender, Class and the Origins of Modern American Office Work, 1900–1930*, Chicago, University of Chicago Press, 1992.

Struik, Dirk, *Yankee Science in the Making*, Boston, Little, Brown and Co., 1948.

Süss, Irmgard, "The Mathematical Research Institute Oberwolfach through Critical Times," in Beckenbach and Walter, pp. 3–17.

Swade, Doron, *The Cogwheel Brain*, London, Abacus, 2000.

Swift, Jonathan, *Gulliver's Travels* (London, 1726), Project Gutenberg, 1997, ftp://ibiblio.org/pub/docs/books/gutenberg/etext97/gltrv10.txt.

Taton, René, and Curtis Wilson, eds., *Planetary Astronomy from the Renaissance to the Rise of Astrophysics, Part B: The Eighteenth and Nineteenth Centuries*, Cambridge, U.K., Cambridge University Press, 1995.

Taussky, Olga, "How I Became a Torchbearer for Matrix Theory," *American Mathematical Monthly*, vol. 95, no. 9, 1988, pp. 801–12.

Taylor, Philip, "Propaganda by Deed—The Greenwich Observatory Bomb of 1894," *Open Space*, no. 9, November 1996, p. 4.

Testimony before the Joint Committee to Consider the Present Organizations of the Signal Service, Geological Survey, Coast and Geodetic Survey and the Hydrographic Office of the Navy Department, New York, Arno Press, 1980.

Theberge, Albert, *History of the Commissioned Corps of the National Oceanic and Atmospheric Administration, Volume I, The Coast Survey, 1807–1867* (unpublished manuscript), Library of the National Oceanic and Atmospheric Administration.

Thompson, Alexander John, *Logarithmetica Britannica, Being a Standard Table of Logarithms to Twenty Decimal Places*, Tracts for Computers, no. 11, Cambridge, U.K., Cambridge University Press, 1924.

Thoreau, Henry David, *Walden*, Project Gutenberg, 1995, ftp://ibiblio.org/pub/docs/books/gutenberg/etext95/waldn10.txt.

Tocqueville, Alexis de, *Democracy in America*, Project Gutenberg, 1997, ftp://ibiblio.org/pub/docs/books/gutenberg/etext97/1dina10.txt and ftp://ibiblio.org/pub/docs/books/gutenberg/etext97/2dina10.txt.

Todd, John, "Applied Mathematical Research in Germany with Particular Reference to Naval Applications," British Intelligence Office Subcommittee, London, His Majesty's Stationery Office, 1945.

Todd, John, "John Hamilton Curtiss, 1909–1977," *Annals of the History of Computing*, vol. 2 no. 2, April 1980, pp. 104–10.

Todd, John, "John von Neumann and the National Accounting Machine," *SIAM Review*, vol. 16, no. 4, October 1974, pp. 526–30.

Todd, John, "Oberwolfach—1945," in Beckenbach and Walter, pp. 19–22.

Todd, John, and D. H. Sadler, "Admiralty Computing Service," *MTAC*, vol. 2, no. 19, July 1947, pp. 289–97.

Tolley, Howard, "Interview of Howard R. Tolley with Dean Albertson," Oral History Research Office of Columbia University, 1972.

Tolley, H. R., and Mordecai Ezekiel, "The Doolittle Method for Solving Multiple Correlation Equations versus the Kelley-Salisbury 'Iteration' Method," *Journal of the American Statistical Association*, vol. 22, no. 160, December 1927, pp. 497–500.

Tolley, Howard, and Mordecai Ezekiel, "A Method of Handling Multiple Correlation Problems," *Journal of the American Statistical Association*, vol. 18, no. 144, December 1923, pp. 933–1003.

Travis, Irven, "The History of Computing Devices," in Campbell-Kelly and Williams, pp. 19–24.

Travis, Irven, Oral History with Nancy Stern, October 21, 1977, OH 36, CBI.

Tuchman, Barbara, *The Guns of August*, New York, Ballantine, 1962.

Tuchman, Barbara, *The Proud Tower*, New York, Ballantine, 1962.

Tuchman, Barbara, *The Zimmerman Telegram*, New York, Ballantine, 1962.

Turner, Frederick Jackson, "The Significance of the Frontier in American History," in *Report of the American Historical Association for 1893*, Madison, Wis., State Historical Society of Wisconsin, 1894, pp. 199–227.

Tyler, H. W., "Biography: John David Runkle," *American Mathematical Monthly*, vol. 10, nos. 8/9, August–September 1903, pp. 183–85.

Upton, Winslow, "Observatory Pinafore" (unpublished manuscript, 1880; typescript ca. 1929), UPTON PAPERS.

U.S. Army, *Ballisticians in War and Peace: A History of the United States Army Ballistic Research Laboratories*, vol. 1, 1914–1956, Aberdeen Proving Ground, 1956.

U.S. Census Commission, *Report of a Commission Appointed by the Honorable Superintendent of Census on Different Methods of Tabulating Census Data*, Washington, D.C., November 30, 1889.

U.S. Coast Survey, *Report of the Superintendent of the Coast Survey*, Washington, D.C., Robert Armstrong.

U.S. Federal Works Agency, *Final Report on the WPA Program, 1935–43*, Washington, D.C., Government Printing Office, 1946.

U.S. Nautical Almanac, "Annual Report of the U.S. Nautical Almanac," Washington, D.C., U.S. Navy.

U.S. Naval Observatory, *Annual Report of the Naval Observatory*, Washington, D.C., Department of the Navy.

U.S. Navy, *Annual Report of the Secretary of the Navy*, Washington, D.C., Government Printing Office.

U.S. Senate, "Testimony of Robert J. Ryan before the Senate Subcommittee on Investigations, January 29, 1953," *Executive Sessions of the Senate Permanent Subcommittee on Investigations of the Committee on Government Operations*, 83rd Cong., 1st sess., 1953, Senate Print 107-84, vol. 1.

Veblen, Thorstein, *The Higher Learning in America*, New York, B. W. Huebach, 1918.

Veysey, Laurence, *The Emergence of the American University*, Chicago, University of Chicago Press, 1965.

Villard, Henry, and James Nagel, *Hemingway in Love and War*, Boston, Northeastern University Press, 1995.

Waff, Craig, "Navigation vs. Astronomy: Defining a Role for an American Nautical Almanac, 1844–1849," in Fiala and Dick, pp. 83–128.

Walkowitz, Judith, "Science, Feminism, and Romance: The Men and Women's Club, 1885–1889," *History Workshop Journal*, vol. 21, 1986, pp. 37–59.

Wallace, Henry A., *Agricultural Prices*, Des Moines, Iowa, Wallace Publishing Company, 1920.

Wallace, Henry A., "What Is an Iowa Farm Worth?" *Wallace's Practical Farmer*, vol. 49, no. 1, January 4, 1924, pp. 1ff.

Wallace, Henry A., "Who Plays the Part of Joseph?" *Wallace's Farmer*, September 27, 1912.

Wallace, Henry A., and George W. Snedecor, *Correlation and Machine Calculation*, Iowa State College of Agricultural and Mechanic Arts, vol. 23, no. 35, Aimes, Iowa State College, 1925.

Wang, Jessica, "Science, Security and the Cold War: The Case of E. U. Condon," *Isis*, vol. 83, 1992, pp. 238–69.

Watson, Thomas J., Jr. and Peter Petre, *Father and Son and Company*, New York, Bantam, 1990.

Weaver, Warren, *Scene of Change*, New York, Charles Scribner's Sons, 1970.

Weber, Gustavus A., *The Naval Observatory: Its History, Activities and Organization*, Baltimore, Johns Hopkins University Press, 1926.

"Weeks Ends Silence in Forecasting Swing of Ax on Deadwood," *Washington Evening Star*, February 13, 1953, p. A3.

"Weeks Is Disputed by Research Group," *New York Times*, April 7, 1953.

"Weeks Sees Ousting of 'Holdovers' Here," *Washington Evening Star*, March 29, 1953, p. A31.

Wells, H. G., "The World of Tomorrow," *New York Times*, May 5, 1939, p. AS4.

Welther, Barbara L., "Pickering's Harem," *Isis*, vol. 73, March 1982, p. 94.

White, Michael, *Isaac Newton*, Reading, Mass., Perseus, 1997.

Whitnah, Donald, *A History of the United States Weather Bureau*, Urbana, University of Illinois Press, 1961.

Whittaker, Edmund T., and G. Robinson, *The Calculus of Observations*, London, Blackie and Son, 1924.

Whyte, William, *Organization Man*, New York, Doubleday, 1956.

Wiener, Norbert, *Ex-Prodigy*, Cambridge, Mass., MIT Press, 1953.

Wilkins, George A., "The History of H.M. Nautical Almanac Office," in Fiala and Dick, pp. 55–82.

Williams, Michael, *A History of Computing Technology*, Los Alamitos, Calif., Computer Society Press, 1997.

Williams, Michael, and Erwin Tomash, "The Sector: Its History, Scales and Uses," *IEEE Annals of the History of Computing*, vol. 25, no. 1, January 2003, pp. 34–47.

Wilson, Curtis, "Appendix: Clairaut's Calculation of the Comet's Return," in *The General History of Astronomy*, ed. Rene Taton and Curtis Wilson, Cambridge, U.K., Cambridge University Press, 1995, pp. 83–86.

Wilson, Curtis, "Clairaut's Calculation of Halley's Comet," *Journal of the History of Astronomy*, vol. 24, 1993, pp. 1–14.

Wilson, Elizabeth Webb, *Compulsory Health Insurance*, New York, National Industrial Conference Board, 1947.

Wilson, Elizabeth Webb, "Health Insurance Costly, Increases Noted in German and English Systems with Policies Involved," *New York Times*, January 1, 1946, p. 26.

Winters, Donald, "The Hoover-Wallace Controversy during World War I," *Annals of Iowa*, vol. 39, no. 8, 1969, pp. 586–97.

Works Progress Administration, *Index of Research Projects*, vol. 3, Washington, D.C., Government Printing Office, 1939.

World's Congress Auxiliary, *The World's Congress Auxiliary of the World's Columbian Exposition of 1893* (circular), WCE.

"WPA Exhibit Opens without Fanfare," *New York Times*, May 23, 1939, p. 18.

"WPA Finds Friends at Its Fair Exhibit," *New York Times*, July 6, 1939, p. 18.

"WPA Will Add 350,000 to Rolls," *Washington Evening Star*, December 10, 1937, p. B5.

Wrench, J. W., "Handbook of Mathematical Functions with Formulas, Graphs and Mathematical Tables," *Mathematics of Computation*, vol. 19, no. 89, April 1965, pp. 147–49.

Wright, Helen, *Sweeper in the Sky*, New York, MacMillan, 1949.

Wright, Helen, "William Mitchell of Nantucket," *Proceedings of the Nantucket Historical Association*, 1949, pp. 33–46.

Yeomans, Donald, "Comet Halley—The Orbital Motion," *Astronomical Journal*, vol. 82, no. 6, 1977, pp. 435–40.

Yeomans, Donald, *Comets: A Chronological History of Observation, Science, Myth, and Folklore*, New York, John Wiley, 1991.

Yule, George Udney, *An Introduction to the Theory of Statistics*, London, Charles Griffin and Co., 1911.

Zabecki, David, *Steel Wind: Colonel Georg Bruchmüller and the Birth of Modern Artillery*, Westport, Conn., Praeger, 1994.

Index

Illustration Credits